Batteries and Supercapacitors Aging

Batteries and Supercapacitors Aging

Special Issue Editors

Pascal Venet
Eduardo Redondo-Iglesias

MDPI • Basel • Beijing • Wuhan • Barcelona • Belgrade • Manchester • Tokyo • Cluj • Tianjin

Special Issue Editors

Pascal Venet
Univ Lyon, Université Claude
Bernard Lyon 1, Ecole Centrale
de Lyon, INSA Lyon, CNRS,
Ampère UMR5005
France

Eduardo Redondo-Iglesias
Univ Gustave Eiffel, AME-Eco7
France

Editorial Office
MDPI
St. Alban-Anlage 66
4052 Basel, Switzerland

This is a reprint of articles from the Special Issue published online in the open access journal *Batteries* (ISSN 2313-0105) (available at: https://www.mdpi.com/journal/batteries/special_issues/Batteries_Supercapacitors_Aging).

For citation purposes, cite each article independently as indicated on the article page online and as indicated below:

LastName, A.A.; LastName, B.B.; LastName, C.C. Article Title. *Journal Name* **Year**, *Article Number*, Page Range.

ISBN 978-3-03928-714-7 (Pbk)
ISBN 978-3-03928-715-4 (PDF)

Contents

About the Special Issue Editors . . . . **vii**

Pascal Venet and Eduardo Redondo-Iglesias
Batteries and Supercapacitors Aging
Reprinted from: *Batteries* **2020**, *6*, 18, doi:10.3390/batteries6010018 . . . . **1**

Christophe Savard, Pascal Venet, Éric Niel, Laurent Piétrac and Ali Sari
Comparison of Battery Architecture Dependability
Reprinted from: *Batteries* **2018**, *4*, 31, doi:10.3390/batteries4030031 . . . . **3**

Peter Kurzweil and Mikhail Shamonin
State-of-Charge Monitoring by Impedance Spectroscopy during Long-Term Self-Discharge of
Supercapacitors and Lithium-Ion Batteries
Reprinted from: *Batteries* **2018**, *4*, 35, doi:10.3390/batteries4030035 . . . . **15**

Nagham El Ghossein, Ali Sari and Pascal Venet
Lifetime Prediction of Lithium-Ion Capacitors Based on Accelerated Aging Tests
Reprinted from: *Batteries* **2019**, *5*, 28, doi:10.3390/batteries5010028 . . . . **29**

Christophe Savard and Emiliia V. Iakovleva
A Suggested Improvement for Small Autonomous Energy System Reliability by Reducing Heat
and Excess Charges
Reprinted from: *Batteries* **2019**, *5*, 29, doi:10.3390/batteries5010029 . . . . **41**

Honorat Quinard, Eduardo Redondo-Iglesias, Serge Pelissier and Pascal Venet
Fast Electrical Characterizations of High-Energy Second Life Lithium-Ion Batteries for
Embedded and Stationary Applications
Reprinted from: *Batteries* **2019**, *5*, 33, doi:10.3390/batteries5010033 . . . . **59**

Tiphaine Plattard, Nathalie Barnel, Loïc Assaud, Sylvain Franger and Jean-Marc Duffault
Combining a Fatigue Model and an Incremental Capacity Analysis on a Commercial
NMC/Graphite Cell under Constant Current Cycling with and without Calendar Aging
Reprinted from: *Batteries* **2019**, *5*, 36, doi:10.3390/batteries5010036 . . . . **77**

Elie Riviere, Ali Sari, Pascal Venet, Frédéric Meniere and Yann Bultel
Innovative Incremental Capacity Analysis Implementation for C/LiFePO$_4$ Cell State-of-Health
Estimation in Electrical Vehicles
Reprinted from: *Batteries* **2019**, *5*, 37, doi:10.3390/batteries5020037 . . . . **93**

George Baure and Matthieu Dubarry
Synthetic vs. Real Driving Cycles: A Comparison of Electric Vehicle Battery Degradation
Reprinted from: *Batteries* **2019**, *5*, 42, doi:10.3390/batteries5020042 . . . . **107**

**Arianna Moretti, Diogo Vieira Carvalho, Niloofar Ehteshami, Elie Paillard, Willy Porcher,
David Brun-Buisson, Jean-Baptiste Ducros, Iratxe de Meatza, Aitor Eguia-Barrio, Khiem Trad
and Stefano Passerini**
A Post-Mortem Study of Stacked 16 Ah Graphite//LiFePO$_4$ Pouch Cells Cycled at 5 °C
Reprinted from: *Batteries* **2019**, *5*, 45, doi:10.3390/batteries5020045 . . . . **123**

Martin Frankenberger, Madhav Singh, Alexander Dinter and Karl-Heinz Pettinger
EIS Study on the Electrode-Separator Interface Lamination
Reprinted from: *Batteries* **2019**, *5*, 71, doi:10.3390/batteries5040071 . . . . **139**

Omonayo Bolufawi, Annadanesh Shellikeri and Jim P. Zheng
Lithium-Ion Capacitor Safety Testing for Commercial Application
Reprinted from: *Batteries* **2019**, *5*, 74, doi:10.3390/batteries5040074 . . . . **151**

Peter Kurzweil and Wolfgang Scheuerpflug
State-of-Charge Monitoring and Battery Diagnosis of NiCd Cells Using
Impedance Spectroscopy
Reprinted from: *Batteries* **2020**, *6*, 4, doi:10.3390/batteries6010004 . . . . **165**

Philipp Teichert, Gebrekidan Gebresilassie Eshetu, Hannes Jahnke and Egbert Figgemeier
Degradation and Aging Routes of Ni-Rich Cathode Based Li-Ion Batteries
Reprinted from: *Batteries* **2020**, *6*, 8, doi:10.3390/batteries6010008 . . . . **179**

About the Special Issue Editors

Pascal Venet received his PhD in electrical engineering in 1993 from Lyon 1 University, France. Shortly after leaving his postdoctoral position he joined Lyon 1 University as an Assistant Professor, from 1995 to 2009. Since 2009, he has been a Professor of Electrical Engineering at Lyon 1 University. His research continues to focus on electrical engineering, mainly in the Electrical Engineering Laboratory (AMPERE). He is the leader of the research team "Safe Systems and Energies" of the laboratory. His current research interests include characterization, modeling, fault diagnostics, reliability, and the aging of energy storage systems such as batteries, supercapacitors, and capacitors. He is also interested in BMSs (Battery Management Systems) and, therefore, the balancing of cells, the determination of state of charge and the state of health of the battery.

Eduardo Redondo-Iglesias was born in Vigo, Spain, in 1981. In 2017 he received the PhD in Electrical Engineering from the University of Lyon (France). Currently, he is in charge of the battery test equipment in IFSTTAR (now Univ Eiffel). His research activities include electrical modeling and characterization of batteries, and their aging for transportation applications.

Editorial

Batteries and Supercapacitors Aging

Pascal Venet [1,2,*] and Eduardo Redondo-Iglesias [2,3,*]

1   Univ Lyon, Université Claude Bernard Lyon 1, École Centrale de Lyon, INSA Lyon, CNRS, Ampère,
    F-69100 Villeurbanne, France
2   ERC GEST (IFSTTAR/Ampère Joint Research Team for Energy Management and Storage for Transport),
    69500 Bron, France
3   Univ Gustave Eiffel, IFSTTAR, AME-Eco7, 69500 Bron, France
*   Correspondence: pascal.venet@univ-lyon1.fr (P.V.); eduardo.redondo@univ-eiffel.fr (E.R.-I.)

Received: 18 February 2020; Accepted: 3 March 2020; Published: 12 March 2020

Electrochemical energy storage is a key element of systems in a wide range of sectors, such as electro-mobility, portable devices, or renewable energy. Energy storage systems (ESS) considered here are batteries, supercapacitors or hybrid components such as lithium-ion capacitors. The durability of ESS determines the total cost of ownership and the global impacts (lifecycle) on a large portion of these applications and thus their viability. Understanding of ESS aging is a key issue to optimize their design and usage towards their applications. Knowledge of the ESS aging is also essential to improve their dependability (reliability, availability, maintainability and safety).

In the call for contributions for this Special Issue, we were looking for contributions helping to understand aging mechanisms, modes and factors, to perform ESS diagnosis and prognosis and innovative solutions to prolong their lifespans.

Topics of interest include, but are not limited to:

- Innovative measurement techniques of ESS aging;
- ESS aging modeling;
- ESS state-of-health (SOH) estimation;
- ESS prognostic and health management;
- Balancing circuits with consideration of the lifetime of ESS;
- Energy management laws taking into account aging;
- Influence of aging on cost and environmental analyses of ESS;
- Multi-objective optimization strategies of ESS including aging consideration;
- Optimal sizing and design of ESS.

In response to this call of papers, 12 research papers [1–12] and one review article [13] from seven different countries have been published. Researchers from academic institutions from France [1,3–7,9], Germany [2,9,10,12,13], United States [8,11], Russia [4], Spain [9], Belgium [9] and Ethiopia [13] participated, sometimes in collaboration with industrial partners of energy or automotive sectors.

The dominant energy storage technology treated in this special issue is without a doubt lithium-ion batteries [1,2,4–10,13]. However, other technologies are of interest as for example supercapacitors [2] or NiCd batteries [12] and the emerging technology of Lithium-ion capacitors [3,11].

Experiments are a very important part of ESS aging studies and most of the papers in this Special Issue included experimental results [2,3,5–12]. Among the different experimental techniques used to measure and detect aging mechanisms taking place in ESS, special attention can be given to Impedance Spectroscopy [2,10,12] and Incremental Capacity [6–8] techniques.

Finally, all of the contributions show that the aging of electrical energy storage systems remains a major problem. It must be studied to improve the dependability of these systems.

Funding: This research received no external funding.

Acknowledgments: We would like to thank all people involved in this special issue, from the authors for their very valuable contributions, to the reviewers for their efforts in analyzing the relevance and quality of the papers that were submitted. Finally, we are grateful to the MDPI publisher for inviting us to be Guest Editors of this Special Issue and the MDPI's editorial team for their availability and their valuable collaboration. Given the success of this special edition, we have decided to open a second special session entitled "Batteries and Supercapacitors Aging II": https://www.mdpi.com/journal/batteries/special_issues/Batteries_Supercapacitors_Aging_2.

Conflicts of Interest: The authors declare no conflict of interest.

References

1. Savard, C.; Venet, P.; Niel, E.; Piétrac, L.; Sari, A. Comparison of Battery Architecture Dependability. *Batteries* **2018**, *4*, 31. [CrossRef]
2. Kurzweil, P.; Shamonin, M. State-of-Charge Monitoring by Impedance Spectroscopy during Long-Term Self-Discharge of Supercapacitors and Lithium-Ion Batteries. *Batteries* **2018**, *4*, 35. [CrossRef]
3. El Ghossein, N.; Sari, A.; Venet, P. Lifetime Prediction of Lithium-Ion Capacitors Based on Accelerated Aging Tests. *Batteries* **2019**, *5*, 28. [CrossRef]
4. Savard, C.; Iakovleva, E.V. A Suggested Improvement for Small Autonomous Energy System Reliability by Reducing Heat and Excess Charges. *Batteries* **2019**, *5*, 29. [CrossRef]
5. Quinard, H.; Redondo-Iglesias, E.; Pelissier, S.; Venet, P. Fast Electrical Characterizations of High-Energy Second Life Lithium-Ion Batteries for Embedded and Stationary Applications. *Batteries* **2019**, *5*, 33. [CrossRef]
6. Plattard, T.; Barnel, N.; Assaud, L.; Franger, S.; Duffault, J.-M. Combining a Fatigue Model and an Incremental Capacity Analysis on a Commercial NMC/Graphite Cell under Constant Current Cycling with and without Calendar Aging. *Batteries* **2019**, *5*, 36. [CrossRef]
7. Riviere, E.; Sari, A.; Venet, P.; Meniere, F.; Bultel, Y. Innovative Incremental Capacity Analysis Implementation for C/LiFePO4 Cell State-of-Health Estimation in Electrical Vehicles. *Batteries* **2019**, *5*, 37. [CrossRef]
8. Baure, G.; Dubarry, M. Synthetic vs. Real Driving Cycles: A Comparison of Electric Vehicle Battery Degradation. *Batteries* **2019**, *5*, 42. [CrossRef]
9. Moretti, A.; Carvalho, D.V.; Ehteshami, N.; Paillard, E.; Porcher, W.; Brun-Buisson, D.; Ducros, J.-B.; De Meatza, I.; Eguia-Barrio, A.; Trad, K.; et al. A Post-Mortem Study of Stacked 16 Ah Graphite//LiFePO4 Pouch Cells Cycled at 5 °C. *Batteries* **2019**, *5*, 45. [CrossRef]
10. Frankenberger, M.; Singh, M.; Dinter, A.; Pettinger, K.-H. EIS Study on the Electrode-Separator Interface Lamination. *Batteries* **2019**, *5*, 71. [CrossRef]
11. Bolufawi, O.; Shellikeri, A.; Zheng, J.P. Lithium-Ion Capacitor Safety Testing for Commercial Application. *Batteries* **2019**, *5*, 74. [CrossRef]
12. Kurzweil, P.; Scheuerpflug, W. State-of-Charge Monitoring and Battery Diagnosis of NiCd Cells Using Impedance Spectroscopy. *Batteries* **2020**, *6*, 4. [CrossRef]
13. Teichert, P.; Eshetu, G.G.; Jahnke, H.; Figgemeier, E. Degradation and Aging Routes of Ni-rich Cathode Based Li-Ion Batteries. *Batteries* **2020**, *6*, 8. [CrossRef]

Article

Comparison of Battery Architecture Dependability

Christophe Savard *, Pascal Venet, Éric Niel, Laurent Piétrac and Ali Sari

Univ Lyon, Université Claude Bernard Lyon 1, École Centrale de Lyon, INSA Lyon, CNRS, Ampère,
F-69622 Villeurbanne, France; pascal.venet@univ-lyon1.fr (P.V.); eric.niel@insa-lyon.fr (É.N.);
laurent.pietrac@insa-lyon.fr (L.P.); ali.sari@univ-lyon1.fr (A.S.)
* Correspondence: christophe.savard@insa-lyon.fr

Received: 14 May 2018; Accepted: 7 June 2018; Published: 3 July 2018

Abstract: This paper presents various solutions for organizing an accumulator battery. It examines three different architectures: series-parallel, parallel-series and C3C architecture, which spread the cell output current flux to three other cells. Alternatively, to improve a several cell system reliability, it is possible to insert more cells than necessary and soliciting them less. Classical RAMS (Reliability, Availability, Maintainability, Safety) solutions can be deployed by adding redundant cells or by tolerating some cell failures. With more cells than necessary, it is also possible to choose active cells by a selection algorithm and place the others at rest. Each variant is simulated for the three architectures in order to determine the impact on battery-operative dependability, that is to say the duration of how long the battery complies specifications. To justify that the conventional RAMS solutions are not deployed to date, this article examines the influence on operative dependability. If the conventional variants allow to extend the moment before the battery stops to be operational, using an algorithm with a suitable optimization criterion further extend the battery mission time.

Keywords: battery; operative dependability; selection algorithm

---

1. Context

An electrical energy storage systems (EESS) may be an electrochemical cell association in a battery [1]. These cells can belong to different technologies and chemistries. The oldest are lead–acid cells. Then, there are the cells using alkaline metals. Finally, for a quarter of a century, lithium cells have been marketed. Lithium-ion cells are more stable than the first lithium cells and have higher energy and power densities than the lead–acid and nickel-based technologies [2]. They also have a higher lifespan, which partly explains their strong growth. The lithium-ion battery market continues to grow, to the point that their unit manufacturing price decreases steadily because of the growing number of units produced [3]. Naturally, in a battery, all cells belong to the same technology.

This paper presents three architectures that can be used in cell batteries: two classics, using permanent series and parallel connections between cells, and one innovative, described later and allowing different connections between the cells. On these architectures, different variants are tested: common solutions (balancing between cells, using of more cells than necessary to provide the nominal power), other conventional but not deployed in best industrial solutions (failure tolerance of some cells with over-solicitation of others, addition of redundant cells to replace the first failing cells) and a new variant (redundant cells management with a control law using cells fundamental states as a choice criterion). The performances of the architectures and their variants are compared by simulation under Matlab, by resorting to a cell model that including aging phenomena.

The cell main electrical characteristic is its maximum storable capacity Q_0. It is well known that this capacity decrease with aging. When a cell is new, this capacity is optimal. It is noted in this paper as Q^*. The battery physical behavior can be modeled by a second order Thevenin model scheme [4] as seen in Figure 1. In this, a cell is represented as a series association of an open-circuit voltage (*OCV*), an equivalent series resistance (*ESR*) and a parallel RC circuit. The *ESR* corresponds to the battery terminals voltage V_{cell}, only if it is measured without an external load connection and with the cell at rest for an adequate time. This time is necessary for balancing internal electrical charges (relaxation phenomenon). A double-layer capacitance (C_w) and a charge transfer resistance (R_w) complete the model by describing relaxation. Other internal phenomena such as a hysteresis voltage can also be modeled by additional subcircuits. Hysteresis consists in a shifting on the open-circuit voltage for the same contained charge between current direction, an upward offset with a negative current I_{cell} in charging phase and a downset offset with a positive current in the discharge phase.

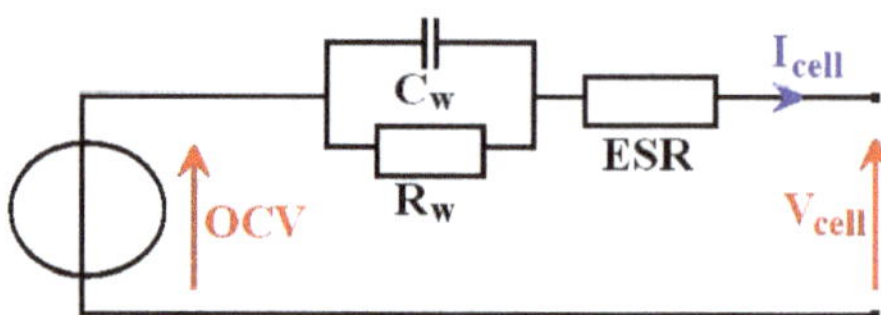

Figure 1. Cell second order Thevenin model.

To express physical state in which a cell is, two indicators are commonly used: *SoC* and *SoH*, respectively, the state of charge and the state of health [5]. The *SoC* describes, in percent, the amount of electrical charge $Q(t)$ it contains at t time, compared to its maximum storable capacity. The *SoC* is determined by Equation (1). Moreover, when a cell ages, it cannot store as much electrical charge than when it is new. The maximum storable capacity Q_0 decreases over time continuously and gradually from its optimal capacity Q^*. So, a cell's State of Health (*SoH*) is defined by the ratio between these two capacity values, as shown by Equation (2). For applications requiring significant power more than a willingness to deliver energy, *SoH* can also be defined relative to the *ESR* [6].

$$SoC(t) = \frac{Q(t)}{Q_0(t)} \tag{1}$$

$$SoH(t) = \frac{Q_0(t)}{Q^*} \tag{2}$$

The ISO 12405-2 standard for electric vehicles [7] specifies test procedures for lithium-ion batteries and electrically propelled road vehicles. It specifies that a cell enters the old age phase when its *SoH* goes down to 80%. Manufacturers communicate a lifetime value for cells that incorporate, as specified in their datasheet, the two aging modes that a cell faces: an age-related calendar mode [8] and a cyclic mode related to cell use [9]. These two phenomena [10] can be described by a combined evolution of a linear aging and an *"aging"* power of time whose principles are described by Equation (3), where the aging parameter is between 0.5 and 2 and A_1 and A_2 are two constants.

$$SoH(t) = 1 - A_1 t - A_2 t^{aging} \tag{3}$$

Cyclic aging is aggravated mainly by three parameters: the operating temperature [11], the amount of electrical charge extracted in a cycle [12] and the current [13]. Despite the cell manufacturing standardization, disparities can occur between cells from the same batch. These are reflected in an optimal capacity Q^* variability and in an equivalent series resistances *ESR* variability [14]. Thus, when cells are associated in an EESS, their disparities should lead to imbalances in currents, cell temperatures and then in their aging. Consequently, a battery may remain operational for a shorter time than a cell lifetime.

Operative dependability O_{dep} is defined as the time, expressed in hours or in number of cycles, during which a multi-cellular battery can meet the external load specifications [15]. In other words,

O_{dep} is the time before a downtime related to a full discharge or a too high age. That is to say, it contains enough operational cells to provide the requested power. For a single cell, this cessation comes from:

- a cell aging failure, resulting in a *SoH* of less than 0.8;
- a sudden random failure (open or short circuit);
- a complete discharge, *SoC* drops down to zero in operating phase.

For a battery, O_{dep} depends on the architecture (how cells are connected); if some cells are in redundancy for the current mission and if failed cells can be isolated or not. It is therefore necessary to monitor the cell states. In order to monitor cells, in particular to control end of charging and to prevent overheating, EESS cells are managed by a BMS (Battery Management System) [16]. Today, multi-cellular EESS are constituted according to two conventional architectures (series-parallel and parallel-series) and sometimes include variants, as explained in the next section, to improve their availability.

From a formal point of view, the operative dependability O_{dep} is defined by the equation set (4), according to *SoC* and *SoH* cells.

$$O_{dep} = min\{SoH(\chi) > 0.8, SoC(\chi) > 0\} \quad with \begin{cases} sufficient \ number \ of \ cells \\ provide \ power > needed \ power; \\ \chi = f(architecture, \ structure) \end{cases} \quad (4)$$

In the next section, different architectures and possible variants to improve the battery operative dependability are presented. Then, part III presents simulations performed on each variant, whatever the cell technology, in order to determine the impact on operative dependability, for LiFePO$_4$ cell example. The next part presents and compares the simulation results, especially for operative dependability.

2. Architectures and Variants

In the literature, to improve the operative dependability, various solutions for reconfiguration of the internal structure are proposed, such as the power tree solution presented in [17] or the DESA architecture (Dependable, efficient, scalable architecture) [18]. The first solution does not allow use of a battery at full power and the second only improves the operative dependability of a similar order to the amount of additional cell number (typically 50% with 50% of additional cells).

Three architectures are compared to determine which has the best operative dependability: two conventionally used and a new one. These architectures are reconfigurables by using a limited number of switches associated with a cell. To increase the battery voltage, the cells are associated in series. To increase the current, they are associated in parallel. Thus, batteries are generally associated in a SP (series-parallel) architecture, as described in Figure 2, which presents a n-rows m-columns structure, noted (n, m). Switches allow to isolate a column, following one of its cell failure, for instance. By duality, the same structure can be inserted into a PS (parallel-series) architecture, as shown in Figure 3. One switch for each cell is needed in order to isolate a failed cell. In an SP architecture, same-column cell voltage disparities can appear. In a PS architecture, the same-row cell currents can be different.

To deliver the same power, by keeping the same structure, another architecture is possible: the C3C [19], depicted in Figure 4. The architecture consists of an element association. Each C3C element comprises a cell and three switches, as shown by Figure 5. Each includes an upstream connection and three downstream. The A indexed switch is placed on the upstream connection. Switch B is on the first downstream connection, C on the third. The middle connection is intended to be connected with the same-column downstream-row cell upstream connection. The C3C architecture combines the advantages of both conventional architectures, as PS' self-balancing and allows a series association of some cells located on different columns. The current that leaves the Cell$_{i,j}$ in the ith row and the jth column can be directed to one of the three following row cells: Cell$_{i+1,j-1}$, Cell$_{i+1,j}$, Cell$_{i+1,j+1}$, in a recharge phase, respectively by the switches $S_{i,j}B$, $S_{i+1,j}A$ and $S_{i,j}C$ or to one of the three upstream row cells: $Cell_{i-1,j-1}$, $Cell_{i-1,j}$, $Cell_{i-1,j+1}$, in a discharge phase. This is made by activating the appropriate switches, respectively $S_{i-1,j-1}C$, $S_{i,j}A$ and $S_{i-1,j+1}B$. Thanks to the architecture, the *m*th column cells are connected with those of the first column.

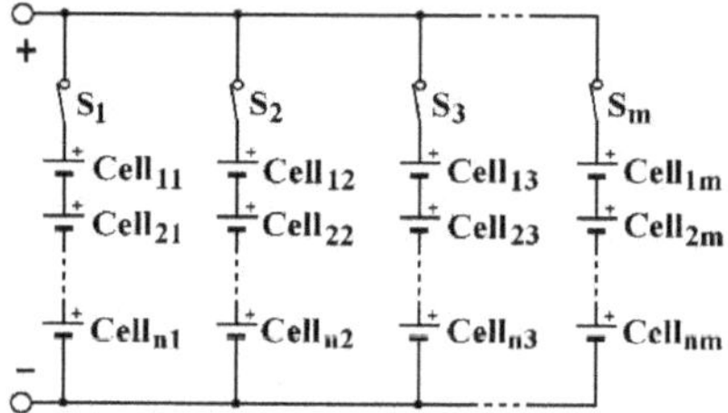

Figure 2. SP (series-parallel) architecture.

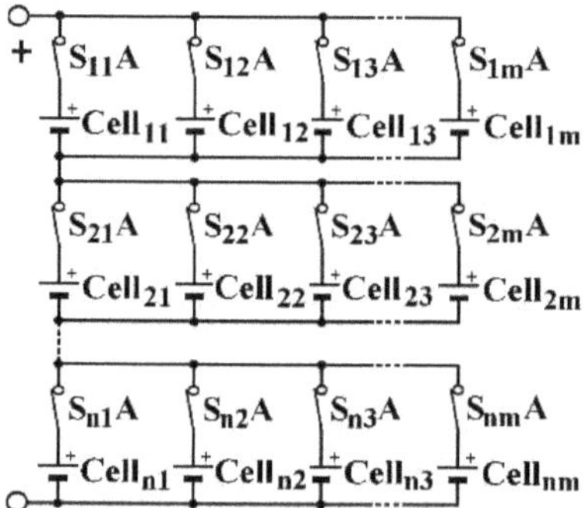

Figure 3. PS (parallel-series) architecture.

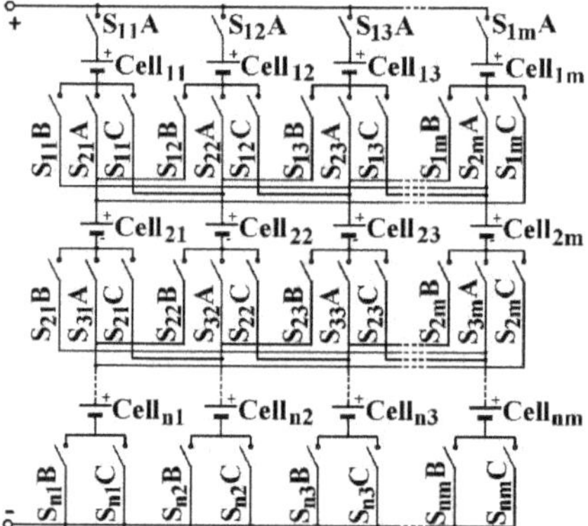

Figure 4. C3C architecture.

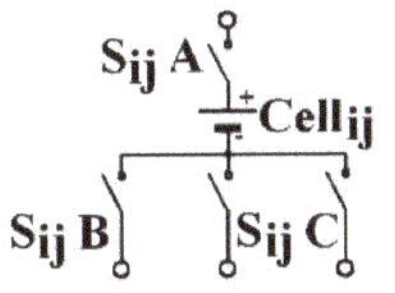

Figure 5. C3C element.

The C3C architecture involves two particularities. Firstly, to take advantage of its very large possible configuration number [20], it must be managed by an algorithm that chooses the best cell combination, with a cell-aging-reducing aim. The cells selection parameters can be chosen to define the optimal combination. Only two are considered in this paper: *SoC* and *SoH*. Secondly, whatever the architecture, to increase battery reliability, a minimum amount of redundancy must be included [21]. This minimal part consists of using a column (especially the *m*th) as a redundant column. Thus, a (*n*, *m*) battery is calibrated to provide a nominal power P_n given by Equation (5).

$$P_n = (n(m-1))V_{cell}I_{cell} \tag{5}$$

As a result, if the conventional architectures include a redundant column, they can also be managed by the same algorithm that chooses the best cell combination. The possible combination number is nevertheless much lower. Therefore, if the battery must supply its nominal current I_{nom}, depict in Equation (6), one cell in each row stays at rest. In PS and C3C architectures, this cell can be any of the *m* cells of each row. In SP, all the same-column cells are placed at rest. To do this, it requires adding a switch in series with each cell in PS and one by column in SP, as shown in Figures 2 and 3.

$$I_{nom} = (m-1)I_{cell} \tag{6}$$

If SP and PS batteries include a redundant column, it becomes possible to use this redundancy classically. That is, the battery can work with its base cells, corresponding to the first $(m-1)$ columns, until one of them stops to be operational. It is thus replaced by the spare cell. In this variant, the *m*th switches in the Figure 3 are off in the beginning for a PS architecture. For an SP architecture, in Figure 2, all cells in the *m*th column are redundant.

Another variant can also be examined, tolerating a cell failure by replacing it with a short circuit in SP architecture and allowing the battery to continue to fulfill its mission, even if it requires the active cells to afford a greater current than their nominal current. To do so, two switches should be placed around each cell: one in series and one in parallel as S$_{i,j}$A is being open and S$_{i,j}$B closed when cell$_{i,j}$ must be isolated, as depicted for cell$_{22}$, as shown by Figure 6. In this example, cell$_{22}$ fails, the current of the second column passes through S$_{22}$B. The cell is marked with a red cross to symbolize its failure. This variant does not require redundancy cells. The same power as in Equation (5) is obtained with a $(n, m-1)$ structure. In a PS architecture, the cell may just be disconnected by only one switch, as shown for cell$_{22}$ in Figure 7, in which cell$_{1m-1}$ and cell$_{22}$ are faulty and marked with a red cross.

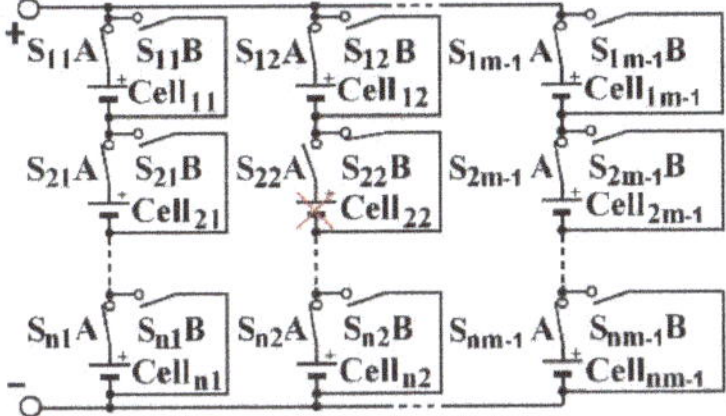

Figure 6. Fault tolerant SP architecture.

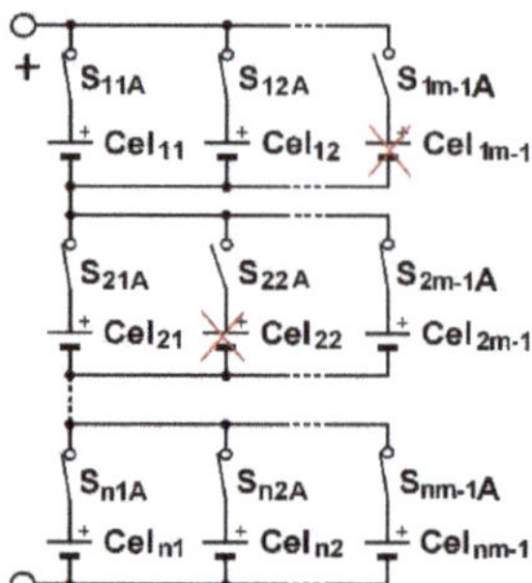

Figure 7. Fault tolerant PS architecture.

It is also possible to use a battery comprising m columns to provide a power corresponding to the nominal power of Equation (5). In this way, the battery has an over-capacity compared to the external load specifications.

Moreover, in order to reduce the disparities between the *SoCs* when cells are associated in series, in an SP architecture, balancing circuits are often used. They enable to homogenize the electrical charges in all cells in the string [22]. These balancing circuits are controlled by a BMS [23]. They allow better stored energy use [24] and a battery operative dependability improvement [25,26]. To evaluate balancing impact, basic SP of Figure 2 (with $(m-1)$ columns) with and without balancing circuits are simulated. All compared batteries must provide the same Equation (5) power. So, this structure only includes $(m-1)$ columns. Several balancing techniques exist in an SP architecture:

- dissipative balancing [27], consisting of balancing the electrical charges from below by removing excess energy by Joule effect;
- redistributive balancing to send excess energy from the most charging cell(s) to the least charging cell(s) in the same column [28]. Its principle is described by the Figure 8 scheme, relating to a single column j in an SP architecture. When $Cell_{a,j}$ is more charged than $Cell_{b,j}$, the intermediate capacitor Cb_j, associate to this column j, is placed in parallel by switching on the $S_{a,j}+$ and $S_{a,j}-$ switches. Then, these switches are switched off and the Sb_j+ and Sb_j- switches are switched on. At the end, $Cell_{aj}$ electric charge have been reduced and $Cell_{aj}$ one have been increased.

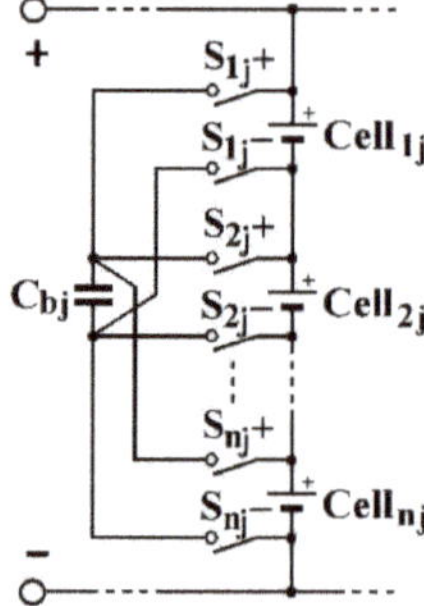

Figure 8. Balancing circuit for an SP architecture's column j.

In this study, only redistributive balancing circuits from one cell to another are considered. This leads to add two switches by cell in SP schemes, each connecting a terminal of the cell to a terminal of an intermediate capacitor, used to temporarily store the energy to be transferred. By this way, the different variants combining architecture and improvement are listed in Table 1. The structure column number m_c and the cell-associated switch number are also specified. For the basic PS and the over-capacity PS variants, the switches in Figure 3 are not useful because cells are not managed individually. The cells are all active or all inactive. In the same way, those of Figure 2 are not useful in the without-balancing SP variant. Apart from the variant reported without balancing, all other SPs include balancing circuits, which add two switches per cell. The variants deployed in present industrial solutions are PS-base, SP-base with or without balancing and over-capacity variants.

Table 1. Simulated architectures and variants.

Architecture	Electric Scheme	Fig.	Variant	Switches per Cell	Column Number (m_c)
PS		3	basic	0	$(m-1)$
			over-capacity	0	m
			With redundancy	1	m
			With *SoC*-based optimization algorithm	1	m
			With *SoH*-based optimization algorithm	1	m
PS		7	Fault tolerant	1	$(m-1)$
SP		2	Without balancing	0	$(m-1)$
SP		2 & 8	Basic, with balancing	2	$(m-1)$
			With redundancy over-capacity	2	m
			With *SoH*-based optimization algorithm	2	m
SP		6	Fault tolerant	4	$(m-1)$
C3C		5	With *SoC*-based optimization algorithm	3	m
			With *SoH*-based optimization algorithm	3	m

3. Simulations

The different variants and architectures were modeled using Matlab. The program simulates cell associations according to each architecture for different (n, m) structures. For each variant, it submits the battery to regular cycling, leading the current shown in Figure 9 for a single 10 Ah cell. The battery is initially full. This cycle consists of firstly, a discharge under a current I equal to the battery nominal current I_{nom} divided by the active column number m_c (that is to say $(m-1)$ or m) for a duration of 2500 s. So, except for over-capacity variants, $I = 10$ A in discharging phase. By this way, the battery is discharged by 70% of its initial capacity. This discharge value is relevant for quantifying cell aging. Indeed, it allows to stop the discharging phase before a complete discharge. The more $Q_0(t)$ decreases, the more the *SoC* at the end of discharging phase decreases. This is because a cell has to provide the same amount of energy. Then, the cells are recharged, for the same duration, to return to full charge. Finally, the battery is placed at rest for an identical duration, allowing the return to internal balance.

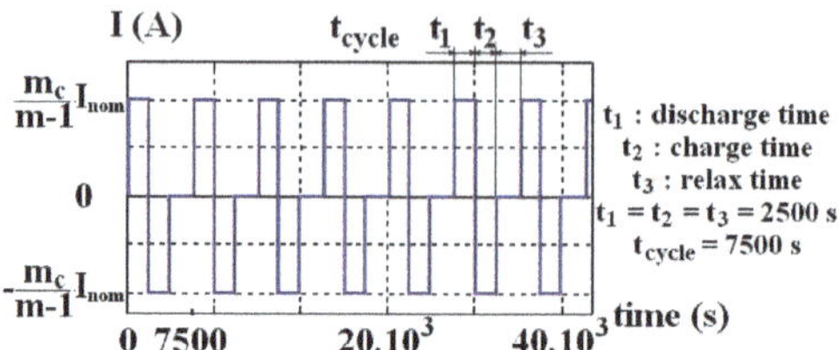

Figure 9. Current profile during a simulation cycle.

To model a cell, a characteristic equation that describes the *OCV* evolution as a function of the *SoC* is used. The typical shape of this curve is described for an amorphous iron phosphate (LiFePO$_4$) cell in Figure 10. Batteries, with an amorphous iron phosphate positive electrode support high intensities in charge and discharge as well as fast charges. These cells have higher power density and lower fire risks than cells with a lithium cobalt oxide (LiCoO$_2$) positive electrode [29]. On this curve, four points are identified by a red dot. These four coordinate points (S_0, E_0), (S_L, E_L), (S_V, E_V) and (S_M, E_M), respectively, delimit three sectors. The curve being continuous, it is so possible to perform partial regression. In the sector where the state of charge is low ($SoC < S_L$) and *OCV* low ($OCV < E_L$), the curve can be described by an exponential function. It is the same for high states of charge ($SoC > S_V$ and $OCV > E_V$). The central sector can be described by a linear function. Ultimately, the characteristic equation can be described by equation set (7) as a function of the *SoC* (explain as a dummy variable x). v and ζ parameters are empirical. v is between 10 and 20. It corresponds to the rapid growth of the voltage when the *SoC* approaches its maximum. ζ parameter is its complement when the voltage collapses, the *SoC* decreasing towards its minimum.

$$OCV(x) = \beta + \alpha x + \gamma\left(1 - e^{\frac{S_0 - x}{\zeta}}\right) + \delta e^{v(x - S_M)} \quad with \begin{cases} \alpha = \frac{E_L - E_V}{S_L - S_V} \\ \beta = \frac{S_V(E_V - E_L - E_0) + S_L E_0}{S_L - S_V} \\ \gamma = E_L - E_0 \\ \delta = E_M - E_V \end{cases} \tag{7}$$

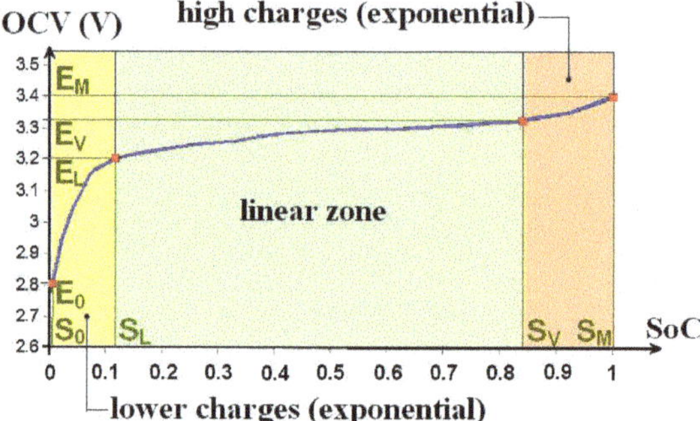

Figure 10. Example of a characteristic curve representation linking the open-circuit voltage to the state of charge for a LiFePO$_4$ cell.

In these simulations, the cells had an initial capacity $Q^* = 10$ Ah and an average *ESR* of 20 mΩ. The ambient temperature was set to 25 °C. Since the $Q_0(t)$ degradation is continuous and progressive, it is logical to consider that the *SoH*, which is a representation of this degradation, also decreases continuously. This degradation is proportional to the using conditions. At each cycle, a cell ages. This translates into an *SoH* decrease.

The cell temperature evolves as a function of the current flowing through it. By convection, radiation and conduction, this heat can spread to other cells. For simplicity, the used model does not integrate thermal coupling phenomena.

Finally, the initial conditions for each cell was given randomly around nominal a value with a variability of more or less 10%. The same $Cell_{i,j}$ is used for all variants in all architectures, so as to ensure a true comparison of the variant intrinsic performance. A cell is considered faulty in two cases. First, when it is completely discharged whereas it should still supply energy ($SoC = 0$). Second, when it is too old and has $SoH = 0.8$. This corresponds, for a single cell, to the operative dependability definition given above. Depending on the architecture, the variant and the failed cell location(s), the battery may or may not continue to provide the requested power. If, at moment, it is no longer able to provide this power, the O_{dep} worth.

Simulation was performed several times with different initial conditions. Only the mean values are reported here, even if the figures illustrating this demonstration are related to a single simulation, among all those performed.

From these different simulations, it is possible to extract operative dependability. O_{dep} corresponds to the time when the battery stops fitting to the specifications: delivery of the battery nominal current I_{nom}. According to the variant, some cells may have failed (by aging, random failure or complete discharge) before this time and be isolated or replaced by others. The simulations presented here were carried out for several structures: with $n = 2$ and m varying from 3 to 10 on the one hand, and on the other hand with $m = 3$ and n varying from 2 to 4.

4. Results and Comparisons

Results from a (3,4) structure simulation performed with a C3C architecture with *SoH*-based optimization algorithm is illustrated in Figure 11, which presents *SoC*, *SoH*, *OCV* and Q_0 evolution. In this example, only the first-row cells are shown because it is in this row that the first two cells fail. $Cell_{14}$ (cyan curve) fails first. O_{dep} is limited by $Cell_{12}$ *aging* (green curve). This simulation was performed willingly in accelerated mode, with an announced cell lifetime of only 50 cycles. In order to read *SoC* evolution on the curve, the aging parameter has been set to its maximum value ($aging = 2$) to better differentiate the *SoH* curves when the battery lifespan approaches. By using this variant, *aging* control is visible on the curve b). All cells age together, their *SoH* decreases together.

Table 2 records the average operative dependability obtained for the different structures. This average is given for cells with lifetimes of 1000 cycles.

In the same way that the reliability of a several identical all-necessary element system decreases if the element number increases, the more a battery contains cells, the lower the operative dependability is. For instance, for a basic PS architecture, with a $n = 2$ structure, O_{dep} reduces when the column number j increases: respectively, to 772, 754, 750, 748, 747 and 674 cycles when j varies from 4 to 10. The lowest operative dependability variant is the PS architecture without redundant cell (basic PS). For instance, $O_{dep} = 772$ cycles for a (2,4) structure. Its performance serves as a reference value. Operative dependability for the variants are relative compared (base 100 for basic PS) in the bar graph in Figure 12, drawn for a (3,4) structure. Thus, the performances can be grouped into three clusters: those that are between 100% and 120% of the basic PS operative dependability, those between 120% and 140%, and those higher [30]. Among the less powerful variants are the classic solutions: balancing in SP architecture, redundancy, fault tolerance and over-capacity. In the second family are the *SoC*-based optimization algorithm variants for the three architectures. Finally, those with the best operative dependability are those that use variant with *SoH* optimization algorithm, regardless of the architecture. Cluster 3 variants show a different O_{dep} improvement by architecture, as summarized in Table 3. On average, this improvement is close to 36%. Unlike the DESA architecture, whose management cannot be deployed in another architecture, with the optimization algorithm, the operative dependability improvement is greater than the share of redundant cells. For example, for a (2,4) structure, O_{dep} is improved by 50% for only 33% of redundant cells. In the same way, for a (2,10), O_{dep} is improved by almost 25% regardless of the architecture for 10% more cells.

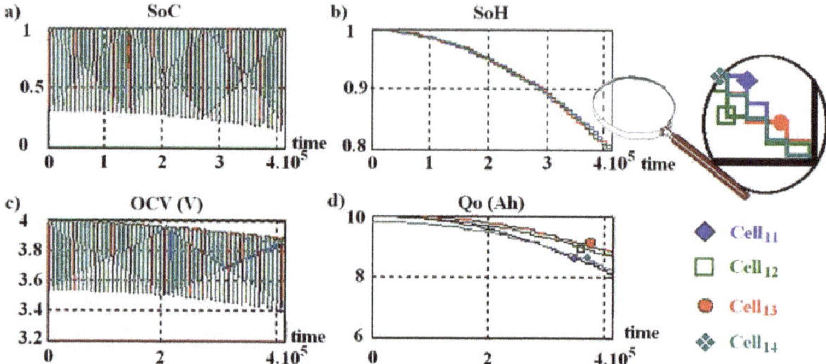

Figure 11. Example of simulation result (*SoC*, *SoH*, *OCV*, Q_0) for a C3C architecture with variant *SoH*-based optimization algorithm: (**a**) *SoC*; (**b**) *SoH*; (**c**) *OCV*, in Volts; (**d**) Q_0, in Ah.

Table 2. Simulation results: operative dependability.

Architecture	Fig	Structure	(2,4)	(2,5)	(2,6)	(2,7)	(2,8)	(2,10)	(3,4)	(4,4)	Cluster
PS	3	basic	772	754	750	748	747	674	748	746	basic
PS	3	With redundancy	868	803	800	799	796	702	818	846	1
PS	3	over-capacity	804	773	772	775	768	768	754	764	1
PS	3	Fault tolerant	796	777	768	773	768	694	760	772	2
PS	3	With *SoC*-based optimization algorithm	1052	892	855	786	874	740	946	1052	2
PS	3	With *SoH*-based optimization algorithm	1172	1043	1002	993	968	832	1104	1254	3
SP	2	Without balancing	808	796	776	765	762	683	771	799	1
SP	2 & 8	Basic with balancing	839	814	780	790	784	708	791	826	1
SP	7	Fault tolerant	744	813	812	808	788	711	786	810	1
SP	2 & 8	With *SoH*-based optimization algorithm	1192	1063	1028	1001	980	830	1080	1228	3
C3C	5	With *SoC*-based optimization algorithm	1076	891	899	884	860	755	981	1080	2
C3C	5	With *SoH*-based optimization algorithm	1172	1044	1001	984	972	843	1108	1234	3

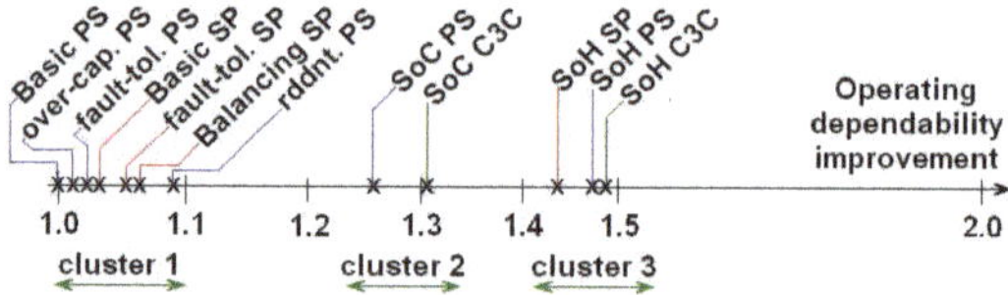

Figure 12. (3,4) structure O_{dep} improvement.

Table 3. Simulation results: operative dependability.

Architecture	(2,4)	(2,5)	(2,6)	(2,7)	(2,8)	(2,10)	(3,4)	(4,4)
PS	52%	38%	34%	33%	30%	23%	48%	68%
SP	54%	41%	37%	34%	31%	23%	44%	65%
C3C	52%	33%	33%	32%	30%	25%	48%	65%

5. Conclusions

In this paper, some variants to improve the operative dependability of EESS are described and compared. For this, a formal model integrating the aging of a cell is used. The different architectures and the possible solutions to improve the performances in duration of use are simulated under Matlab.

Whatever the architecture, classical variants as over-capacity and balancing only improve O_{dep} by up to 20%. By using a minimal portion of redundant cells and an *SoH*-based optimization algorithm, it is possible to improve, on average, the battery operative availability by more than 35%, regardless of its architecture.

When the current flowing through the battery is below nominal current, in a C3C architecture, it is possible to perform specific cell-to-cell balancing while using other cells to meet the current demand. No other architecture allows this differentiated use of cells. It is necessary to continue this work by comparing the cost of adding the redundant cells, the switches and the over-cost induced on the BMS in terms of computing capacity with the gain in additional mission time.

Author Contributions: C.S. wrote the paper; P.V., L.P., A.S. and É.N. have made corrections; All authors discussed results data and decided on next steps.

Funding: This research received no external funding.

Conflicts of Interest: The authors declare no conflicts of interest.

References

1. Pellegrino, G.; Armando, E.; Guglielmi, P. An Integral Battery Charger with Power Factor Correction for Electric Scooter. *IEEE Trans. Power Electron.* **2010**, *25*, 751–759. [CrossRef]
2. Sinkaram, C.; Rajakumar, K.; Asirvadam, V. Modeling Battery Management System Using the Lithium-Ion Battery. In Proceedings of the 2012 IEEE International Conference on Control System, Computing and Engineering, Penang, Malaysia, 23–25 November 2012; pp. 50–55.
3. Lahyani, A.; Sari, A.; Lahbib, I.; Venet, P. Optimal hybrid and amortized cost study of battery/supercapacitor system under pulsed loads. *J. Energy Storage* **2016**, *6*, 222–231. [CrossRef]
4. He, H.; Xiong, R.; Zhang, X.; Sun, F.; Fan, J. State-of-Charge Estimation of the Lithium-Ion Battery Using an Adaptive Extended Kalman Filter Based on an Improved Thevenin model. *IEEE Trans. Veh. Technol.* **2011**, *60*, 1461–1469.
5. Le, D.; Tang, X. Lithium-ion Battery State of Health Estimation Using Ah-V Characterization. In Proceedings of the 2011 Annual Conference of the Prognostics and Health Management Society, Montreal, QC, Canada, 25–29 September 2011.
6. Lievre, A.; Sari, A.; Venet, P.; Hijazi, A.; Ouattara-Brigaudet, M.; Pelissier, S. Practical Online Estimation of Lithium-Ion Battery Apparent Series Resistance for Mild Hybrid Vehicles. *IEEE Trans. Veh. Technol.* **2016**, *65*, 4505–4511. [CrossRef]
7. *ISO 12405: Electrically Propelled Road Vehicles, Test Specification for Lithium-Ion Traction Battery Packs and Systems, Part 2: High-Energy Applications*; ICS: 43.120 Electric road vehicles Std. International Organization for Standardization: Geneva, Switzerland, 2012.
8. Xu, B.; Ondalov, A.; Ulbig, A.; Andersson, G.; Kirschen, D.S. Modeling of Lithium-Ion Battery Degradation for Cell Life Assessment. *IEEE Trans. Smart Grid* **2018**, *9*, 1131–1140. [CrossRef]
9. Broussely, M.; Biensan, P.; Bonhomme, F.; Blanchard, P.; Herreyre, S.; Nechev, K.; Staniewicz, R.J. Main aging mechanisms in Li ion batteries. *J. Power Sources* **2005**, *146*, 90–96. [CrossRef]
10. Zhang, Y.; Xiong, R.; He, H.; Pecht, M. Lithium-ion battery remaining useful life prediction with Box-Cox transformation and Monte Carlo simulation. *IEEE Trans. Ind. Electron.* **2018**, *99*. [CrossRef]
11. Chen, Z.L.; Lee, J. Review and recent advances in battery health monitoring and prognostics technologies for electric vehicle (EV) safety and mobility. *J. Power Sources* **2014**, *256*, 110–124. [CrossRef]
12. Urbain, M. Modelisation Electrique et Energetique des Accumulateurs Lithium-ion, Estimation en Ligne du SoH. Ph.D. Thesis, Institut National Polytechnique de Lorraine, Vandœuvre-lès-Nancy, France, June 2009.
13. Ning, G.; Haran, B.; Popov, B.N. Capacity fade study of lithium-ion batteries at high discharge rates. *J. Power Sources* **2003**, *117*, 160–169. [CrossRef]
14. Wen, S. *Fast Cell Balancing Using External MOSFET*; Application Report SLUA420A; Texas Instruments: Dallas, TX, USA, May 2007; Revisited January 2009.
15. Savard, C. Amelioration de la Disponibilite Operationnelle des Systemes de Stockage de L'energie Electrique Multicellulaires. Ph.D. Thesis, University Lyon, INSA Lyon, Villeurbanne, France, November 2017.

16. Yang, B.G.; Zhang, B.G.; Feng, Z.K.; Han, G.S. An algorithm on branches number of a tree based on extended fractal square root law. *Int. Arch. Photogramm. Remote Sens. Spat. Inf. Sci.* **2008**, *37*, 551–554.

17. Jin, F.; Shin, K. Pack sizing and reconfiguration for management of large-scale batteries. In Proceedings of the 2012 IEEE/ACM Third International Conference Cyber-Physical Systems (ICCPS), Beijing, China, 17–19 April 2012; pp. 138–147.

18. Kim, H.; Shin, K.G. Dependable, efficient, scalable architecture for management of large-scale batteries. *IEEE Trans. Ind. Inform.* **2012**, *8*, 406–417. [CrossRef]

19. Savard, C.; Sari, A.; Venet, P.; Niel, E.; Pietrac, L. C3C: A structure for high reliability with minimum redundancy for batteries. In Proceedings of the 17th IEEE International Congres of Industrial Technology, Taipei, Taiwan, 14–17 March 2016; pp. 281–286.

20. Savard, C.; Niel, E.; Venet, P.; Pietrac, L.; Sari, A. Modelisation par un graphe de flots d'une architecture alternative pour les systemes de stockage multi-cellulaire de l'energie electrique. In Proceedings of the SEEDS Days—JCGE 2017, Arras, France, 30 May–1 June 2017.

21. Savard, C.; Niel, E.; Pietrac, L.; Venet, P.; Sari, A. Amelioration de la fiabilite des structures matricielles de batteries. In Proceedings of the 20eme Congres de Maitrise des Risques et de Surete de Fontionnement, St-Malo, France, 11–13 October 2016.

22. Lu, L.; Han, X.; Li, J.; Hu, J.; Ouyang, M. A review on the key issues for lithium-ion battery management in electric vehicles. *J. Power Sources* **2013**, *226*, 272–288. [CrossRef]

23. Lorentz, V.R.H.; Wenger, M.M.; Grosch, J.L.; Giegerich, M.; Jank, M.P.M.; März, M.; Frey, L. Novel cost-efficient contactless distributed monitoring concept for smart battery cells. In Proceedings of the Industrial Electronics (ISIE) 2012 IEEE International Symposium, Hangzhou, China, 28–31 May 2012; pp. 1342–1347.

24. Einhorn, M.; Roessler, W.; Fleig, J. Improved Performance of Serially Connected Li-Ion Batteries With Active Cell Balancing in Electric Vehicles. *IEEE Trans. Veh. Technol.* **2011**, *60*, 2248–2457. [CrossRef]

25. Wei, C.L.; Huang, M.F.; Sun, Y.; Cheng, A.; Yen, C.W. A method for fully utilizing the residual energy in used batteries. In Proceedings of the 2016 IEEE International Conference on Industrial Technology (ICIT), Taipei, Taiwan, 14–17 March 2016; pp. 230–233.

26. Dost, P.; Sourkounis, C. Sensor minimal cell monitoring with integrated direct active cell balancing. In Proceedings of the 2017 Twelfth International Conference on Ecological Vehicles and Renewable Energies (EVER), Monte Carlo, Monaco, 11–13 April 2017.

27. Shili, S.; Hijazi, A.; Sari, A.; Lin-Shi, X.; Venet, P. Balancing circuit new control for supercapacitor storage system lifetime maximization. *IEEE Trans. Power Electron.* **2017**, *32*, 4939–4948. [CrossRef]

28. Shili, S.; Hijazi, A.; Sari, A.; Bevilacqua, P.; Venet, P. Online supercapacitor health monitoring using a balancing circuit. *J. Energy Storage* **2016**, *7*, 159–166. [CrossRef]

29. Minakshi, M. Lithium intercalation into amorphous $FePO_4$ cathode in aqueous solutions. *Electrochim. Acata* **2010**, *55*, 9174–9178. [CrossRef]

30. Savard, C.; Venet, P.; Pietrac, L.; Niel, E.; Sari, A. Increase lifespan with a cell management algorithm in electric energy storage systems. In Proceedings of the 2018 IEEE International Conference on Industrial Technology (ICIT), Lyon, France, 20–22 February 2018; pp. 1748–1753.

Article

State-of-Charge Monitoring by Impedance Spectroscopy during Long-Term Self-Discharge of Supercapacitors and Lithium-Ion Batteries

Peter Kurzweil [1,*] and Mikhail Shamonin [2]

[1] Electrochemistry Laboratory, University of Applied Sciences (OTH), Kaiser-Wilhelm-Ring 23, 92224 Amberg, Germany

[2] Faculty of Electrical Engineering and Information Technology, University of Applied Sciences (OTH), 93053 Regensburg, Germany; mikhail.chamonine@oth-regensburg.de

* Correspondence: p.kurzweil@oth-aw.de; Tel.: +49-9621-482-3317

Received: 11 June 2018; Accepted: 10 July 2018; Published: 1 August 2018

Abstract: Frequency-dependent capacitance $C(\omega)$ is a rapid and reliable method for the determination of the state-of-charge (SoC) of electrochemical storage devices. The state-of-the-art of SoC monitoring using impedance spectroscopy is reviewed, and complemented by original 1.5-year long-term electrical impedance measurements of several commercially available supercapacitors. It is found that the kinetics of the self-discharge of supercapacitors comprises at least two characteristic time constants in the range of days and months. The curvature of the Nyquist curve at frequencies above 10 Hz (charge transfer resistance) depends on the available electric charge as well, but it is of little use for applications. Lithium-ion batteries demonstrate a linear correlation between voltage and capacitance as long as overcharge and deep discharge are avoided.

Keywords: capacitance; state-of-charge monitoring; self-discharge; supercapacitor; aging

---

1. Introduction

Batteries and supercapacitors [1–3] have conquered compact electronic systems for portable applications. As reversible short-time power sources, supercapacitors are used for light emitting diode torches (LED), computer memory backup, actuators and fire protection drive units. Current lithium-ion batteries already reach specific energies above 140 Wh/kg. Integrated electronic systems require life-cycle monitoring of these energy storage devices.

The definition of the state-of-charge (SoC) is based either on the momentary value of open-circuit voltage $U(t)$ or the available electric charge Q; for capacitors, the capacitance value C is employed as well.

$$\text{SoC}(t) = \frac{Q(t)}{Q_{max}} = \frac{\int I(t)\mathrm{d}t}{C\,U_{max}} \approx \frac{U(t)}{U_{max}} \tag{1}$$

Alternatively, SoC measurements [4] use

1. The relative voltage until 'zero' or the end-point voltage U_e is reached:

$$\text{SoC} = \frac{U(t)}{U_0} \quad \text{or} \quad \frac{U(t) - U_e}{U_0 - U_e} \tag{2}$$

2. Ampere-hour counting of the consumed electric charge Q due to the load current $I(t)$, starting from the known initial charge or nominal capacity Q_0, whereby a coulombic efficiency $\eta \approx 1$ is estimated:

$$Q(t) = Q_0 - \int \eta\, I(t)\, \mathrm{d}t \tag{3}$$

3.   The change of resistance or impedance, which does not yield absolute SoC values.
4.   Model-based estimation methods using electrochemical theories, equivalent-circuits, Kalman filters, machine learning applications and so on.

Unfortunately, electric charge and voltage correlate in a different way depending on the cell chemistry. SoC values based on voltage or capacity are hardly comparable. The widely used reference tables connect the momentary open-circuit voltage and actual SoC measured by different methods. External charge/discharge currents, temperature changes, aging, voltage drift, and hysteresis disturb SoC measurements. Therefore, a simple measuring method is required for the determination of the actual residual charge of the storage device.

This work addresses the design of a simple SoC control technique for electrochemical storage devices such as supercapacitors and batteries. As a first step towards a more general application, we investigated time-dependent self-discharge of capacitive interfaces because this process excludes faradaic reactions related to external charging and discharging.

2. State-of-the-Art

The more recent literature reflects the growing interest in SoC estimation of energy storage devices. Research gaps exist with respect to SoC prediction, which is essential to safe application, optimizing energy management, extending the life cycle, and reducing the cost of battery systems [5–7]. Near-future electric vehicles require some comprehensive model-based battery health and lifetime monitoring applications including predictive functions [8–10]. As a matter of fact, batteries suffer from nonlinear characteristics and are further influenced by operational conditions, driving loads, and further random factors of the application. Appropriate models for accurate SoC estimation of lithium-ion batteries are still being researched. No general approach for the best equivalent circuit model has yet been found even for the same cell chemistry [11] The first- and second-order RC models seem to be sufficiently accurate and reliable, whereas the higher-order RC model provide better robustness with variation in model parameters (R resistance, C capacitance). Considerable research has been focusing on universal modeling tools for rechargeable batteries [12,13] as well as the real-time evaluation of impedance data along the charge-discharge characteristics [14]. This evaluation is generally difficult in the area of diffusion impedance due to the relatively long measurement duration at low frequencies.

This work mainly considers commercial supercapacitors and hybrid capacitors and does not focus on the many interactions between the electrode structure and charge capacity [15]. Therefore, we refer to promising new materials such a carbon dots and similar nanostructures [16,17], as well as composites using conducting polymers [18] which might give future insights into the molecular dimensions of self-discharge.

2.1. Self-Discharge of Supercapacitors

According to the literature, little is known about the mechanisms that cause the gradual loss of voltage in charged double-layer capacitors stored at open-circuit for several months. The origin of self-discharge is believed to be caused by different processes in different time scales (Conway [19]).

1.   The rapid redox reactions at the phase boundary between electrode and electrolyte (charge transfer reaction), especially with overcharging and electrolyte decomposition, change in time according to a logarithmic relation: $U \sim \ln t$ (voltage U, time t).
2.   The diffusion limited voltage relaxation in the porous electrodes in the course of several hours, especially in the presence of traces of water, seems to obey a square root law: $U \sim t^{1/2}$. The diffusion current density by excess ions i from the electrolyte, which cause the self-discharge, follows Fick's 2nd law ($dc/dt = D \, d^2c/dx^2$; molar concentration c, layer thickness $x \approx 60$ μm, diffusion coefficient $D \approx 4 \times 10^{-13}$·m^2·s^{-1}) at given temperature and starting voltage $U_0 > 1$ V [20]:

$$i_d = \frac{z_i e c_i}{2} \sqrt{\frac{D}{\pi t}} \cdot e^{-\frac{x^2}{Dt}} = -C_{12} \frac{dU}{dt} \tag{4}$$

$$U \approx U_0 - \sqrt{\frac{c_i^2 D}{C_{12}^2 \pi}} \cdot \sqrt{t} = U_0 - m(U_0) \cdot \sqrt{t} \tag{5}$$

3. The genuine self-discharge in the electric field between the microporous electrodes within some days follows an exponential law:

$$U = a_1 e^{-\frac{t}{\tau_1}} + a_2 e^{-\frac{t}{\tau_2}} \overset{t \to \infty}{\longrightarrow} a_2 e^{-\frac{t}{\tau_2}} \tag{6}$$

The leakage current by self-discharge, in the range of microamperes depending on the ambient temperature, can roughly be estimated with the assistance of the rated capacitance C of the supercapacitor, and the gradual drop of the momentary open-circuit voltage in time [21]:

$$I_L = -C(dU/dt). \tag{7}$$

The mechanisms of self-discharge are speculative. The ions, which adsorb at polar groups in the carbon material (C–O, C=O, COOH), seem to lose their mobility so that consequently the ionic conductivity of the electrode/electrolyte interface drops [22]. A concentration-driven reorganization of the multilayer adsorbed ions into the deeper pores of the electrodes might be possible. The thicker the blocking layer for the charge-transfer reaction, the greater the leakage resistance. Tevi's model [23] treats self-discharge with the assistance of the Butler-Volmer equation at given rate constant $k(E,d) = -(dE/dt)/E$, electrode potential E, and barrier layer thickness d. The empirical coefficients α, β, and m are found by a comparison of electrodes both with and without additional polymer layer.

$$\frac{dE}{dt} + m \cdot e^{\frac{\alpha FE}{RT}} \cdot e^{-\beta d} \cdot E = 0 \tag{8}$$

4. The long-term trend of self-discharge after months and years is largely unknown [24]. It was reported that the leakage resistance R_L continuously increases. However, it slows in time, until a maximum is reached [25]. The $R_L(t)$ curve was described to show three slopes: after 24 h, after some weeks, and after some months; then it drops again.

The voltage-time curve of a supercapacitor, starting from different states-of-charge, is not simply an exponential function; rather it follows a fifth-order polynomial [26]. There is no clear equivalent circuit that completely displays the self-discharge processes. The $C\|R_L$ parallel network with the leakage resistance R_L appears to be insufficient. In a further alternative, ladder networks comprising three and more R–C elements might be fruitful.

2.2. Frequency Response of Capacitance

We propose to utilize capacitance as a measure of the SoC. The definition of capacitance, $C = Q/U$, suggests a practical correlation between electric charge Q and voltage window ΔU. The capacitive voltage drop in time is useful for constant-current discharge measurements. In addition, a given scan rate v allows capacitance determination by cyclic voltammetry, thus, evaluating both charge and discharge currents.

$$C = \frac{dQ}{dU} = \frac{I}{dU/dt} = \frac{I}{v} \tag{9}$$

For capacitance determination using impedance spectroscopy, we employ the frequency-dependent quantity $C(\omega)$ [27].

$$C(\omega) = \mathrm{Re}\,\underline{C} = \frac{\mathrm{Im}\,\underline{Y}}{\omega} = = \frac{-\mathrm{Im}\,\underline{Z}}{\omega\,|\underline{Z}|^2} = \frac{-\mathrm{Im}\,\underline{Z}}{\omega\left[(\mathrm{Re}\,\underline{Z})^2 + (\mathrm{Im}\,\underline{Z})^2\right]} \tag{10}$$

Equation (10) might be interpreted as the parallel combination of a frequency-dependent capacitor and a resistor, although no specific model is required, and can, therefore, be used for in-depth evaluation of complex impedance without any previous knowledge on reaction kinetics [28]. A perfect capacitor does not show any self-discharge or leakage current. A real capacitor, however, tends to discharge during several weeks when stored at open terminals. Therefore, the equivalent circuit provides a leakage resistance in parallel to the capacitance (Figure 1). Under the assumption that this leakage resistance is infinitely large, Equation (10) becomes Equation (11).

$$C(\omega) \approx -\frac{1}{\omega \cdot \mathrm{Im}\,\underline{Z}} \quad \text{at}\ \ \omega \to \infty \tag{11}$$

At frequencies above 1 kHz, the phase boundary capacitance between electrode and electrolyte turns towards the geometric double-layer capacitance of the active surface. At low frequencies, ions from the electrolyte penetrate the porous electrode material, and battery-like faradaic reactions play a role. The resulting pseudo capacitance, $C(\omega,T,U)$, depends on frequency, temperature, and voltage, and cannot be modeled by intuitive and simply plausible equivalent-circuits [29].

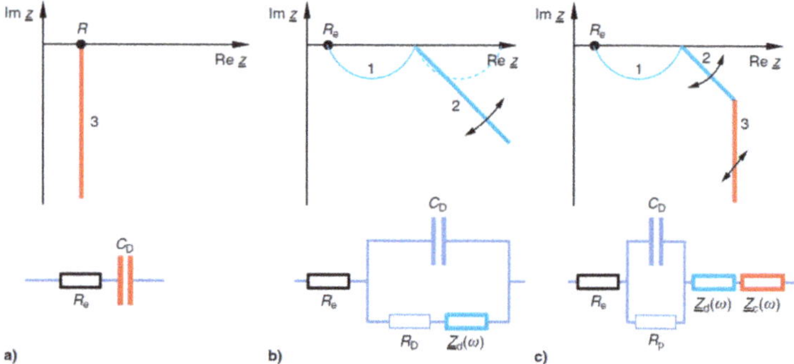

Figure 1. General equivalent circuits and impedance spectra of electrochemical storage devices (mathematical convention): (**a**) charge-transfer, (**b**) diffusion in mesopores and micropores (grain boundaries), (**c**) diffusion impedance in macro pores at a blocking interface.

2.3. State-of-Charge of Batteries

At present, there is a need for a reliable impedance method for the SoC determination for automotive applications. Earlier approaches [30,31] concentrated on the cell resistance. The internal resistance R of a lithium-ion battery was reported to reach an unclear minimum around SoC $\approx$ 50%. The curvature of the U-shaped R(SoC) characteristics [32] was found to grow with the age of the battery and corresponds to the gradually increasing slope of the line of R/SoC against SoC, which might be used as a measure of the battery state-of-health (SoH).

The charge transfer resistance (first arc in the impedance spectrum) drops with increasing C-rate (discharge current), according to Ohm's law: $R \sim I^{-1}$. Depending on the cell chemistry, the resistance at frequencies above 100 Hz is usually reduced with recharging the battery. Unfortunately, the relative change of cell resistance strongly depends on temperature [33]. The passivation and charge-transfer resistance of cobalt-manganese and nickel-cobalt-aluminum batteries are significantly determined by the graphite anode [34]. Lithium iron phosphate (LFP) provides a small internal resistance although the diffusion resistance in the solid phase is large.

The diffusion resistance (second arc in the impedance spectrum) gets slightly smaller at high C-rates. With repeated charging the battery, the slope of diffusion impedance (third section in the complex plane) appears less steep.

Deep discharge dramatically alters both the charge transfer and the diffusion impedance. The growing imaginary parts in the diffusion arc at low frequencies (0.01 Hz) indicate that the battery has been damaged by deep discharge, for example, SoC < 27% of a Ni-Mn-Co oxide/graphite system [31]. The minimum frequency f_2 of the complex plane plot, and, less significantly, the transition frequency f_{23} near the real axis between charge transfer and diffusion arc, trend towards small values (SoC < 40%). Cell resistance seems to reflect the SoC in an unreliable way. Hitherto, the value of capacitance has not been considered in detail with respect to SoC measurements in batteries.

3. Results and Discussion

3.1. Self-Discharge of Supercapacitors

The goal of this study was to find a correlation between the capacitance and the SoC. We tried to avoid the impact of electrochemical aging by not applying any current to the supercapacitor, i.e., we observed the self-discharge at open terminals. Our previous experience has shown that fully charged supercapacitors can be stored for several years without showing significant changes of the impedance spectra. Aging does require either thermal stress or overvoltage [29].

We charged carbon-based supercapacitors at rated voltage, U_0, for at least 1 h until the leakage current dropped below 2 mA. The devices were stored with open terminals at room temperature and 80 °C for 600 days. The cell voltage was measured every day for some seconds using a high-impedance voltmeter. Impedance spectra were recorded in the frequency range between 10 kHz to 0.1 Hz using a sinusoidal voltage (amplitude: 50 mV) superimposed on the momentary open-circuit voltage. Several tens of repeated impedance measurements of fully charged capacitors did not show any significant impact on the SoC, because of the bias control during the measurement.

The influence of the impedance measurement on the discharge of the capacitor can be neglected because the duration of measurement (approximately 4 min) is much less than the characteristic time constants as determined in Table 1. The total connection time of the impedance spectrometer is also less than these time constants and the uncertainty arising from the fitting procedure. Compared with the nearly 600 days of the self-discharge experiment, the total duration of all impedance measurements (that were recorded once a week or less) is negligible. The tiny AC currents during the impedance measurements may be neglected as well, because, they periodically charge and discharge the capacitor for a couple of minutes without changing the current SoC.

3.1.1. Impedance Spectra

In the long-term trend, the complex plane plots in Figure 2 show qualitatively the same shape. With increasing self-discharge, the Nyquist plot gets narrower, $R(t)$ is proportional to $U(t)$. The cell resistance, corrected by the electrolyte resistance, $R - R_e$, drops rather than increases in time. Arbitrary shifts on the real axis in the milliohm range are caused by the contact leads. The electrolyte resistance, $R_e = \mathrm{Re}\,\underline{Z}(\omega \to \infty)$, i.e., the zero passage in the complex plane at about 200 Hz, does not change on the average during self-discharge within the measuring accuracy. In addition, the nearly constant equivalent series resistance (ESR) at 1 kHz and 1 Hz have little meaning for SoC determination, even if the cable resistance $R(\omega \to \infty)$ is properly subtracted.

The resistance at low frequencies, $R(\leq 1 \text{ Hz})$, does not significantly change with self-discharge. The quarter circle at high frequencies is caused by the grain-boundary resistance of the active carbon particles in the electrolyte solution, which is superimposed by the charge transfer reaction. The more or less linear section at medium frequencies reflects the diffusion of charge carriers between the wet carbon particles. With aged capacitors especially, the resistance at high frequencies is determined by the adhesive that holds together the powder composite on the aluminum current collector.

Capacitance, in contrast to resistance, shows a clear logarithmic correlation with storage time. In general, capacitance at medium frequencies is determined by the charge transfer of electrons at the electrode-electrolyte interface.

$$C(t) = C_0 - A \ln t \tag{12}$$

$$C(\alpha) = C_0 + B \cdot \alpha \tag{13}$$

where $\alpha = U/U_0$ denotes the SoC. At low frequencies, the frozen mass transport of charge carriers in the porous electrodes leads to a more or less constant capacitance. According to Equation (11), the value $C = [\omega\,\mathrm{Im}\,\underline{Z}]^{-1}$, at any given frequency, changes roughly linearly with the state-of-charge (SoC, α), as shown in Figure 2.

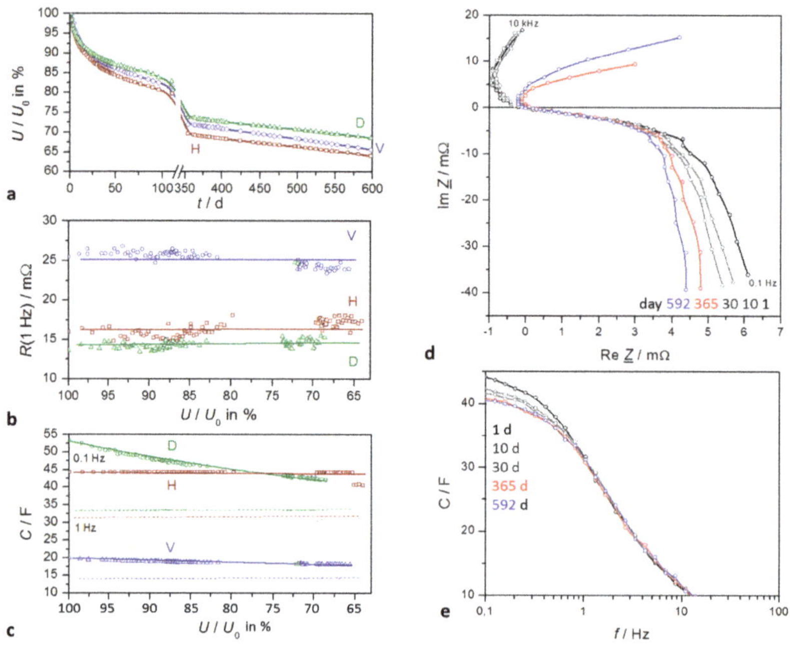

Figure 2. Self-discharge of supercapacitors for a time period of about 600 days: $\underline{H}$ VINA Hy-Cap, 2.7 V, 50 F; $\underline{D}$ VINA Hy-Cap, 3 V, 50 V; $\underline{V}$ Vitzrocell 2.7 F, 25 F. (**a**) State-of-charge (SoC) versus time (in days) emphaszising the early exponential decay and the late linear behavior; (**b**) Internal resistance at 1 Hz as measured, without correction of cable resistances; (**c**) Capacitance at 0.1 Hz (solid) and 1 Hz (dashed); (**d**) Impedance spectra in the complex plane, corrected by the contact and electrolyte resistance R_e; mathematical convention; (**e**) frequency response of capacitance during self-discharge.

3.1.2. Self-Discharge Characteristics

In the course of self-discharge, the open-circuit voltage drops exponentially in time according to $-\log U \sim t$, as shown in Figure 3. One might expect a first-order kinetics, whereby $a_0 \approx 0$ is the limiting value of the 'empty' capacitor after several years of self-discharge.

$$\frac{U(t)}{U_0} = a_0 + a_1\,e^{-\frac{t}{T_1}} \tag{14}$$

The time constant T_1 lies in the range of two weeks for the first 100 days of self-discharge. However, after two months a second time constant T_2 becomes obvious.

$$\frac{U(t)}{U_0} = a_0 + a_1\,e^{-\frac{t}{T_1}} + a_2\,e^{-\frac{t}{T_2}} \tag{15}$$

For the first 400 days, fit quality reaches more than 99.9%. However, different supercapacitors of the same type show some divergence of the parameters, so that no universal values valid for all devices can be given. Table 1 compiles the self-discharge after nearly 600 days.

With respect to the approximation that $e^{-t} \approx 1 + t^{-1}$, we can explain the reciprocal relationship between cell voltage and time, $U \sim t^{-1}$, which is obvious by the asymptote in the first hours of self-discharge, and as well after some weeks ($t \gg 20$ d) in Figure 3c.

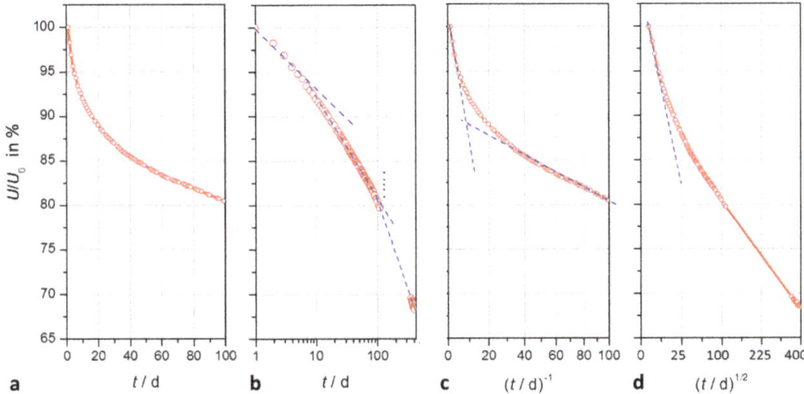

Figure 3. Modeling of the self-discharge of a supercapacitor (2.7 V, 50 F) within 100 days and 400 days, respectively: (**a**) quasi-exponential decay of open-circuit voltage versus time on a linear scale; (**b**) logarithmic scale showing three time domains; (**c**) voltage versus the reciprocal of time with asymptotes for short and long times; (**d**) voltage versus the square root of time; a linear relation holds for diffusion processes.

Table 1. Fitting of the time-dependent self-discharge U/U_0 of different supercapacitors by Prony series after approximately 600 days: Coefficients a_i are dimensionless (V/V), time constants T_i in days (1 d = 24 h). Important time constants are printed in bold.

Capa-Citor	Number of Time Constants	a_0	$+a_1 e^{-\frac{t}{T_1}}$		$+a_2 e^{-\frac{t}{T_2}}$		$+a_3 e^{-\frac{t}{T_3}}$		$+a_4 e^{-\frac{t}{T_4}}$		Residuum
Hy-Cap 50 F, 2.7 V	1	0.66	—	—		—	0.28	129		—	0.019
	2	0.61	—	—	0.11	12	0.27	302		—	0.00069
	3	0.59	0.064	**3.2**	0.085	**20**	0.28	**356**		—	0.000091
	4	0.55	0.099	2.9	0.085	19	0.062	299	0.22	1195	0.000083
	1	0	—			—		—	0.89	1591	0.084
	2	0	—		0.13	22		—	0.85	1965	0.0048
	3	0	0.14	**8.2**	0.14	59		—	0.76	3510	0.00028
	4	0	0.14	8.2	0.10	54	0.02	(2332)	0.78	3469	0.00028
Hy-Cap 50 F, 3 V	1	0.70	—	—	—	—	—	—	0.24	146	0.01292
	2	0.65	—	—	—	—	0.09	11	0.25	315	0.00064
	3	0.63	0.062	**3.1**	—	—	0.061	**22**	0.26	**389**	0.0000840
	4	0.63	0.060	3.0	(0.002)	(5.14)	0.063	22	0.26	386	0.0000838
	1	0	—			—	—	—	0.91	1880	0.054
	2	0	0.11	23			—	—	0.87	2278	0.0039
	3	0	0.12	6.8	0.09	77	—	—	0.80	3798	0.00019
	4	0	0.08	6.8	0.12	118	<0.01	(1666)	0.82	3724	0.00019
Vitzrocell 25 F, 2.7 V	1	0.66	—	—	—	—	0.28	129	—	—	0.019
	2	0.61	0.11	12	—	—	0.27	300	—	—	0.00069
	3	0.46	0.10	**8.1**	0.11	107	—		0.33	**1030**	0.00028
	4	0.33	0.10	7.7	0.08	77	0.25	508	0.25	1077	0.00027
	1	0	—			—		—	0.90	1690	0.00252
	2	0	0.12	22		—		—	0.87	2064	0.00252
	3	0	0.087	8.0	0.10	84		—	0.82	2717	0.00016
	4	0	0.087	8.0	0.10	84	<0.001	(1400)	0.82	2705	0.00016

Nonlinear least squares fitting does not generally exclude a square root law, $U \sim t^{1/2}$, which explains diffusion processes. However, fit quality is worse than with exponential functions.

In the transition time between 10 and 100 days, cell voltage follows $U \sim -\log t$, which suggests a self-discharge reaction determined by faradaic charge transfer.

$$\frac{U}{U_0} = 1 - B \ln t \tag{16}$$

The slope roughly equals $B \approx 0.04$ for supercapacitors of different manufacturers.

We conclude that the self-discharge of supercapacitors is described best by an exponential function comprising two time constants in the range of several days and weeks, respectively. A third time constant affects self-discharge not earlier than after about ten months. A fourth time constant in the range of five years is not important (Table 1).

3.1.3. Temperature Dependence

At elevated temperatures, the pseudo-exponential self-discharge characteristics (Figure 4) drop markedly steeper than in the cold. For example: 2.7 V $\rightarrow$ 2.36 V (13% change) at 20 °C, but 2.7 V $\rightarrow$ 1.0 V (63%) at 80 °C within four days.

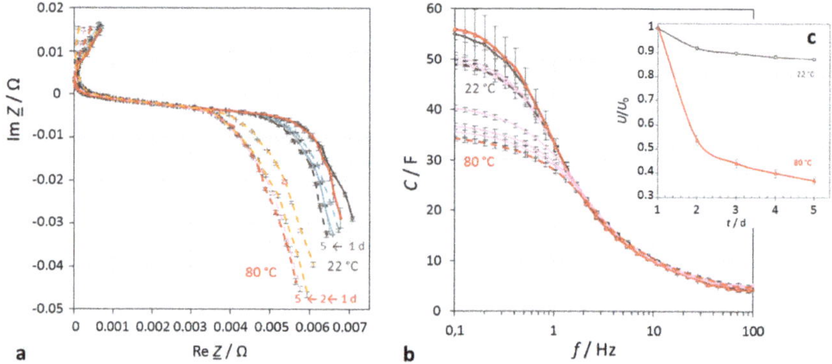

Figure 4. Impact of temperature on the self-discharge of a supercapacitor (Vina Hy-Cap 3 V, 50 F) in the course of five days. (**a**) Impedance spectra in the complex plane at 22 °C and 80 °C ambient temperature, mathematical convention; (**b**) frequency response of capacitance; (**c**) self-discharge.

Self-discharge depends moderately on the initial voltage (SoC). With fully charged supercapacitors, the voltage drops more strongly than at incomplete charge. For example, at 50 °C: 2.4 V $\rightarrow$ 2.25 V (6% change), in contrast to 1.8 V $\rightarrow$ 1.75 V (3%).

In addition, our earlier investigations [35] indicated that self-discharge increases with rising temperature, following first-order kinetics according to Arrhenius's law.

$$k = k_0 e^{-\frac{E_A}{RT}} \tag{17}$$

The activation energy in activated carbon supercapacitors amounts to about $E_A = 50$ Kj·mol^{-1}, which is typical for sorption processes in heterogeneous catalysis. Hence, slow electrochemical reactions drive a small leakage current across the double-layer at the electrode-electrolyte interface. The mechanisms of self-discharge are speculative at the moment. Redox reactions by impurities and dissolved gases might play a role, as well as the relaxation of overcharged states to a lower potential.

3.2. State-of-Charge of Lithium-Ion Batteries

According to our measurements, the self-discharge of lithium-ion batteries stored at open terminals proceeds substantially slower than with supercapacitors. For comparison purposes, a hybrid supercapacitor was measured, too.

The impedance spectra show two regions (Figure 5):

1. Electrolyte resistance R_e and the subsequent vague quarter-circle, which is related to the lithium-ion conducting passive layer on the electrode (solid-electrolyte interphase, <1000 Hz), semicircle of the double-layer and charge transfer reaction (<100 Hz),
2. Diffusion impedance of the porous electrodes (<2 Hz).

In this study, the quantity U/U_0 appeared to be a good and simple measure of the SoC, because we aimed to find a general correlation with capacitance. In practical systems, OCV-SoC reference tables are most useful for the estimation of the real SoC and further SoH statements (Zou et al. [8,9]).

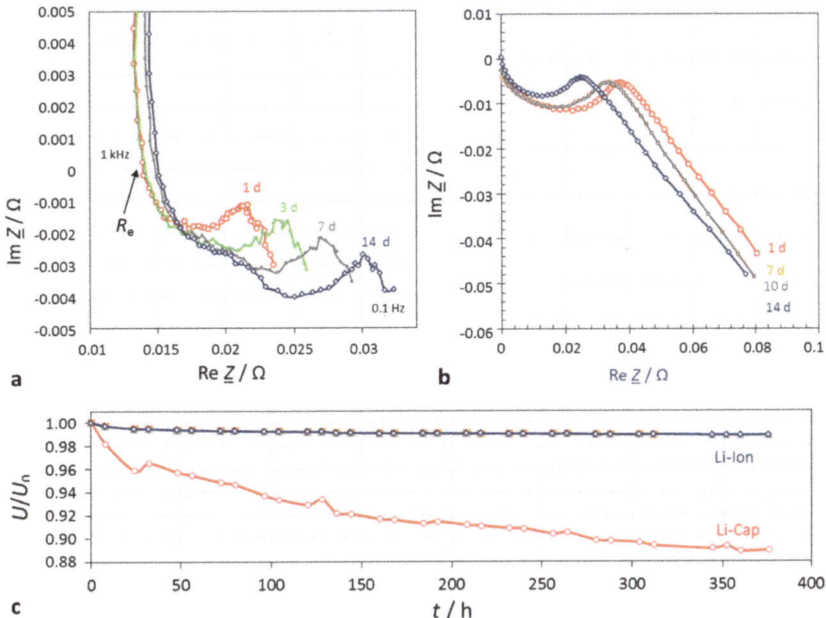

Figure 5. Self-discharge within 14 days. (**a**) Lithium-ion battery (Samsung INR18650-25R, 1.5 Ah, 3.6 V); (**b**) lithium-ion hybrid capacitor (Taiyo Yuden 3.8 V, 100 F). The electrolyte resistance was corrected; (**c**) SoC characteristics.

3.3. Capacitice Charge Status

Fully charged lithium-ion batteries were discharged with the help of an external load by defined partial charged dQ until predefined SoC states were reached.

The impedance spectra in Figure 6 get the narrower, the higher is the SoC. This means that the charge transfer and diffusion resistance increase with the depth-of-discharge of the battery. The electrolyte resistance is more or less constant during charging and discharging.

With lithium-ion batteries, electric charge and SoC correlate excellently:

$$Q(U) \sim e^U \quad \Leftrightarrow \quad \lg Q \sim U \tag{18}$$

Capacitance appears to be a roughly linear measure of the cell voltage: $U \sim C$. Overcharge and deep discharge cause deviations, so that rather a sigmoidal relation is assumed. Using double-logarithmic axes, a sufficient linearity of capacitance and cell voltage is observed.

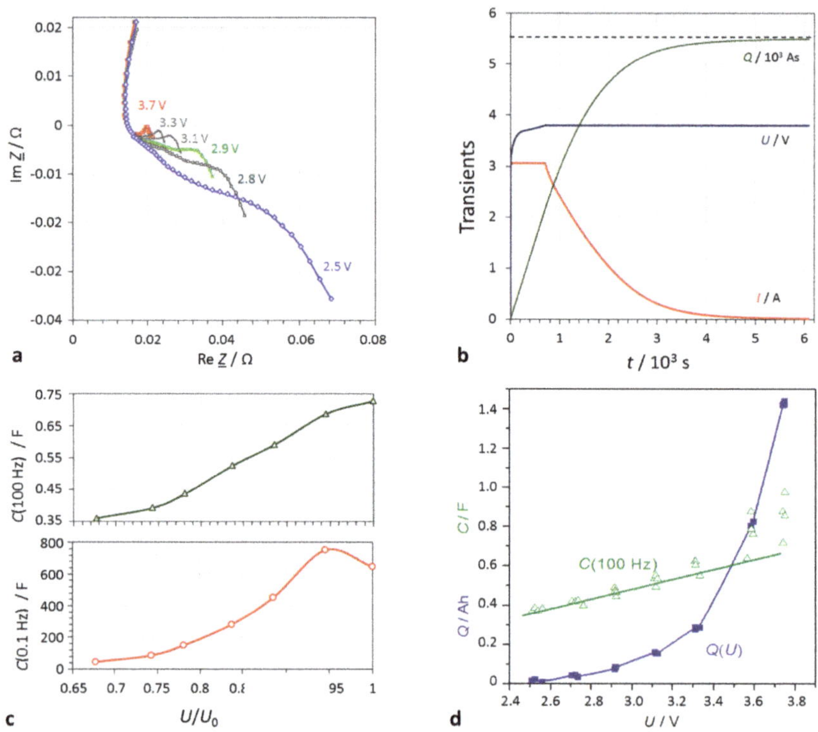

Figure 6. Lithium-ion battery (Samsung INR18650-25R, 1.5 Ah, 3.6 V) at defined SoC states. (**a**) Impedance spectra in the complex plane (mathematical convention): The cell resistance drops with increasing terminal voltage; (**b**) charge characteristics: applied current I, and voltage U, stored charge Q; (**c**) capacitance at a given frequency (e.g., 100 Hz, and 0.1 Hz) indicates the SoC; (**d**) quasi-linear correlation of capacitance and quasi-exponential correlation of electric charge with cell voltage.

Table 2 compiles resistances and capacitances of a lithium-ion battery at different SoC values. Capacitance at 0.1 Hz multiplied by the cell voltage yields the approximate actually available electric charge, $Q = CU$. This means that DC capacitance reflects the true SoC. The capacitance at 100 Hz directly indicates the SoC: $\alpha = (C + 1.27)/2.59$, in the example in Figure 6. However, a useful scaling has yet to be found to provide absolute charge values, when AC capacitance is measured at high frequencies. Moreover, it becomes obvious that the state-of-voltage values U/U_0 differ from the real SoC Q/Q_0.

Table 2. Correlation between state-of-charge (SoC) and capacitance for a lithium-ion battery (Samsung NR18650-25R).

SoC		68%	74%	81%	88%	94%	100%
$U\rightarrow$input	(V)	2.522	2.747	2.992	3.245	3.488	3.703
$R_e\rightarrow$measured	(mΩ)	14.8	14.9	14.9	14.7	14.5	14.5
R (0.1 Hz)$-R_e$	(mΩ)	99	82	61	37	7.7	5.5
$C_{100}\rightarrow$measured	(F)	0.540	0.609	0.784	0.993	1.23	1.30
$C_{0.1}\rightarrow$measured	(F)	18.3	25.2	46.8	119	593	674
$Q\rightarrow$measured	(As)	–	27.6	95.9	265	1366	2889
$CU\rightarrow$calculated	(As)	46.2	69.2	137	3878	1883	2497

Curve Fitting

The Nyquist plot of a lithium-ion battery can be modeled by a capacitance (constant phase element, CPE) in parallel to the solid-electrolyte interphase (SEI) and charge-transfer resistance. The diffusion arc obeys a Warburg impedance W, supplemented by a serial capacitance C_0, which considers the periodic intercalation of lithium ions during the impedance measurement. In additional, an inductance L shifts the impedance spectrum towards positive imaginary parts (Huang et al. [31]).

$$L - R_e - (CPE_1 || R_1) - (CPE_2 || R_2) - W - C_0 \tag{19}$$

where $-$ denotes a series combination, $||$ a parallel combination. The frequency at the minimum of the semicircle (most negative imaginary part in mathematical convention) reflects the reciprocal of the time constant of the charge-transfer process, $\tau = (2\pi \cdot f)^{-1} = RC$.

$$Z_{CPE}(\omega) = \frac{1}{Q(j\omega)^\alpha}, \quad \omega = 2\pi f = \frac{1}{(RQ)^{\frac{1}{\alpha}}}, \text{ and } C = Q^{\frac{1}{\alpha}} \cdot R^{\frac{1-\alpha}{\alpha}} \tag{20}$$

The Warburg coefficient roughly equals $\alpha \approx 0.9$ for the SEI and $\alpha \approx 0.6$ for the charge transfer process of a 15 Ah-NCM/graphite battery (Huang et al. [31]).

In recent approaches, the impedance spectrum is transformed to the time domain to find out the statistical distribution of the time constants of a hypothetical transmission line model consisting of a ladder network of incremental $dR || dC$ elements [34,36].

As far as the curve fitting efforts are concerned, we conclude that the simple correlation of capacitance and SoC fits for the purpose of practical use.

4. Conclusions

To summarize, the cell resistance appears to be an inappropriate measure of the available electric charge in a storage device. The capacitance value determined at a frequency of about 100 Hz allows a straightforward SoC control of supercapacitors and lithium-ion batteries. There is no need for any model descriptions or equivalent circuits, which are often unclear and complicate the system analysis during operation.

For lithium-ion batteries, capacitance extracted from impedance spectra reliably indicates the available electric charge in the working range between full charge (without overload) and cut-off voltage (without deep discharge). An approximately linear correlation was found between capacitance at medium SoC values:

$$\frac{U}{U_0} \sim \log \frac{Q}{Q_0} \sim \frac{C}{C_0} \tag{21}$$

Log denotes the decadic logarithm. This relation is obstructed by external charge/discharge currents so that a sigmoidal function is more reasonable in the regimes of overcharge and deep discharge. A correlation that directly yields absolute SoC values from impedance spectra is still missing.

Author Contributions: Writing—Original Draft Preparation, both authors; Writing—Review & Editing, both authors.

Funding: This work was by financially supported by the Federal State of Bavaria, Germany, in the framework of internal cluster funding (OTH Amberg-Weiden and OTH Regensburg).

Acknowledgments: We thank our students Anton Udalzow and Thierry Assopguimya for some experimental contributions according to our instructions.

Conflicts of Interest: The authors declare no conflict of interest. The funders had no role in the design of the study; in the collection, analyses, or interpretation of data; in the writing of the manuscript, and in the decision to publish the results.

References

1. Conway, B.E. *Electrochemical Supercapacitors: Scientific Fundamentals and Technological Applications*; Kluwer Academic/Plenum Publishers: New York, NY, USA, 1999.
2. Beguin, F.; Frackowiak, E. *Supercapacitors: Materials, Systems, and Applications*; Wiley-VCH: Weinheim, Germany, 2013.
3. Kurzweil, P. Electrochemical Double Layer Capacitors. In *Electrochemical Energy Storage for Renewable Sources and Grid Balancing*; Moseley, P.T., Garche, J., Eds.; Elsevier: Amsterdam, The Netherlands, 2015; Chapter 19; pp. 346–407.
4. Xiong, R.; Cao, J.; Yu, Q.; He, H.; Sun, F. Critical Review on the Battery State of Charge Estimation Methods for Electric Vehicles. *IEEE Access* **2017**, *6*, 1832–1843. [CrossRef]
5. Barre, A.; Deguilhem, B.; Grolleau, S.; Gerard, M.; Suard, F.; Riu, D. A review on lithium-ion battery ageing mechanisms and estimations for automotive applications. *J. Power Sources* **2013**, *241*, 680–689. [CrossRef]
6. Weigert, T.; Tian, Q.; Lian, K. State-of-charge prediction of batteries and battery-supercapacitor hybrids using artificial neural networks. *J. Power Sources* **2011**, *196*, 4061–4066. [CrossRef]
7. Yang, H. Estimation of Supercapacitor Charge Capacity Bounds Considering Charge Redistribution. *IEEE Trans. Power Electron.* **2018**, *33*, 6980–6993. [CrossRef]
8. Zou, C.; Manzie, C.; Nesic, D.; Kallapur, A.G. Multi-time-scale observer design for state-of-charge and state-of-health of a lithium-ion battery. *J. Power Sources* **2016**, *335*, 121–139. [CrossRef]
9. Zou, C.; Hu, X.; Wei, Z.; Tang, X. Electrothermal dynamics-conscious lithium-ion battery cell-level charging management via state-monitored predictive control. *Energy* **2017**, *141*, 250–259. [CrossRef]
10. Zou, C.; Zhang, L.; Hu, X.; Wang, Z.; Wik, T.; Pecht, M. A review of fractional-order techniques applied to lithium-ion batteries, lead-acid batteries, and supercapacitors. *J. Power Sources* **2018**, *390*, 286–296. [CrossRef]
11. Lai, X.; Zheng, Y.; Sun, T. A comparative study of different equivalent circuit models for estimating state-of-charge of lithium-ion batteries. *Electrochim. Acta* **2018**, *259*, 566–577. [CrossRef]
12. Dubarry, M.; Liaw, B.Y. Development of a universal modeling tool for rechargeable lithium batteries. *J. Power Sources* **2007**, *174*, 856–860. [CrossRef]
13. Dubarry, M.; Truchot, C.; Liaw, B.Y. Synthesize battery degradation modes via a diagnostic and prognostic model. *J. Power Sources* **2012**, *219*, 204–216. [CrossRef]
14. Cui, Y.; Zuo, P.; Du, C.; Gao, Y.; Yang, J.; Cheng, X.; Ma, Y.; Yin, G. State of health diagnosis model for lithium ion batteries based on real-time impedance and open circuit voltage parameters identification method. *Energy* **2018**, *144*, 647–656. [CrossRef]
15. Yang, H. Effects of supercapacitor physics on its charge capacity. *IEEE Trans. Power Electron.* 2018. [CrossRef]
16. Genc, R.; Alas, M.O.; Harputlu, E.; Repp, S.; Kremer, N.; Castellano, M.; Colak, S.G.; Ocakoglu, K.; Erdem, E. High-Capacitance Hybrid Supercapacitor Based on Multi-Colored Fluorescent Carbon-Dots. *Sci. Rep.* **2017**, *7*, 11222. [CrossRef] [PubMed]
17. Repp, S.; Harputlu, E.; Gurgen, S.; Castellan, M.; Kremer, N.; Pompe, N.; Wörner, J.; Hoffmann, A.; Thomann, R.; Emen, F.M.; et al. Synergetic effects of Fe^{3+} doped spinel $Li_4Ti_5O_{12}$ nanoparticles on reduced graphene oxide for high surface electrode hybrid supercapacitors. *Nanoscale* **2018**, *10*, 1877–1884. [CrossRef] [PubMed]
18. Lehtimäki, S.; Suominen, M.; Damlin, P.; Tuukkanen, T.; Kvarnström, C.; Lupo, D. Preparation of Supercapacitors on Flexible Substrates with Electrodeposited PEDOT/Graphene Composites. *ACS Appl. Mater. Interfaces* **2015**, *7*, 22137–22147. [CrossRef] [PubMed]
19. Conway, B.E.; Pell, W.G.; Liu, T.C. Diagnostic analyses for mechanisms of self-discharge of electrochemical capacitors and batteries. *J. Power Sources* **1997**, *65*, 53–59. [CrossRef]
20. Ricketts, B.W.; Ton-That, C. Self-discharge of carbon-based supercapacitors with organic electrolytes. *J. Power Sources* **2000**, *89*, 64–69. [CrossRef]
21. Bohlen, O.; Kowal, J.; Sauer, D.U. Ageing behaviour of electrochemical double layer capacitors, Part II. Lifetime simulation model for dynamic applications. *J. Power Sources* **2007**, *173*, 626–632.
22. Zhang, Q.; Cai, C.; Qin, J.; Wei, B. Tunable self-discharge process of carbon nanotube based supercapacitors. *Nano Energy* **2014**, *4*, 14–22. [CrossRef]
23. Tevi, T.; Takshi, A. Modeling and simulation study of the self-discharge in supercapacitors in presence of a blocking layer. *J. Power Sources* **2015**, *273*, 857–862. [CrossRef]

24. Yang, H.; Zhang, Y. Self-discharge analysis and characterization of supercapacitors for environmentally powered wireless sensor network applications. *J. Power Sources* **2011**, *196*, 8866–8873. [CrossRef]

25. Kowal, J.; Avaroglu, E.; Chamekh, F.; Senfelds, A.; Thien, T.; Wijaya, D.; Sauer, D.U. Detailed analysis of the self-discharge of supercapacitors. *J. Power Sources* **2011**, *196*, 573–579. [CrossRef]

26. Shen, J.F.; He, Y.J.; Ma, Z.F. A systematical evaluation of polynomial based equivalent circuit model for charge redistribution dominated self-discharge process in supercapacitors. *J. Power Sources* **2016**, *303*, 294–304. [CrossRef]

27. Kurzweil, P.; Ober, J.; Wabner, D.W. Method for Correction and Analysis of Impedance Spectra. *Electrochim. Acta* **1989**, *34*, 1179–1185. [CrossRef]

28. Kurzweil, P.; Fischle, H.J. A new monitoring method for electrochemical aggregates by impedance spectroscopy. *J. Power Sources* **2004**, *127*, 331–340. [CrossRef]

29. Kurzweil, P.; Frenzel, B.; Hildebrand, A. Voltage-dependent capacitance, aging effects, and failure indicators of double-layer capacitors during lifetime testing. *ChemElectroChem* **2015**, *2*, 6–13. [CrossRef]

30. Fernández Pulido, Y.; Blanco, C.; Ansean, D.; García, V.M.; Ferrero, F.; Valledor, M. Determination of suitable parameters for battery analysis by Electrochemical Impedance Spectroscopy. *Measurement* **2017**, *106*, 1–11. [CrossRef]

31. Huang, Q.A.; Shen, Y.; Huang, Y.; Zhang, L.; Zhang, J. Impedance Characteristics and Diagnoses of Automotive Lithium-Ion Batteries at 7.5% to 93.0% State of Charge. *Electrochim. Acta* **2016**, *219*, 751–765. [CrossRef]

32. Hung, M.H.; Lin, C.H.; Lee, L.C.; Wang, C.M. State-of-charge and state-of-health estimation for lithium-ion batteries based on dynamic impedance technique. *J. Power Sources* **2014**, *268*, 861–873. [CrossRef]

33. Skoog, S.; David, S. Parameterization of linear equivalent circuit models over wide temperature and SOC spans for automotive lithium-ion cells using electrochemical impedance spectroscopy. *J. Energy Storage* **2017**, *14*, 39–48. [CrossRef]

34. Manikandan, B.; Ramar, V.; Yap, C.; Balaya, P. Investigation of physico-chemical processes in lithium-ion batteries. *J. Power Sources* **2017**, *361*, 300–309. [CrossRef]

35. Kurzweil, P.; Hildebrand, A.; Weiß, M. Accelerated Life Testing of Double-Layer Capacitors: Reliability and Safety under Excess Voltage and Temperature. *ChemElectroChem* **2015**, *2*, 150–159. [CrossRef]

36. Itagaki, M.; Ueno, M.; Hoshi, Y.; Shitanda, I. Simultaneous Determination of Electrochemical Impedance of Lithium-ion Rechargeable Batteries with Measurement of Charge-discharge Curves by Wavelet Transformation. *Electrochim. Acta* **2017**, *235*, 384–389. [CrossRef]

 *batteries*

Article

Lifetime Prediction of Lithium-Ion Capacitors Based on Accelerated Aging Tests

Nagham El Ghossein *, Ali Sari and Pascal Venet

Univ Lyon, Université Claude Bernard Lyon 1, Ecole Centrale de Lyon, INSA Lyon, CNRS, Ampère, F-69100 Villeurbanne, France; ali.sari@univ-lyon1.fr (A.S.); pascal.venet@univ-lyon1.fr (P.V.)
* Correspondence: naghamghossein@gmail.com

Received: 18 December 2018; Accepted: 26 February 2019; Published: 5 March 2019

Abstract: Lithium-ion Capacitors (LiCs) that have intermediate properties between lithium-ion batteries and supercapacitors are still considered as a new technology whose aging is not well studied in the literature. This paper presents the results of accelerated aging tests applied on 12 samples of LiCs. Two high temperatures (60 °C and 70 °C) and two voltage values were used for aging acceleration for 20 months. The maximum and the minimum voltages (3.8 V and 2.2 V respectively) had different effects on capacitance fade. Cells aging at 2.2 V encountered extreme decrease of the capacitance. After storing them for only one month at 60 °C, they lost around 22% of their initial capacitance. For this reason, an aging model was developed for cells aging at the lowest voltage value to emphasize the huge decrease of the lifetime at this voltage condition. Moreover, two measurement tools of the capacitance were compared to find the optimal method for following the evolution of the aging process. It was proved that electrochemical impedance spectroscopy is the most accurate measurement technique that can reveal the actual level of degradation inside a LiC cell.

Keywords: lithium-ion capacitor; aging model; langmuir isotherm; lifetime prediction; aging mechanisms; calendar aging; floating aging

1. Introduction

The performance of Energy Storage Systems (ESSs) depends on several factors such as current, voltage, and temperature. The operating and environmental conditions have an important impact on their behavior. Accelerated aging tests are usually applied to these devices for the aim of developing aging law. These laws can be used in management systems that are integrated with ESSs to predict the remaining time before their failure.

The accelerated aging tests can be divided into two categories: calendar aging and cycling aging [1–4]. The behavior of an ESS during its storage is tested in calendar aging. The impacts of the temperature and the voltage on the lifetime are studied throughout this type of tests. However, ESSs during charge or discharge suffer from a different kind of aging, especially when the applied currents are very high. This is cycling aging whose corresponding tests could be done also at very low or high temperatures to further accelerate the process.

Lifetime of Lithium-ion Batteries (LiBs) and Supercapacitors (SCs) was frequently studied in the literature [5–9]. SCs have a longer lifespan than LiBs and a better behavior at low and higher temperatures. Nonetheless, their energy density is much lower than that of LiBs. The new technology of SCs, Lithium-ion Capacitors (LiCs), possesses an improved energy density, around triple the one of conventional SCs. Since they combine the negative electrode of LiBs with the positive electrode of SCs, they have intermediate electrical characteristics [10]. Several combinations of electrodes have been evaluated for ensuring the most optimal features of the hybrid device [11,12]. The majority of commercial LiCs are composed of a negative electrode of carbon pre-doped with lithium and a positive electrode of activated carbon. The pre-lithiation of the negative electrode can be done using multiple methods such as ensuring a direct contact between lithium metal and the carbon electrode, short-circuiting the carbon electrode with a lithium metal electrode or adding lithium excess in the positive electrode [13]. Commercial cells developed by JM Energy and JSR Micro include a sacrificial lithium electrode that serves as a lithium source for carbon pre-lithiation. Others developed by General Capacitors comprise lithium stripes that are in direct contact with the negative electrode at which pre-lithiation occurs [14]. The negative electrode in LiCs is oversized with respect to the positive electrode in order to benefit from the total capacitance of the positive electrode [11,15].

Since LiCs are a new technology, a limited number of publications had studied their aging. An advantage of over-sizing the negative electrode remains in reducing the effects of the Solid Electrolyte Interface (SEI), which is the main destructive aging mechanism of this type of electrodes as found in LiBs [16,17]. However, the impacts of the SEI could not be completely suppressed. For example, in [18], during continuous cycling of a LiC, the capacity decreased because of the growth of the SEI at the surface of the negative electrode, which induced the increase of its resistance. As a result, an important number of cyclable lithium ions was lost and the potential of the positive electrode drifted to high positive values. Usually, the potential of the positive electrode should not exceed 4 V vs. Li/Li^+ [19]. Otherwise, a formation of Lithium Fluoride (LiF) at the negative electrode could happen [19]. Moreover, accelerated aging tests were applied in [20] on commercial LiCs that belong to three different manufacturers. A calendar test at 60 °C and 3.8 V resulted in a 10% decrease of the capacity after 5000 hours while another at 0 °C and 3.8 V caused a decrease of less than 2%. Therefore, it was found that the highest the temperature is, the greater the degradation becomes. In addition, cycle aging tests showed that the depth of discharge does not affect the degradation of the capacity of a LiC. However, another study proved that calendar aging of specific commercial LiCs extremely depends on their state of charge [21].

A LiC has a particular operating principle. Each potential of the cell reflects a different chemical composition [15]. At 3 V, which is the open circuit voltage of a commercial LiC right after its assembly, the positive electrode of activated carbon is at the approximate neutral state. When discharging the cell from 3 to 2.2 V, some of the lithium ions that are pre-intercalated into the negative electrode des-intercalate from carbon layers and migrate towards the positive electrode where they accumulate. However, when charging the cell from 3 to 3.8 V, the anions forming the salt of the electrolyte accumulate at the surface of the activated carbon and the cations of the salt intercalate into the negative electrode. Therefore, at 2.2 and 3.8 V, the double layer formed at the activated carbon positive electrode includes different ions and the degree of lithiation of the negative electrode increases from 2.2 to 3.8 V. For this reason, calendar aging at both voltage values will be analyzed in this paper so the effects of both chemical compositions could be compared.

Accelerated aging tests were applied to several cells as it will be described in Section 2 of the paper. To track the evolution of the aging process, the capacitance and the equivalent series resistance of the cells can be measured using frequency or time domain measurements. Section 3 will explain why the frequency domain measurements were chosen to follow the properties of the LiCs. In addition, Section 4 will present the results of the capacitance decrease and the equivalent series resistance increase found from frequency domain measurements over an aging time of 20 months. Based on the evolution of these parameters with respect to the aging time, an aging model will be developed in Section 3 for cells that aged at the lowest voltage value.

2. Experimental Setup

Twelve prismatic LiCs were divided into two equal groups and mounted in two climatic chambers at 60 °C and 70 °C. In each chamber, the voltage of three cells was maintained constant at 2.2 V while the voltage of the rest of the cells was maintained at 3.8 V. Cells used during these tests belong to the "Ultimo" prismatic series developed by JM Energy and JSR Micro. The capacitance of the cells is equal to 3300 F and they can operate over a large temperature window, from $-30\,°C$ to $70\,°C$. Three similar cells were tested per aging condition to guarantee accurate and reproducible results.

3. Comparison of Frequency and Time Domain Measurements

After placing the LiCs in the climatic chambers at high temperatures, their capacitance and equivalent series resistance were measured using electrochemical impedance spectroscopy (EIS). This measurement tool in the galvanostatic mode consists on fixing the potential of a cell, applying a small sinusoidal current at different frequencies and then measuring the impedance using the corresponding voltage response. The amplitude of the sinusoidal current was chosen to be 5 A. Moreover, prior to each measurement process, the voltage of the cell was maintained constant for 30 min. These periodic measurements were done at high temperatures without getting the LiCs off from the climatic chambers. To check whether additional measurements should be done, the LiCs were taken off the climatic chambers after 109 days and their properties were measured using frequency and time domain measurements at $25\,°C$. In [15], these measurement techniques are explained in details.

3.1. Frequency Domain Measurements

The impedance of a LiC depends on the voltage V and the frequency f [15,22]. The capacitance $C(f, V)$ can be extracted from the imaginary part of the impedance $Img(f, V)$ as follows:

$$C(f, V) = \frac{-1}{2\pi \times f \times Img(f, V)} \tag{1}$$

The series resistance of a LiC at a specific frequency is equal to the real part of the impedance. These parameters were measured at 100 mHz, at the aging voltage and at $25\,°C$ and then normalized with respect to the value found before aging (also at $25\,°C$). The frequency of 100 mHz was chosen for its accuracy in showing the evolution of the parameters during aging. Figures 1 and 2 show the evolution of the normalized capacitance at $25\,°C$ after 109 days of aging at $60\,°C$ and $70\,°C$, respectively. The parameters of the three samples per voltage condition are represented in these figures.

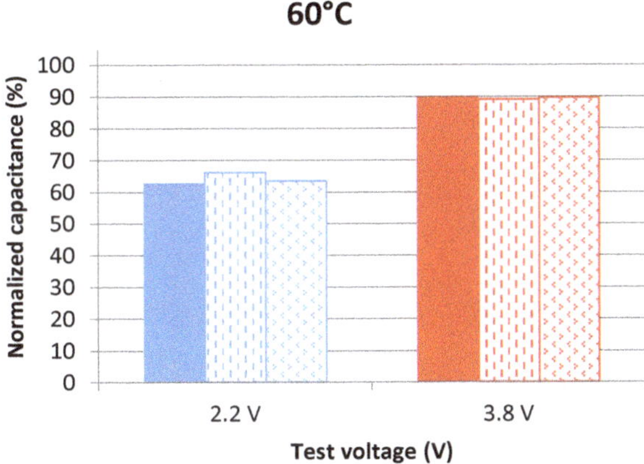

Figure 1. Capacitance evolution of cells aging at 2.2 V and 3.8 V at $60\,°C$ after 109 days of calendar aging. The results are found from measurements in the frequency domain at $25\,°C$.

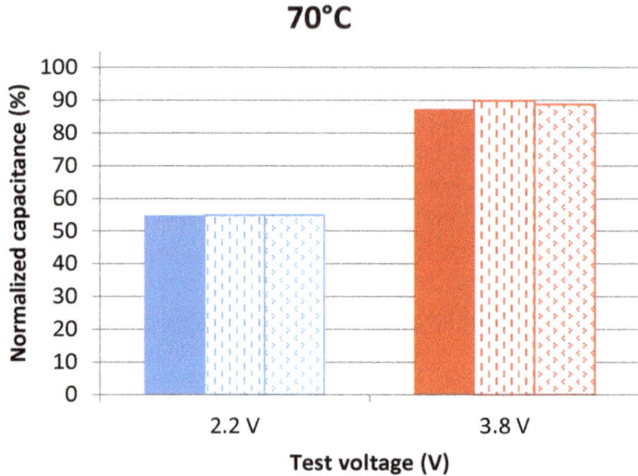

Figure 2. Capacitance evolution of cells aging at 2.2 V and 3.8 V at 70 °C after 109 days of calendar aging. The results are found from measurements in the frequency domain at 25 °C.

LiCs aging at 3.8 V obviously have better capacitance retention than the ones aging at 2.2 V. The average capacitance decrease of cells aging at 3.8 V is 11% at 60 °C and 12% at 70 °C. On the other hand, the average capacitance decrease of cells aging at 2.2 V is 36% at 60 °C and 46% at 70 °C. In [21], the best storage voltage of LiCs was found to be 3 V since the capacitance decrease through accelerated aging tests at this voltage value was insignificant. To validate the results mentioned previously, additional measurements in the time domain were done.

3.2. Time Domain Measurements

Measurements in the time domain are based on using the voltage response of a LiC during its discharge with a DC current as explained in the traditional protocol detailed in [15]. The capacitance is calculated from the difference between the maximum and the minimum voltages during a full discharge, the time of discharge, and the value of the current. This technique evaluates the total discharge capacitance while frequency domain measurements assess the capacitance at a specific voltage value. The normalized capacitance is shown in Figure 3 for cells aging at 60 °C and in Figure 4 for the others aging at 70 °C.

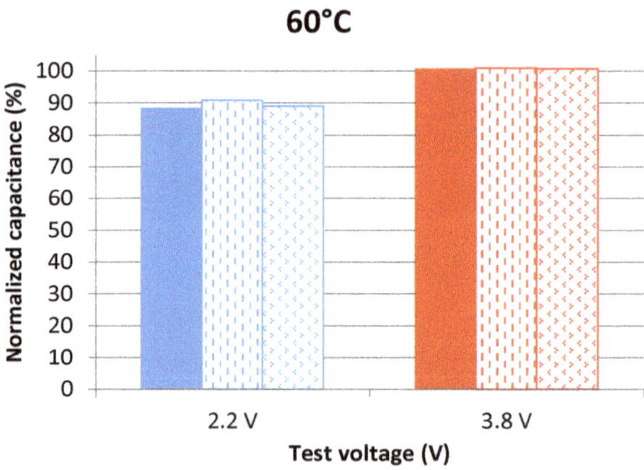

Figure 3. Capacitance evolution of cells aging at 2.2 V and 3.8 V at 60 °C after 109 days of calendar aging. The results are found from measurements in the time domain at 25 °C.

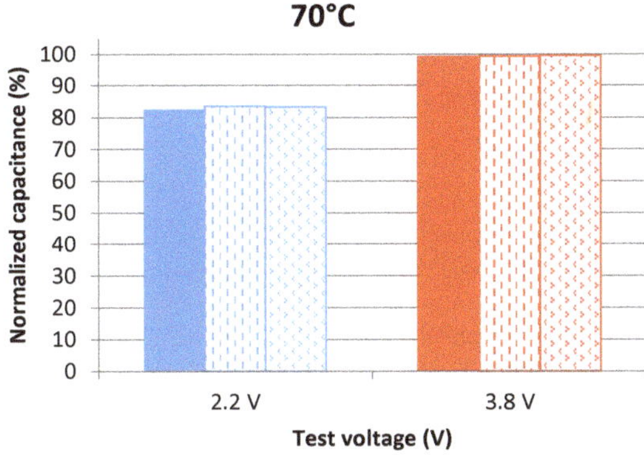

Figure 4. Capacitance evolution of cells aging at 2.2 V and 3.8 V at 70 °C after 109 days of calendar aging. The results are found from measurements in the time domain at 25 °C.

Results of time domain measurements show that LiCs aging at 3.8 V barely age since their capacitance does not decrease even after being for 109 days under calendar stresses. The capacitance decrease of cells aging at 2.2 V is also much less than the one found in frequency domain measurements. For this reason, the capacitance of a cell aging at 3.8 V at each voltage value, during its discharge with a DC current, was measured according to the method explained in [15]. A comparison between the evolution of the capacitance with respect to the voltage before and after aging is presented in Figure 5.

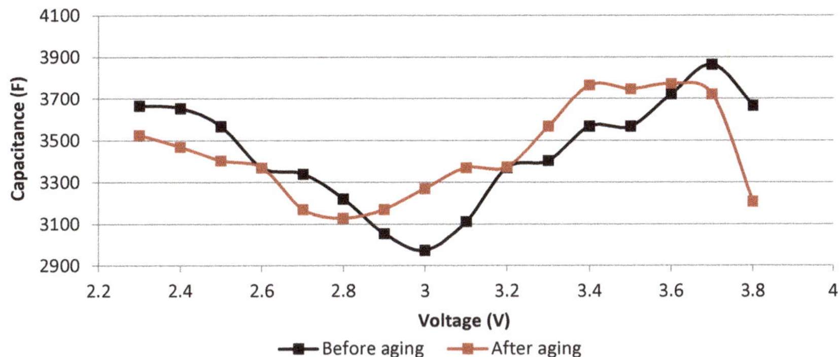

Figure 5. Effects of aging at 3.8 V and 60 °C for 109 days on the capacitance evolution with respect to the voltage, deduced from time domain measurements at 25 °C.

The minimum capacitance is usually an indicator of the neutral state of the positive electrode whose potential is around 3.1 V vs. Li/Li$^+$ at this state [15,22]. The corresponding voltage value of the complete cell shifts from 3 to 2.8 V due to aging (cf. Figure 5). This voltage is the difference between the potentials of the positive electrode and of the negative electrode. Therefore, since the potential of the negative before aging is around 0.1 V vs. Li/Li$^+$ when the total voltage is equal to 3 V, one can conclude that its potential drifts to a more positive value that may be around 0.3 V vs. Li/Li$^+$ when the new total voltage is equal to 2.8 V. This can be caused by the growth of the SEI at the negative electrode and the loss of cyclable lithium ions [23]. Therefore, the potential window at which the double layer at the positive electrode is formed by the anions extends from 3–3.8 V to 2.8–3.8 V. Since the anions have a small size compared to the cations [24], the associated double layer generates a higher capacitance. As a result, the total capacitance measured over the total voltage window may not show the real aging state of the cell. In fact, it may benefit from the increase of the capacitance produced by the accumulation of the anions at the positive electrode over an extended voltage window. This stability of the total capacitance may particularly happen at the beginning of aging when the only aging effect is the drift of the potential of the negative electrode. Therefore, since the measurements in the time domain did not reflect the actual aging state compared to the measurements in the frequency domain, this latter was used to follow the evolution of the parameters throughout the whole aging period.

4. Interpretation of Calendar Aging

LiCs under accelerated aging tests at 60 °C and 70 °C degraded in different ways. The evolution of their parameters was the tool for predicting the aging mechanisms at each test condition.

4.1. Results of Periodic Frequency Domain Measurements

The capacitance decrease and the equivalent series resistance increase of different cells that are induced by the calendar aging are shown in Figures 6 and 7 at 60 °C and 70 °C, respectively. The values were extracted from impedance measurements at the test voltage and at 100 mHz. Initial capacitance and resistance values of cells aging at 2.2 V are the following: 2586 F, 0.8 mΩ at 60 °C and 2600 F, 0.7 mΩ at 70 °C. Cells aging at 3.8 V had the following initial capacitance and resistance values: 3200 F, 0.8 mΩ at 60 °C and 3283 F, 0.8 mΩ at 70 °C.

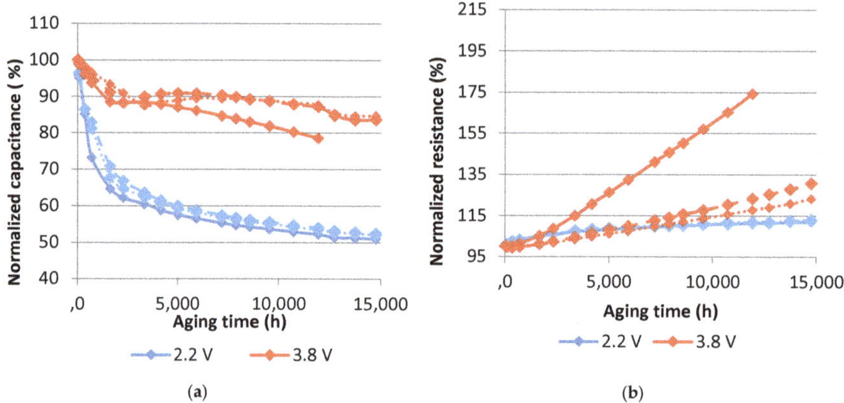

Figure 6. Evolution of the normalized capacitance (**a**) and the normalized resistance (**b**) with aging time at 60°C. Each curve represents one sample.

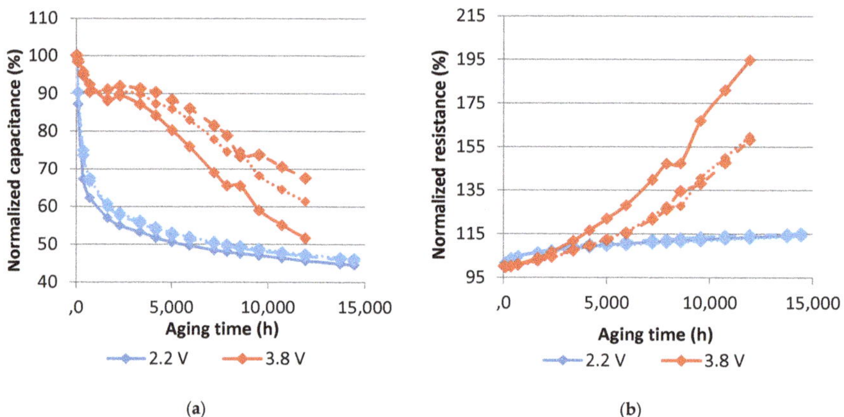

Figure 7. Evolution of the normalized capacitance (**a**) and the normalized resistance (**b**) with aging time at 70°C. Each curve represents one sample.

Accelerated aging tests at 70 °C obviously provoke more severe degradations than the ones at 60 °C. The capacitance decrease at 2.2 V reaches 55% at 70 °C while the resistance increase is around 14% after 20 months. At 60 °C, less damage is found in this voltage condition. The capacitance decreases by around 49% and the equivalent series resistance increases by around 12%. The effects of calendar aging on cells that were maintained at 3.8 V are more pronounced in their equivalent series resistance. In fact, the resistance noticeably rises especially for two cells originating from a particular production batch (95% at 70 °C and 74% at 60 °C after 17 months of aging). The rest of the cells have around 26% of resistance increase after 20 months at 60 °C and 60% after 17 months at 70 °C. The capacitance of these cells decreases at a rate less than the one at 2.2 V. The corresponding drop is equal to around 18% at 60 °C and 40% at 70 °C.

According to [16], during the calendar aging of a lithium-ion battery, the degradation of its negative electrode depends on its degree of lithiation. When the negative electrode of graphite is highly lithiated, the SEI layer at its surface would grow causing the increase of the equivalent series resistance of the cell. This phenomenon can be the reason behind the high increase of the resistance of LiCs aging at 3.8 V since at this state of charge, their negative electrode is highly lithiated. Moreover, at this voltage value, the potential of the positive electrode would drift to more positive values. Consequently, redox reactions would happen between some parasitic groups present at the surface of activated carbon and the components of the electrolyte [5,25]. As a result, solid or gaseous products will appear in the cell inducing the increase of the overall resistance.

As for cells aging at 2.2 V, aging mechanisms are different due to the different chemical states of both electrodes at each voltage value [15]. Since the capacitance of LiCs that aged at 2.2 V decreases with a very high rate, an aging model for this voltage condition will be developed in the next paragraph, as well as an interpretation of the corresponding aging mechanisms.

4.2. Aging Model

Aging mechanisms in ESSs are mainly attributed to parasitic chemical reactions. These reactions can generate gases that become irreversibly adsorbed at the surface of the electrodes. Langmuir assumes that the surface consists of several equivalent sites where species can be adsorbed physically or chemically. The adsorption and desorption processes are considered to be dynamic. A law of speed can then be defined for each process and when the rates become equal, a state of equilibrium can be characterized by a reduced surface of the solid. Admitting that the double layer at the positive electrode is no longer formed where the adsorption sites are saturated with these products, the storage area decreases with their accumulation during aging [26,27]. The Langmuir isotherm defines the lost area (ΔS) using the concentration of gases ([C]) and the Langmuir adsorption coefficient (a), as follows:

$$\Delta S = \frac{a \times [C]}{1 + a \times [C]} \tag{2}$$

The mass of products generated by parasitic reactions, M, can be expressed by the following equation based on Faraday's law [26,27]:

$$M = \frac{1}{F} \times Q \times \frac{M_{at}}{z} \tag{3}$$

where, F is the Faraday constant, Q is the total electric charge passed through the electrolyte, M_{at} is the molar mass of the substance and z is the valance number of ions enrolled in the reaction. The current I required for maintaining the voltage constant during the calendar aging is considered constant. Therefore, the number of moles $N(t)$ of a gas in function of time can be extracted from Equation (3):

$$N(t) = \frac{I \times t}{F \times z} \tag{4}$$

As a result, the concentration of gases can be described by a time dependent function. The decrease of the storage area leads to a decrease of the capacitance of the electrode. Therefore, considering the amount of produced gases constant with respect to time, the loss of capacitance in Farad (ΔC) can be derived from the above equations [26,27]:

$$\Delta C = \frac{a_c \times t}{1 + b_c \times t} \tag{5}$$

The irreversible adsorption at the surface of the activated carbon electrode could block the pores of the material. As a result, the equivalent series resistance increase in ohm (ΔR) may also be modeled by Langmuir isotherm that depends on the aging time:

$$\Delta R = \frac{a_R \times t}{1 + b_R \times t} \tag{6}$$

As a result, the models of the normalized capacitance and resistance at an aging time t, $C(\%)(t)$ and $R(\%)(t)$ respectively, can be expressed by the following equations:

$$C(\%)(t) = \frac{C(0) - \frac{a_c \times t}{1 + b_c \times t}}{C(0)} \times 100 \tag{7}$$

$$R(\%)(t) = \frac{R(0) - \frac{a_R \times t}{1 + b_R \times t}}{R(0)} \times 100 \tag{8}$$

where, $C(0)$ and $R(0)$ are the initial values of the capacitance and the resistance before aging. The values of $C(\%)(t)$ and $R(\%)(t)$ deduced from the impedance measurements at 100 mHz at different aging times were used to identify the parameters of Equations (5) and (6). The averages of capacitance loss and equivalent series resistance increase of the three tested samples were taken into consideration at 60 °C and 70 °C. The errors between the predicted values of the normalized capacitance and equivalent series resistance and the measured values were also evaluated (e_c and e_R respectively). Table 1 combines the identified parameters and Figures 8 and 9 compare the simulated model and the measured values at 60 °C and 70 °C, respectively.

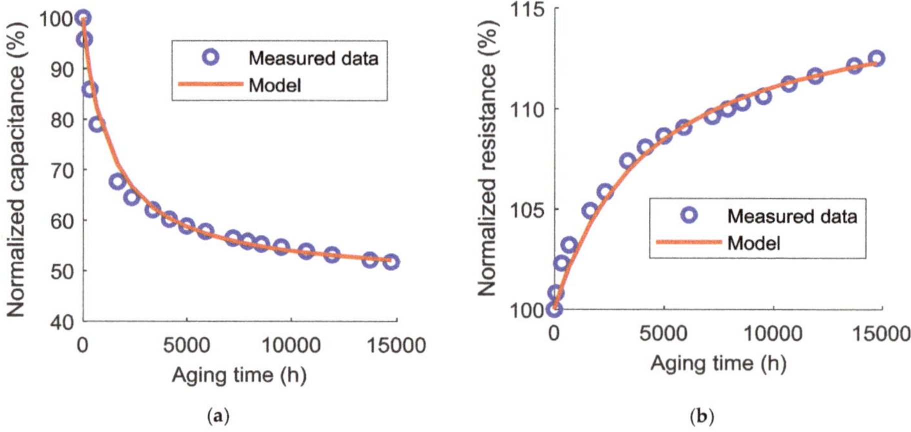

Figure 8. Comparison between the measured data and the model for the capacitance (**a**) and the equivalent series resistance (**b**) evolutions at 60 °C.

Table 1. Identified parameters of the Langmuir model at 60 °C and 70 °C.

Temperature (°C)	$a_c(F{\cdot}h^{-1})$	$b_c(h^{-1})$	$e_c(\%)$	$a_R(m\Omega{\cdot}h^{-1})$	$b_R(h^{-1})$	$e_R(\%)$
60	1.2	0.001	0.7	3.5×10^{-5}	3.2×10^{-4}	0.3
70	1.9	0.0014	3	4.1×10^{-5}	3.1×10^{-4}	0.7

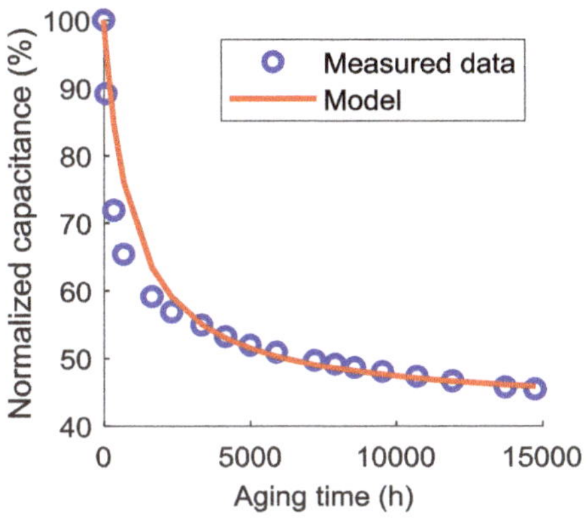

Figure 9. Comparison between the measured data and the models of the capacitance (**a**) and the equivalent series resistance (**b**) evolution at 70 °C.

As can be seen in these figures, the Langmuir model is able to accurately predict the evolution of the capacitance and the equivalent series resistance with the calendar aging time. Therefore, when cells are mounted at a fully discharged state, their lifetime can be now estimated according to the end-of-life criteria. Usually, a normalized capacitance of 80% or a normalized resistance of 200% is a sufficient indicator that the ESS has reached its end-of-life.

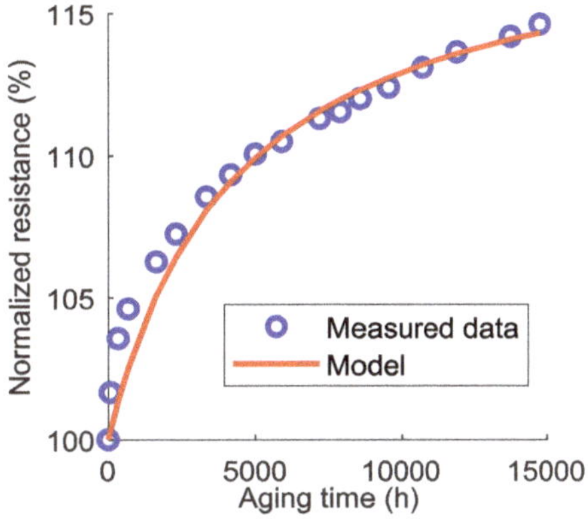

Based on the operating principle of a brand new LiC, the positive electrode at 2.2 V attracts the lithium ions (Li^+) that form the double layer at its surface. An important reason behind the degradation of supercapacitors concerns the parasitic groups present at the surface of activated carbon after the activation treatments [28]. Therefore, these groups, which reside at the surface of the positive electrode in a LiC, might irreversibly adsorb the cations Li^+ and block the pores of the activated carbon. As a result, additional lithium ions should be deintercalated from the negative electrode to compensate the loss of lithium ions at the positive electrode. Accordingly, the overall capacitance of the device diminishes. The aging model based on the Langmuir isotherm can then describe with a small error the evolution of the capacitance and the equivalent series resistance with respect to the aging time.

5. Conclusions

The hybrid configuration of LiCs has a huge effect on their performance at different voltage values. Calendar aging tests were applied at the minimum and maximum voltages of the total voltage window and at 60 °C and 70 °C. During the aging process, periodic measurements were applied to the cells for the aim of following the decrease of their capacitance and the increase of their equivalent series resistance. Time domain measurements were not able to reveal the real aging status. The potential drift of the negative electrode at the beginning of the aging influenced the value of the measured capacitance during a full discharge. Frequency domain measurements were then chosen as periodic measurement tool since they reflect the actual aging state of the cells.

During the calendar aging, cells that aged at the minimum voltage value had a huge capacitance decrease. They lost around 55% of their initial capacitance at 70 °C and 49% at 60 °C after 20 months of calendar aging. An aging model was developed for these cells based on Langmuir isotherm. The model can accurately predict the capacitance increase and the equivalent series resistance decrease at a specific aging time. Additional analysis should be conducted to check whether this model can be applied during the cycle aging of LiCs or not. This will also depend on the aging mechanisms generated throughout the lifetime of LiCs on duty. In fact, the model was unable to predict the capacitance decrease and the resistance increase of cells aging at 3.8 V since the corresponding aging mechanisms generated non-monotonous change of properties during the whole aging period due to different aging mechanisms.

Author Contributions: Writing and original draft preparation, N.E.G; Review & Editing, A.S, P.V.; Results discussion, all authors.

Funding: This research received no external funding.

Conflicts of Interest: The authors declare no conflict of interest.

References

1. Uddin, K.; Perera, S.; Widanage, W.D.; Somerville, L.; Marco, J. Characterising Lithium-Ion Battery Degradation through the Identification and Tracking of Electrochemical Battery Model Parameters. *Batteries* **2016**, *2*, 13. [CrossRef]
2. Stroe, D.-I.; Swierczynski, M.; Stroe, A.-I.; Knudsen Kær, S. Generalized Characterization Methodology for Performance Modelling of Lithium-Ion Batteries. *Batteries* **2016**, *2*, 37. [CrossRef]
3. Kandasamy, N.K.; Badrinarayanan, R.; Kanamarlapudi, V.R.K.; Tseng, K.J.; Soong, B.-H. Performance Analysis of Machine-Learning Approaches for Modeling the Charging/Discharging Profiles of Stationary Battery Systems with Non-Uniform Cell Aging. *Batteries* **2017**, *3*, 18. [CrossRef]
4. Ossai, C.I. Prognosis and Remaining Useful Life Estimation of Lithium-Ion Battery with Optimal Multi-Level Particle Filter and Genetic Algorithm. *Batteries* **2018**, *4*, 15. [CrossRef]
5. German, R.; Venet, P.; Sari, A.; Briat, O.; Vinassa, J.M. Improved Supercapacitor Floating Ageing Interpretation Through Multipore Impedance Model Parameters Evolution. *IEEE Trans. Power Electron.* **2014**, *29*, 3669–3678. [CrossRef]
6. German, R.; Sari, A.; Briat, O.; Vinassa, J.M.; Venet, P. Impact of Voltage Resets on Supercapacitors Aging. *IEEE Trans. Ind. Electron.* **2016**, *63*, 7703–7711. [CrossRef]
7. Redondo-Iglesias, E.; Venet, P.; Pelissier, S. Eyring acceleration model for predicting calendar ageing of lithium-ion batteries. *J. Energy Storage* **2017**, *13*, 176–183. [CrossRef]
8. Redondo-Iglesias, E.; Venet, P.; Pelissier, S. Global Model for Self-discharge and Capacity Fade in Lithium-ion Batteries Based on the Generalized Eyring Relationship. *IEEE Trans. Veh. Technol.* **2018**. [CrossRef]
9. Savoye, F.; Venet, P.; Millet, M.; Groot, J. Impact of Periodic Current Pulses on Li-Ion Battery Performance. *IEEE Trans. Ind. Electron.* **2012**, *59*, 3481–3488. [CrossRef]
10. Ronsmans, J.; Lalande, B. Combining energy with power: Lithium-ion capacitors. In Proceedings of the 2015 International Conference on Electrical Systems for Aircraft, Railway, Ship Propulsion and Road Vehicles (ESARS), Aachen, Germany, 3–5 March 2015; pp. 1–4.
11. Zuo, W.; Li, R.; Zhou, C.; Li, Y.; Xia, J.; Liu, J. Battery-Supercapacitor Hybrid Devices: Recent Progress and Future Prospects. *Adv. Sci.* **2017**, *4*, 1600539. [CrossRef] [PubMed]
12. Adelowo, E.; Baboukani, A.R.; Chen, C.; Wang, C. Electrostatically Sprayed Reduced Graphene Oxide-Carbon Nanotubes Electrodes for Lithium-Ion Capacitors. *C* **2018**, *4*, 31. [CrossRef]
13. Holtstiege, F.; Bärmann, P.; Nölle, R.; Winter, M.; Placke, T. Pre-Lithiation Strategies for Rechargeable Energy Storage Technologies: Concepts, Promises and Challenges. *Batteries* **2018**, *4*, 4. [CrossRef]
14. Cao, W.J.; Luo, J.F.; Yan, J.; Chen, X.J.; Brandt, W.; Warfield, M.; Lewis, D.; Yturriaga, S.R.; Moye, D.G.; Zheng, J.P. High Performance Li-Ion Capacitor Laminate Cells Based on Hard Carbon/Lithium Stripes Negative Electrodes. *J. Electrochem. Soc.* **2017**, *164*, A93–A98. [CrossRef]
15. Ghossein, N.E.; Sari, A.; Venet, P. Nonlinear Capacitance Evolution of Lithium-Ion Capacitors Based on Frequency- and Time-Domain Measurements. *IEEE Trans. Power Electron.* **2018**, *33*, 5909–5916. [CrossRef]

16. Keil, P.; Schuster, S.F.; Wilhelm, J.; Travi, J.; Hauser, A.; Karl, R.C.; Jossen, A. Calendar Aging of Lithium-Ion Batteries I. Impact of the Graphite Anode on Capacity Fade. *J. Electrochem. Soc.* **2016**, *163*, A1872–A1880. [CrossRef]

17. Savoye, F. Impact des impulsions périodiques de courant sur la performance et la durée de vie des accumulateurs lithium-ion et conséquences de leur mise en oeuvre dans une application transport. Ph.D. Thesis, Université Claude Bernard Lyon 1, Lyon, France, 2012.

18. Sun, X.; Zhang, X.; Liu, W.; Wang, K.; Li, C.; Li, Z.; Ma, Y. Electrochemical performances and capacity fading behaviors of activated carbon/hard carbon lithium ion capacitor. *Electrochim. Acta* **2017**, *235*, 158–166. [CrossRef]

19. Aida, T.; Murayama, I.; Yamada, K.; Morita, M. Analyses of Capacity Loss and Improvement of Cycle Performance for a High-Voltage Hybrid Electrochemical Capacitor. *J. Electrochem. Soc.* **2007**, *154*, A798–A804. [CrossRef]

20. Uno, M.; Kukita, A. Cycle Life Evaluation Based on Accelerated Aging Testing for Lithium-Ion Capacitors as Alternative to Rechargeable Batteries. *IEEE Trans. Ind. Electron.* **2016**, *63*, 1607–1617. [CrossRef]

21. Ghossein, N.E.; Sari, A.; Venet, P. Effects of the Hybrid Composition of Commercial Lithium-Ion Capacitors on their Floating Aging. *IEEE Trans. Power Electron.* **2018**. [CrossRef]

22. Ghossein, N.E.; Sari, A.; Venet, P. Interpretation of the Particularities of Lithium-Ion Capacitors and Development of a Simple Circuit Model. In Proceedings of the 2016 IEEE Vehicle Power and Propulsion Conference (VPPC), Hangzhou, China, 17–20 October 2016; pp. 1–5.

23. Sivakkumar, S.R.; Pandolfo, A.G. Evaluation of lithium-ion capacitors assembled with pre-lithiated graphite anode and activated carbon cathode. *Electrochim. Acta* **2012**, *65*, 280–287. [CrossRef]

24. Shellikeri, A.; Hung, I.; Gan, Z.; Zheng, J. In Situ NMR Tracks Real-Time Li Ion Movement in Hybrid Supercapacitor–Battery Device. *J. Phys. Chem. C* **2016**, *120*, 6314–6323. [CrossRef]

25. Azaïs, P.; Duclaux, L.; Florian, P.; Massiot, D.; Lillo-Rodenas, M.-A.; Linares-Solano, A.; Peres, J.-P.; Jehoulet, C.; Béguin, F. Causes of supercapacitors ageing in organic electrolyte. *J. Power Sources* **2007**, *171*, 1046–1053. [CrossRef]

26. German, R.; Sari, A.; Venet, P.; Zitouni, Y.; Briat, O.; Vinassa, J.M. Ageing law for supercapacitors floating ageing. In Proceedings of the 2014 IEEE 23rd International Symposium on Industrial Electronics (ISIE), Istanbul, Turkey, 1–4 June 2014; pp. 1773–1777.

27. Chaari, R. Evaluation et Modélisation du Vieillissement des Supercondensateurs pour des Applications Véhicules Hybrides. Ph.D. Thesis, Université Bordeaux 1, Bordeaux, France, 2013.

28. Zhang, T.; Fuchs, B.; Secchiaroli, M.; Wohlfahrt-Mehrens, M.; Dsoke, S. Electrochemical behavior and stability of a commercial activated carbon in various organic electrolyte combinations containing Li-salts. *Electrochim. Acta* **2016**, *218*, 163–173. [CrossRef]

 **batteries**

Article

A Suggested Improvement for Small Autonomous Energy System Reliability by Reducing Heat and Excess Charges

Christophe Savard [1],* and Emiliia V. Iakovleva [2]

[1] Mainate Labs, 77430, 25 rue de Sens, Champagne-sur-Seine, 77430, France & GEII Department,
    IUT Gratte-Ciel, 17 rue de France, 69627 Villeurbanne Cedex, France
[2] Electromechanical Department, Saint-Petersburg Mining University, 21st line 2,
    Saint-Petersburg 199106, Russia; em88mi@gmail.com
* Correspondence: cjs@mainate.com

Received: 30 January 2019; Accepted: 6 March 2019; Published: 11 March 2019

Abstract: Devices operating in complete energy autonomy are multiplying: small fixed signaling applications or sensors often operating in a network. To ensure operation for a substantial period, for applications with difficult physical access, a means of storing electrical energy must be included in the system. The battery remains the most deployed solution. Lead-acid batteries still have a significant share of this market due to the maturity of their technology. However, even by sizing all the system elements according to the needs and the available renewable energy, some failure occurs. The battery is the weak element. It can be quickly discharged when the renewable energy source is no longer present for a while. It can also be overloaded or subjected to high temperatures, which affects its longevity. This paper presents a suggested improvement for these systems, systematically adding extra devices to reduce excess charges and heat and allowing the battery use at lower charges. The interest of this strategy is presented by comparing the number of days of system failure and the consequences for battery aging. To demonstrate the interest of the proposed improvement track, a colored Petri net is deployed to model the battery degradation parameters evolution, in order to compare them.

Keywords: autonomous devices; lead-acid batteries; Petri nets

---

1. Context

Some devices operate autonomously by drawing and harvesting available energy in their environment. These kinds of devices are often used for signaling or for compilation and information data storage [1]. In this case, the sensors can have a network connection. Having an opportunity to provide renewable energy provides these small applications with sustainability in the achievement of their mission, which is not always the case for networks of sensors located in inaccessible places and subject to strong constraints that thus must optimize their consumption and their communications [2]. This paper looks at devices profiting daylight, but for which systematic forecast maintenance visits are spaced out over time. These stand-alone systems are often designed according to the schematic diagram in Figure 1. This is the case for devices that collect solar energy and convert it into electrical energy to power a device that operates continuously [3,4]. To take into account the weather conditions of the implantation site, it is necessary to associate a source of electrical energy storage [5]. For this, one or more batteries are included in the system. These batteries are often still lead-acid batteries because their mature technology, proven for many decades; their high mass and volume capacity, still superior to the present Li-ion batteries [6,7]; their apparent robustness; and their low production cost relative to the stored kilowatt-hour, make them competitive. They retain a significant market share, particularly since they can operate under low temperatures. This is especially true since it is necessary to oversize the battery in order to operate in the most degraded conditions: in winter, in rainy weather, or when the energy sensors are hidden under accumulated snow. Right now, however, power management units (PMUs), even though they are responsible for energy security and optimization, are not functioning optimally. Thus, the battery is often overcharged in summer, which causes rapid aging, shortening the system's useful operating time before failure. The failure consists of a current providing termination to the autonomous device. It is possible to add a complementary energy storage system to the autonomous device, for example, with a flywheel or in the form of thermal storage [8]. This avoids short feeding breaks, but is not enough in winter. Indeed, the low level of brightness associated with short periods of sunshine can also lead to battery discharges, such as the PMU disconnecting the autonomous device to protect the battery by discharging too much. The periods of low light are often long in winter and a second remote energy storage system should then be sized taking into account the probable duration of power failure. As a matter of principle, the PMU only works in two modes: Safe and Functional. A good quality PMU monitors the battery voltage to switch modes and also to set the battery charge mode, sending power from the energy harvesting system (EHS) when it is surplus to the device requirements, respecting as much as possible the principle of a charge in continuous current (CC) mode followed by a continuous voltage (CV) mode when the charge is close to its maximum. However, this does not guarantee that the battery lifespan is optimized. This paper proposes a simple approach to modeling battery aging and uses a Petri net to simulate the influence of the proposed adaptations on the aging aggravation. It suggests improvement to optimize energy management and reduce premature aging, even if it sometimes means current-shedding the device.

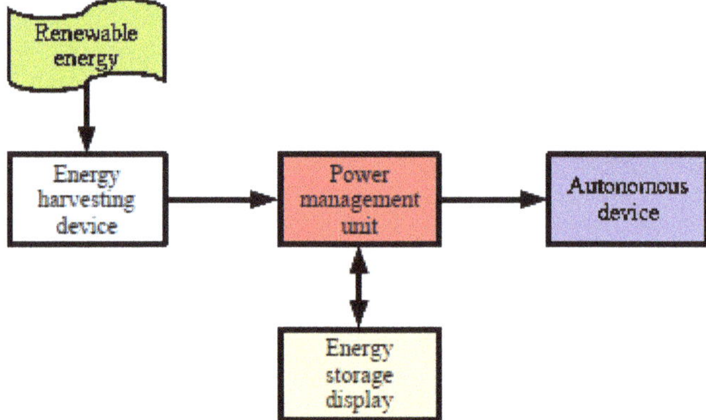

Figure 1. Autonomous system supply principle.

2. Battery Aging Issue

Lead-acid batteries used in these systems consist of several cells connected in series to provide the required voltage. The degradation of performance that results from aging cells is the consequence of three main phenomena, which are related to each other [9]. In the first place, the electrodes corrode, mainly during excessive recharges and when the battery remains in full charge for a long time. The positive electrode loses some of the active mass at each recharge-discharge cycle, but not linearly and continuously throughout the electrode. This reduction in mass decreases the amount of electrical charge that can be stored in the battery. In the second place, the electrodes cover themselves with lead sulphate. This sulphation is mainly related to a high depth of discharge (DoD) [9]. Typically, when the battery contains less than 25% of its maximum charge, it is considered in deep discharge. PMU manufacturers consider that $SoC = 25\%$ is a threshold below which the battery should not be discharged. Actually, below this threshold, the OCV decline accelerates because under low charges, the electrolyte contains only a low acid concentration. During a discharge, it is strongly diluted in the electrolyte [10]. The larger the DoD, the lower the acid concentration. In the third place, the amount of electrolyte decreases with use. This is the phenomenon of drying out, following the evaporation of water appearing during the chemical reactions [11]. This can be a result of high charges or high ambient temperature.

Obviously, even occasional battery short-circuiting aggravates its degradation. The PMU makes sure to remove this risk. A battery in deep discharge can be permanently damaged by sulphation. If the sulphate of lead formed during a weak discharge has a very fine crystalline structure, which will be easily decomposed by the charging current, the lead sulphate crystals created during the deep discharges are too big to be dissolved by the charging current [12]. In addition, the larger the crystals, the faster the voltage between the electrodes increases during recharge and decreases during discharge, reducing the amount of energy stored and restored [10]. When a battery remains discharged for too long, the lead particles dissolve in the electrolyte, and the solubility increases. They then form a lead hydrate, which crystallizes in the separator [9]. If these crystals come into electrical contact, pure lead dendrites (growth in branching) form and grow progressively, increasing self-discharge. This hydration phenomenon leads to internal short circuits. To reduce the formation of crystals, it is possible to resort to the pulsed-current technique [13,14] during fast refills and thus to reduce the effects on the aging aggravation. The pulse-current charging strategy is employed to recharge the battery with a direct current to which a middle frequency alternating current is added. This pulsed-current technique increases the battery lifespan [14] because it reduces the amount of lead crystals in the active material and minimizes the development of lead hydrate [12,13]. Corrosion also occurs due to self-discharge reactions between the lead and the sulfuric acid of the electrolyte. Under normal charging conditions, a protective layer of lead oxide is formed. It is unfortunately dissolved if the battery is discharged and the acid concentration is low, causing an acceleration of corrosion. Corrosion is irreversible.

The PMU measures the voltage across the battery [15]. It uses an internal model based on the relation that exists between the electric charge $Q(t)$ contained in the battery and the terminal voltage V_{bat}, even if it is not identical for the same charge, depending on whether the battery is recharging or discharging. Depending on V_{bat}, the system is placed in one of two operational modes. Any type of model allows the battery lifespan to be determined after calculation provided that it is based on a sufficient quantity of parameters and that it includes a suitable description of aging processes [11]. However, since Sauer and Al. [16] recall that no model perfectly correlates the aging processes of lead-acid batteries and their impact on performance, it is superfluous to resort to a perfectly precise model.

One of the advanced-PMU missions is to limit conditions that will accelerate degradation, such as maintaining a small size for sulphate crystals to avoid deep discharges, operating at a not too high ambient temperature to reduce internal evaporation, or finally reducing the time during which the battery is fully charged (cause of accelerated corrosion). In a contradictory manner, in order to limit the sulphation, it is necessary to regularly carry out a complete recharge, typically under a temperature oscillating around 45 °C, at least once every month. On the other hand, when the cell voltage falls below 2.1 V, the corrosion already present accelerates. Optimally, a cell voltage close to 2.25 V, called the float voltage, reduces corrosion. This voltage is often recommended in power supply systems for which corrosion is a major factor in aging. For a classic 12 V battery, consisting of six cells with a nominal voltage of 2.1 V, this leads to an ideal float voltage of 13.5 V. In applications such as autonomous photovoltaic panel power systems (A3PS), the battery continually discharges in the absence of sun, even when it does not provide power, due to its self-discharge. Typically, this self-discharge is close to 10% per month.

The battery conditions of use contribute to its aging and the acceleration of it. Cycling is another phenomenon that reduces the battery lifespan used in an A3SP. Since charge regularly oscillates between a maximum (often the full charge) and a minimum (close to the deep discharge), it combines the corrosion and its aggravation. In addition, the batteries must not be placed on undercharging for long periods of time, otherwise sulphation will occur. The temperature has a positive influence on the battery capacity, which increases by just under 1% per degree Celsius. Since sulphate crystals are more readily dissolved at an elevated temperature, it appears preferable to perform regular full charges at a high temperature while keeping the battery at a lower temperature outside these times.

This paper is organized as follows: after recalling the different causes of lead-acid batteries aging, a simple cell model based on different parameters is proposed in order to control the aging. To achieve this, Part IV will present an implantable model in the PMU to track battery operation, as well as a suggested improvement track. The model is based on Petri nets, whose principles are recalled in the previous chapter. Finally an example of A3SP is presented before its operation is analyzed by comparing the impact on aging of the suggested improvement.

3. Battery Parameters

In a conventional approach, the battery can be modeled as a voltage generator according to the first or second order model of Thevenin. The open circuit voltage (OCV) is related to the amount of electric charge in the battery at time t, denoted $Q(t)$. The battery can store a maximum amount of charge, noted as Q_0, which will decrease with aging, so with time t. Therefore, to quantify the charge contained in the battery, it is defined as a state of charge (SoC) at any time t, defined by Equation (1).

$$SoC(t) = \frac{Q(t)}{Q_0(t)} \tag{1}$$

The equivalent series resistance (ESR) of the first order Thevenin model reflects the electrical consequences of heat loss in the electrolyte and in the connections, as well as the ohmic resistance of the electrodes and connections. The resistance value grows with aging since it represents a brake on the current flow in the physical structure of the battery.

Aging is directly related to the battery use: cycling, charges and deep discharges, operating, and storage temperature. In cells, the lead (Pb) and lead oxide (PbO_2) conversion to $PbSO_4$ during the reaction with electrolyte sulfuric acid induce high mechanical stress, and the volumes of $PbSO_4$ are about double. The weak charge fluctuation only leads to small variations in the electrolyte volume and thus to lower mechanical stresses that will have less impact on cell aging. Deep discharges cause corrosion and sulphation. We consider that a battery that operates under optimal conditions must age in the same way as when it was tested and that its lifespan was determined by its manufacturer. Optimally, the charge varies around a float capacity at each cycle. The float voltage which ensures the least possible aging for a battery, also when it is not stressed [17], is thus close to 2.25 V. This value corresponds to $SoC = 0.85$. Subject to different operating regimes, its aging will worsen. Thus, beyond a charge fluctuation of 30%, the impact on aging can be considered to increase linearly. On the other hand, to model the temperature impact, it is possible to use the Arrhenius law, because of the electrochemical reaction process temperature dependence. In a simplified manner, it is generally considered that the service life is halved for a device operating at a temperature of 10 °C and above. For a lead-acid battery, the optimum operating temperature is close to 30 °C, as specified by Gauri and Al [18].

Some articles, such as [19], are based on a neural network to model a lead-acid battery operation and integrate aging, so as to overcome the parameter variability through self-learning and model parameter correction. Indeed, the estimate lead-acid battery SoC is more delicate than for more modern technologies because many side reactions occur in the electrolyte and interfaces, in addition to the losses incurred during the charging process. Haddad and Al. propose a model differentiating the series resistance according to the current direction, in parallel with an over-voltage capacity simulating the return to the thermodynamic equilibrium after a recharge or a discharge. The cell voltage is different, depending on whether it is recharging (higher voltage) or discharging (lower voltage), because of relaxation phenomena [20]. A voltage generator controlled by the SoC is completed in parallel by a resistor representing the self-discharge. In [21], the SoC calculation precision is dependent on the number of equations, allowing the authors to integrate the variable parameters such as the temperature, the quantity of electrolyte, and the electrode sizes, as well as the aging. However, in a PMU, the *SoC* is often determined from the simple OCV measurement, with corrective measures made when the battery is placed at rest or following heuristic methods. The present work does not retain the principle of system operation real-time monitoring, but focuses on the state in which it is at dusk. If during the daytime, the autonomous device has not been fed continuously, it is considered that the system has failed. Thus, the parameters to be followed for the battery are the *SoC* and the OCV voltage, and for the system, the parameter is if the mission has been fulfilled or not. Finally, based on the operating conditions influencing the battery lifespan, we define a fourth parameter: aging aggravation, noted as *Agg*. If the battery operated under optimum conditions (temperature close to 30 °C, average OCV close to the 2.25 V float voltage, and daily charge variation less than 30%), *Agg* would be worth 1. Reference [22] shows that the battery lifespan can be considerably reduced when it does not work under these conditions. We will then consider that the following three phenomena, when in convolution, lead to aggravation *Agg* = 8: high temperature, full charge, and low *SoC*. In the latter case, we consider two thresholds: a strong degradation threshold for SoCs less than 25% (inducing a factor 2) and average degradation for SoCs less than 50% (inducing a factor of 1.5). This is the reason why the current PMUs shed the autonomous device as soon as the cell OCV falls below 2.1 V, i.e., 12.6 V for a six-cell battery. This voltage corresponds to SoC = 50%, as shown schematically in Figure 2, showing the relation between SoC and OCV as a yellow dotted line, expressed in mV, for a 24 Ah lead-acid battery cell. When the cell is charged at half of its capacity (*SoC* = 0.5), its OCV reaches the nominal value of 2.1 V. The more the cell is charged, the higher its OCV is. When the cell approaches full charge, its OCV increases faster with *SoC*. As previously indicated, the OCV decreases more quickly when SoC is less than 25%.

Since the model describing the relationship between electric charge and OCV does not need to enter either the chemical parameters evolution trend or the electrical parameters evolution, but must follow the four previously defined parameters, the fundamental relation between the SoC and the OCV can be simplified according to the red line in Figure 2. This allows the voltage to be expressed with respect to the state of charge according to three linear zones. The limits between these zones correspond to *SoC* = 0.85 (float capacity), *SoC* = 0.5 (voltage of 2.1 V serving as a threshold for current PMUs), and *SoC* = 0.25 (threshold before a deep discharge causing sulphation). Finally, the battery self-discharge must be integrated into the A3PS energy balance.

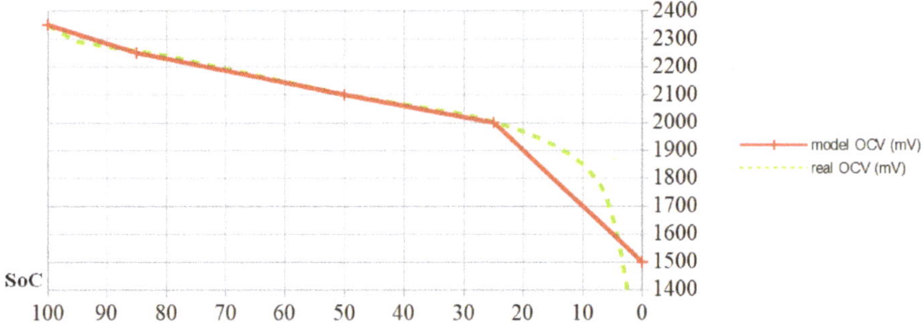

Figure 2. Linearization of the OCV (SoC) curve (SoC) by parts for a 24 Ah lead-acid cell.

4. Petri Net Tool

In order to realize simulation of cell behavior and parameters evolution, a Colored Petri network (PN) is proposed [23]. Figure 3 shows an example of a PN, which associates formal semantics and visual representation with precise syntax and a graphical language, unlike a state machine with its own alphabet. A PN is represented by a graph, consisting of states, transitions, and arcs connecting these states and transitions. It belongs to the world of discrete events. An event is embodied by the firing of a transition. The operation of the system described by an PN is thus visualized by a synthetic structured and compact representation. Formally, a PN has properties such as liveliness, boundary, persistence, and reset [24]. Dedicated mathematical tools ensure that it is analyzed.

In a PN, ovoids represent the states, for example, state "A". Transitions are represented by rectangles, i.e., transition "a". Oriented arcs link them, implying a consecutive relationship. The states may contain a marking of one or more tokens, as one token in state "A" and two in state "B". When at least one token is in a state, this state is active. On the contrary, state "F" is not active. When all upstream states are marked (contain at least one token), the transition is firable. If the associated guard is respected, it is shot. A guard is a logical condition on upstream markings and/or on external variables. For instance, "d" transition guard "[x > 3]" means that state "D" has to contain at least three tokens. Arcs can be weighted by a number of needed tokens greater than unity. Then, the net is designated as a generalized Petri net. The upstream place token(s) are then transferred to the downstream state(s). In the Figure 3 example, if a token is present in state "A" and three tokens are in state "B", transition "a" can be passed. In this case, state "C" acquires two additional tokens. On the other hand, if a token is present in state "E" and state "D" contains at least three tokens, transition "d" is fired. A function weights the arc, leading to state "F". So, here it receives $10x$-xy-y tokens.

Tokens can have different colors. A color assigned to a token may, for example, be a set of positive integers (color INT, for integer); boolean (BOOL); a restriction of some listed elements, such as integers from 0 to 10; or a list of predefined colors, such as the days of the week or a list of day codes. A token may consist of a different color token set. It is then a muticolor token. Thus, formally, a PN is defined by a 9-tuple given by relation (2).

$$R = \{S, \Theta, A, \Sigma, W, C, G, E, Y\} \tag{2}$$

It consists of the following fields:

S: finite set of states;

Θ: finite set of transitions, dissociated from states;

A: finite set of oriented arcs connecting a state to a transition or vice versa;

Σ: non-empty finished set of color sets;

W: set of colored variables such as $\forall; \in W, \exists$ color() $\in \Sigma$; $\tag{3}$

C: $S \rightarrow \Sigma$: function assigning a color to each state;

G: set of guard functions, associated with each transition, of Boolean type;

E: function set associated with each arc, of the state type connected to the arc;

Y: set of initialization functions, associated with each state, of the state type.

Using a formal model makes it easy to visualize physical phenomena such as, for instance, a cell electrical voltage, V_{cell}. [25] gives the example of an association on photovoltaic panels in a micro-grid. Wind turbines and batteries were modeled by a Petri Net, with the states being associated with the source and battery states, so as to optimize the electrical flows on the network. Previously, PNs have already been used to model single-source system operation modes, depending on the demand and energy stored amount in [26]. An algorithm has been implemented to control the network so as to modulate the energy stored in the battery.

We use this tool to determine the evolution of parameters associated with a model in order to determine system performance with a PMU operating in a conventional manner and with the suggested improvements.

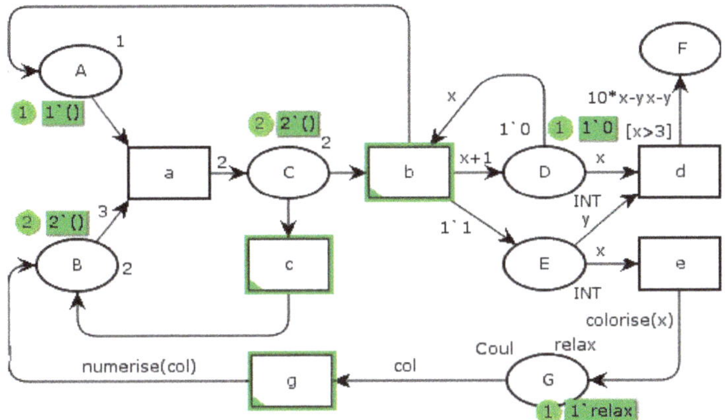

Figure 3. Colored Petri net example.

5. Model

To control the battery aging, it is necessary to add a cooling system and a temperature sensor to the A3PS. Indeed, often due to lack of space reasons, the PMU, the battery, and the autonomous device control unit are inserted into the same box. If possible, this box should not be exposed to the sun. However, to reduce the temperature impact, it appears necessary to add a ventilation system, so as to limit the temperature in the box, at most, to a few degrees above the outside temperature. In the same way, depending on the sensitivity of the control unit and the minimum winter temperatures, it may be necessary to add a heating system. In the example presented here, only the high temperature control will be taken into consideration.

A conventional PMU meets the following specifications:

- When the current I_{pv} from the EHD is greater than the current I_{spe} requested by the autonomous device, the excess current is sent to the battery. Otherwise, the battery provides the missing current. The system then switches from a Direct mode with possible recharging of the battery to a battery supply mode, with a discharge of the battery under a current less than or equal to the demand. The system operates in two modes, as shown in Figure 4;
- PMU uses the CC-CV procedure to recharge the battery, with a switching threshold of 2.3 V per cell and a CV mode voltage of 2.35 V.

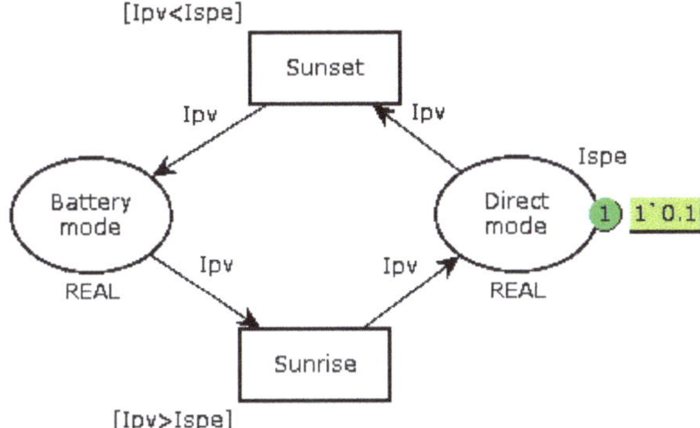

Figure 4. PMU operating modes according to the current coming from the photovoltaic panel.

To lessen the aging of the battery, we propose to add the following constraints:

- If possible, keep the temperature around 30 °C, and otherwise limit it to a value close to the ambient temperature;
- To control the battery charge:
- perform complete recharges with a temperature between 40 and 50 °C;
- If a battery has not been fully charged for 30 days and the I_{pv} current is sufficient, shed the device so as to perform a full charge;
- If a full charge has occurred less than 15 days ago and there is a risk of doing another full charge, redirect the excess current to an auxiliary device. By default, this auxiliary device should be a cold lighting system, implanted inside the box;
- Replace the bi-modal load-shedding when the battery voltage is less than 12.6 V (6 cells of 2.1 V) by a three-mode operation, as shown in Figure 5. When the voltage is lower than a first threshold V_1, the system goes into Intermediate mode, which consists of prioritizing battery recharging and reducing the current supplied to the autonomous device if it can operate in this way. It is also possible to add a mechanical storage system, such as a small flywheel, so as to continue the mission if it requires the nominal current I_{spe}. When the battery voltage drops below the second threshold V_2, a break-contact mechanical relay directly connects the EHD to the battery. The value of these thresholds must be slightly different in recharge and in discharge (V_{1d}, V_{1c} and V_{2d}, V_{2c}) so as to take into account internal relaxation phenomena;
- Monitor the full charge (voltage across a cell greater than 2.35 V). If the voltage reaches this threshold, affect the excess current to the auxiliary device.

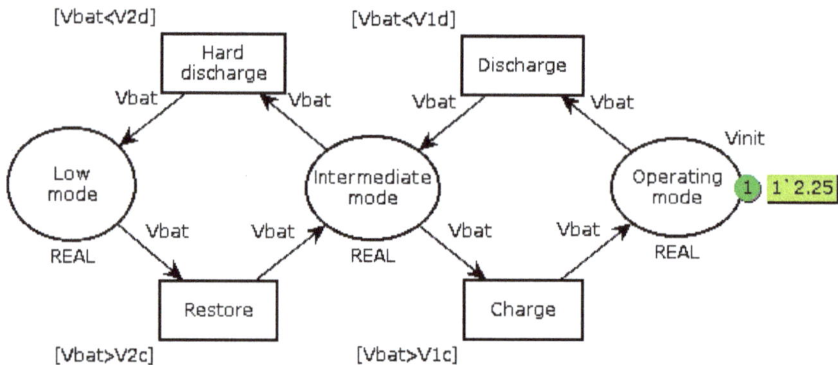

Figure 5. Proposed operation of the system in three modes, depending on the cell voltage.

In the simulation presented below, the usual functions performed by the PMU, such as the CC-CV charging, are considered effective and do not result in simulation. The complementary improvement of including a small storage system in the autonomous device, such as a flywheel, is not taken into account. We will consider the worst case, which is the one where the device always asks I_{spe}, including when the system is running in Intermediate mode.

6. Sizing an Example

The battery capacity must be calculated so as to avoid a winter failure. Sizing must be made according to the A3PS location and include the usual weather conditions. We take into consideration the coverage of the solar panels by snow for five consecutive days to size our system. During this time, the *SoC* should not have fallen below 30%, starting from the float capacity. So, we add a safety margin of 5% before the battery goes into deep discharge, corresponding to a cell *SoC* = 25%. We consider an isolated device, having to work in total autonomy, implanted in the North of France (for example, in Ardennes), which continuously consumes I_{spe} = 100 mA. This implies a battery capacity greater than 262 Wh. A battery of 12 V–24 Ah satisfies this need. The monthly output of a 20 W-peak PV panel can be determined via the European Commission's Photovoltaic Geographical Information System (PVGIS) website [27]. For the chosen location, the monthly distribution of the energy collected by an EHV associated with a photovoltaic panel of 20W is specified in Figure 6. Simulation is made for a 30° inclination of a south-facing panel, optimal in France. In December, average production will only be 14 Wh per day. Two panels of this type must be installed in conjunction with a device that disconnects one of the two panels from April to October to reduce the excess energy sent by the EHV in sunny periods. In order for the system to operate with a fan on hot days, it is necessary for the EHV to provide sufficient current. To cool the inside temperature to 2 °C above the outside temperature, it is necessary to have a fan delivering at least 35 m^3/h. Any professional fan, operating at a DC voltage of 12 V, 0.6 W, and 87 m^3/h of flow is suitable. In the Ardennes, the average maximum temperature is 28 °C and the record is close to 40 °C. To study the worst case, we consider that the box is not completely implanted in the shade and that it requires strong ventilation, which is only reduced in winter.

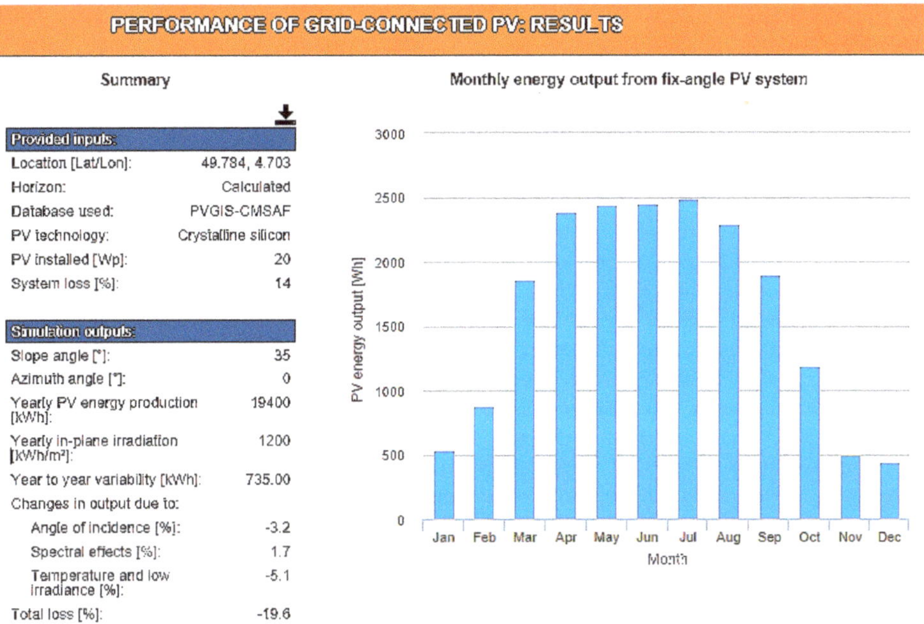

Figure 6. Monthly energy captured by a 20W solar panel installed in the North of France simulation, from the European Commission PVGIS. [27].

The system thus fitted makes it possible to operate, even in winter, as shown by Table 1, which includes the self-discharge in the daily energy requirements. The table details the EHV's monthly and daily average energy production and compares them with the monthly autonomous device associated with the fan consumption. The calculations are made assuming a warm-up all year long during the days of strong sunshine and with a second panel feeding the system the four winter months.

Table 1. Energy balance between production and consumption in the system.

Date	Monthly Energy Product (Wh) [27]	Daily Energy Product (Wh)	Daily Consumption (Wh) Device	Daily Consumption (Wh) Fan	Needs	Balance
January	1068	36	28.8	0.0	29.52	6.1
February	1742	58	28.8	0.7	30.24	27.8
March	1860	62	28.8	2.9	32.40	29.6
April	2380	79	28.8	10.8	40.32	39.0
May	2440	81	28.8	13.0	42.48	38.9
June	2451	82	28.8	14.4	43.92	37.8
July	2491	83	28.8	14.4	43.92	39.1
August	2290	76	28.8	14.4	43.92	32.4
September	1900	63	28.8	12.2	41.76	21.6
October	1190	40	28.8	7.9	37.44	2.2
November	992	33	28.8	2.2	31.68	1.4
December	900	30	28.8	0.2	29.68	0.3

The balance sheet shows two extremes. The first is in winter, when the days have the least amount of sunshine. It is possible that on some days, the battery will not be charged enough to complement the autonomous device needs, which causes a system failure. The second extremum is in July. Hot temperatures associated with high charges will lead to early aging of the battery. Thus, it is sufficient to test the system behavior in December and July and compare the number of failure days for conventional operation in two modes and for operation with the suggested improvement. Table 2 shows the change in luminous flux, expressed in lux, as a function of weather conditions. A cloudy sky reduces the value of sunshine by about half, as well as the production of electrical energy for a solar EHV, since it is directly related to the sunshine degree. Snow accumulated on the ground reduces the energy collected to almost zero. A thunderstorm can be considered as masking the energy collection for three day hours. The A3PS is not established in a tropical zone, so it is also considered that during a rainy day in summer, it is not necessary to ventilate the box. The same principle of summer and winter cycling is included in the standard cycling test IEC 61427 of the International Energy Agency's IEA PVPS T3-11: 2002, Implementing Agreement on Photovoltaic Power Systems [28], which aims to evaluate the battery lifespan, expressed in terms of capacity loss, by simulating a typical use. It consists of testing the battery in winter conditions, with an SoC varying between 5 and 35% and in summer conditions, with an SoC varying between 75 and 100%. In the simulation carried out here, the constraints imposed lead to an operation, on the one hand, with an SoC between 50% and 100% (conventional PMU) and with an SoC always less than 100% and greater than 25%.

Table 2. Energy production rate according to the sunshine.

Weather Conditions	Brightness (lux)	Energy Production Rate
Sunrise	20,000	100.00%
Clouds	10,000	50.00%
Rain	40	0.20%
Snow	4	0.02%
Storm	10	82.39% (3 h of rain)

The Figure 7 PN is used to model the system behavior according to whether it works in a conventional manner or with the proposed improvements. The automaton is interested in the cell behavior. It can be resumed to be implanted in the PMU to estimate the cell parameters. "Cell" is a multicolor token, including a color for the electric charge, the OCV, the aggravation rate, the complementary device state, and the mission success. The "Cells" state has an initial marking corresponding to the ideal conditions: float voltage, neutral aggravation rate ($Agg = 1$), complementary light device extinguished, and mission fulfilled (autonomous device powered by I_{spe}). The "Weather conditions" state contains a token list "CycleF" simulating the sunshine conditions of each day of the tested month. The days are coded according to the relation (4) coding:

1 - day of full sun in summer;

2 - summer day with a thunderstorm, storm lasting on average three hours;

3 - cloudy day in summer;

4 - rainy day in summer; (4)

5 - day of full sun in winter;

6 - cloudy day in winter;

7 - rainy day in winter;

8 - snow day.

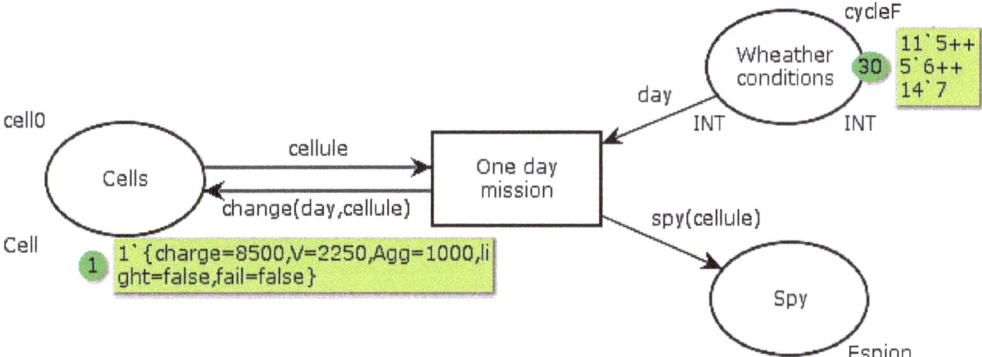

Figure 7. PN simulating the daily operation and listing the selected parameters.

The function "change" modifies the marking in the state "Cells" after the transition "One day mission" fire. It is given by algorithm 1. The values introduced in the sub-functions used are those determined in Tables 1 and 2, expressed in thousandths of units. The "Spy" state is used to display the token list corresponding to the daily cell parameters. The main colors and statements of this PN are given in relations (5). The color "light" is "true" when the complementary device is powered, to keep the battery slightly below the full charge (SoC = 99%). In conventional operation, "light" is always "false".

colset Cell = record charge:INT * V:INT * Agg:INT * light:BOOL * fail:BOOL;

colset Espion = record V:INT * Agg:INT * light:BOOL * fail:BOOL;

val cell0 = {charge = 8500, V = 2250, Agg = 1000, light = false, fail = false}; (5)

val s1 = 2465: val s2 = 1791; val s3 = 553; val s4 = 850; val s5 = 2086; val s6 = 528;

val s7 = 1024;

var cellule:Cell; var day:INT;

Algorithm 1: Influence of a Mission Day on System Parameters

```
fun change(jour:INT,c:Cell) =
if (jour = 1) then {{charge = (Downch((#charge(c)),s1,(#fail(c)))),
V = Vdown((#charge(c)),s1,(#fail(c))), Agg = (Bas((#charge(c)),s1)),
light = Allume((#charge(c)),s1),fail = failure(#charge(c))}
else if (jour = 2) then {charge=(Downch((#charge(c)),s2,(#fail(c)))),
V = Vdown((#charge(c)),s2,(#fail(c))), Agg = (Bas((#charge(c)),s2)),
light = Allume((#charge(c)),s2),fail = failure(#charge(c))}
else if (jour = 3) then {charge = Highch((#charge(c)),s3,(#fail(c))),
V = Vup((#charge(c)),s3,(#fail(c))), Agg=(Haut((#charge(c)),s3)),
light = Allume((#charge(c)),s3),fail = failure(#charge(c))}
else if (jour = 4) then {charge = (Downch((#charge(c)),s4,(#fail(c)))),
V = Vdown((#charge(c)),s4,(#fail(c))), Agg = (Bas((#charge(c)),s4)),
light=Allume((#charge(c)),s4),fail=failure(#charge(c))}
else if (jour = 5) then {charge = (Downch((#charge(c)),s5,(#fail(c)))),
V = Vdown((#charge(c)),s5,(#fail(c))), Agg = (Bas((#charge(c)),s5)),
light = Allume((#charge(c)),s5),fail = failure(#charge(c))}
else if (jour = 6) then {charge = Highch((#charge(c)),s6,(#fail(c))),
V = Vup((#charge(c)),s6,(#fail(c))), Agg = (Haut((#charge(c)),s6)),
light = Allume((#charge(c)),s6),fail = failure(#charge(c))}
else {charge = (Downch((#charge(c)),s7,(#fail(c)))),
V = Vdown((#charge(c)),s7,(#fail(c))), Agg = (Bas((#charge(c)),s7)),
light = false,fail = failure(#charge(c))};
```

The values S_i correspond to the daily charge variation of the type i (from 1 to 8) day. It is considered that the weather is invariant on a full day. In the calculation of the aging aggravation parameter, we have considered that, as previously stated, an *SoC* of less than 0.5 implies an aggravation factor *Agg* of 1.5 and 2 if *SoC* < 0.25. The high battery charges also lead to an aggravation factor of 1.5, and then 2, 3, and 4, when the SoC is above the 90%, 95, 99, and 100% thresholds, respectively, in the absence of thermal regulation, so that the convolution of corrosion and evaporation phenomena are taken into account. When the temperature is kept below the outside temperature by more than 2 °C, *Agg* is 1.5 and 2 beyond the thresholds of 90% and 95%, respectively. The example proposed here exists and has shown two big faults. In the first summer after installation, the temperature in the cabinet was enough to literally melt the PMU. Then, in winter, despite the cessation of the mission for a *SoC* < 0.5, the battery went into deep discharge and could not be recharged without external intervention.

7. Analysis and Comparisons

The weather of the two extreme months is synthesized in the "weather" columns of Table 3, which also gives the energy produced and the energy balance, in Wh, as well as the variation that it represents with respect to the battery Q_0 capacity. The monthly amount of solar energy corresponds to the average indicated by Table 1. In the example used, because of the similarity in daily variations in the battery charge, the type of day 8 (snow) is merged with the type 7 (rain). The initial value "CellF" returns in order the 30 or 31 values of the column "day code" for the month considered.

Table 3. Daily data of simulations on extreme months.

Day	Weather	Energy Product (Wh)	Balance (Wh)	Charge Change	Day Code	Weather	Energy Product (Wh)	Balance (Wh)	Charge Change	Day Code
			December					July		
1	sunshine	89.76	60.08	20.86%	5	sunshine	110.1	71.0	24.65%	1
2	snow	0.02	−29.66	−10.30%	8	clouds	55.0	15.9	5.53%	3
3	snow	0.02	−29.66	−10.30%	8	sunshine	110.1	71.0	24.65%	1
4	snow	0.02	−29.66	−10.30%	8	sunshine	110.1	71.0	24.65%	1
5	snow	0.02	−29.66	−10.30%	8	storm	90.7	51.6	17.91%	2
6	snow	0.02	−29.66	−10.30%	8	rain	0.2	−24.5	−8.50%	4
7	rain	0.18	−29.56	−10.24%	7	rain	0.2	−24.5	−8.50%	4
8	rain	0.18	−29.56	−10.24%	7	clouds	55.0	15.9	5.53%	3
9	rain	0.18	−29.56	−10.24%	7	sunshine	110.1	71.0	24.65%	1
10	rain	0.18	−29.56	−10.24%	7	sunshine	110.1	71.0	24.65%	1
11	clouds	44.88	15.20	5,28%	6	sunshine	110.1	71.0	24.65%	1
12	rain	0.18	−29.56	−10.24%	7	clouds	55.0	15.9	5.53%	3
13	clouds	44.88	15.20	5,28%	6	clouds	55.0	15.9	5.53%	3
14	sunshine	89.76	60.08	20.86%	5	sunshine	110.1	71.0	24.65%	1
15	sunshine	89.76	60.08	20.86%	5	sunshine	110.1	71.0	24.65%	1
16	rain	0.18	−29.56	−10.24%	7	sunshine	110.1	71.0	24.65%	1
17	rain	0.18	−29.56	−10.24%	7	sunshine	110.1	71.0	24.65%	1
18	clouds	44.88	15.20	5,28%	6	sunshine	110.1	71.0	24.65%	1
19	sunshine	89.76	60.08	20.86%	5	storm	90.7	51.6	17.91%	2
20	clouds	44.88	15.20	5,28%	6	rain	0.2	−24.5	−8.50%	4
21	rain	0.18	−29.56	−10.24%	7	clouds	55.0	15.9	5.53%	3
22	rain	0.18	−29.56	−10.24%	7	sunshine	110.1	71.0	24.65%	1
23	rain	0.18	−29.56	−10.24%	7	sunshine	110.1	71.0	24.65%	1
24	sunshine	89.76	60.08	20.86%	5	sunshine	110.1	71.0	24.65%	1
25	clouds	44.88	15.20	5.28%	6	storm	90.7	51.6	17.91%	2
26	rain	0.18	−29.56	−10.24%	7	storm	90.7	51.6	17.91%	2
27	rain	0.18	−29.56	−10.24%	7	clouds	55.0	15.9	5.53%	3
28	sunshine	89.76	60.08	20.86%	5	clouds	55.0	15.9	5.53%	3
29	sunshine	89.76	60.08	20.86%	5	sunshine	110.1	71.0	24.65%	1
30	clouds	44.88	15.20	5.28%	6	storm	90.7	51.6	17.91%	2
31	clouds	44.88	15.20	5.28%	6					

The OCV and the aging aggravation evolution are given, respectively, in Figures 8 and 9 for the months of December and July. In December, the battery OCV with the suggested improvements is permanently lower or equal to that obtained with a conventional PMU. In both cases, due to weather conditions, the mission is often unfilled. This is particularly the case when snow covers the photovoltaic panels or in a durably rainy period. With a dual-mode PMU, the number of failure days is much greater since it stops feeding the autonomous device when the SoC becomes less than 50%. The suggested improvements allow us to divide this number of unavailability days of almost three, in this example. This is at the cost of a lower OCV voltage, which reflects a greater restored energy. To preserve the battery, it is necessary to perform a full charge at least once a month, to the detriment of the autonomous device supplying power. The simulation giving the dashed curves respects this constraint (V_{cell_adapt} and $Fail_{adapt}$). To achieve this full charge, from December 28, the mission success is set aside to affect all the current I_{pv} to recharge the battery. When it reaches the full charge, the current is reassigned to the autonomous device. This forced full charge induces an additional three days of failure, but the number of failure days does not reach the number in the conventional solution. If this constraint was not applied, the voltage would follow the violet curve $V_{cell\_improve}$ and the failure according to the cyan curve $Fail_{improve}$.

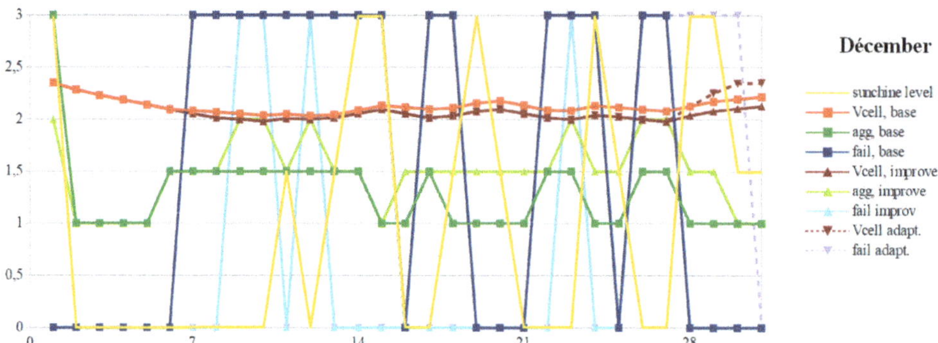

Figure 8. Simulated variations in OCV, failure, and aging aggravation, with or without the suggested improvements slopes, in December.

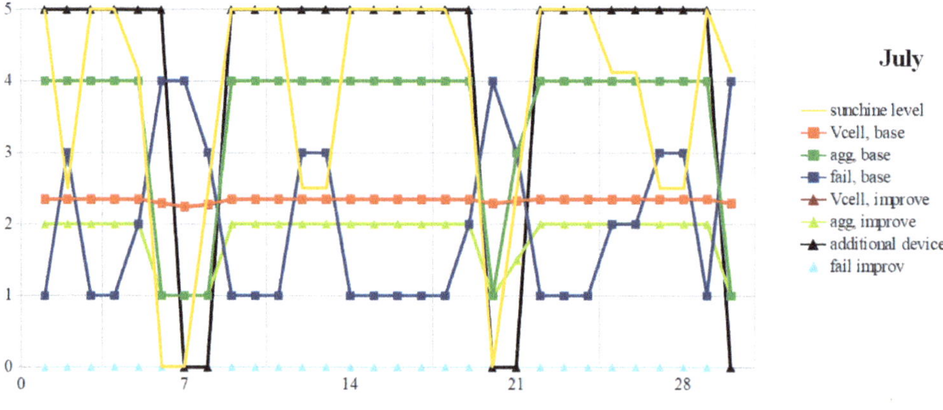

Figure 9. Same curves, in July.

Since the battery charge is lower in winter with the suggested improvements, the cells will suffer more from sulphation. Their aging will worsen, by a little more than two thirds in the example. To reduce its impact, the pulsed-current technique should be used during full forced charging. As the battery is not overcharged in winter, the electrodes will not corrode. Advanced techniques such as the dynamic definition of the pulsed-current frequency have been described in [13]. It consists of minimizing the battery impedance, consisting of the ESR in series with the parallel combination of charging resistance with the over-voltage capacity. With this technique, it is necessary to test the full charge and to vary the frequency, depending on the *SoC*. As a result, corrosion will not be accelerated by these deep discharges.

In summer, the battery is often carried in full charge, even if the weather remains rainy for a few days. With a conventional PMU, the battery is almost always overcharged, which accelerates corrosion and aggravates aging. Thanks to the additional device, the battery is never overcharged. It works almost continuously, except during rainy periods. Thus, the battery ages three times slower in summer with the proposed tracks.

A two-dimensional performance indicator can be established by adding, on the one hand, the December and July daily aging aggravations, and on the other hand, the failure days. It is given by relation (6):

$$P = \begin{bmatrix} \sum\limits_{i=1st\ of\ july}^{july\ 31} Agg_i a + a \sum\limits_{j=1st\ of\ december}^{december\ 30} Agg_j \\ \sum\limits_{i=1st\ of\ july}^{july\ 31} Fail_i a + a \sum\limits_{j=1st\ of\ december}^{december\ 30} Fail_j \end{bmatrix} \tag{6}$$

The value of this indicator is given in Equation (7):

$$P_{standardPMU} = \begin{bmatrix} 74 \\ 16 \end{bmatrix} a; \ aP_{improvedPMU} = \begin{bmatrix} 24.5 \\ 6 \end{bmatrix} \tag{7}$$

Overall, over the year, with the suggested improvements, aging aggravation is halved and the number of failure days is reduced more than twice. The aging aggravation due to deeper discharges in winter is offset by the removal of summer full charges, which results in less electrolyte evaporation and more electrode corrosion.

8. Conclusions

Isolated devices can operate in autarky as long as they have a system to capture and store renewable energy. The battery and the energy sensor sizes are sometimes significantly oversized compared to the autonomous device daily energy requirements. Their sizing is often the result of a compromise between the equipment size, failure acceptance (non-power supply of the device), and battery aging. To preserve the batteries, the PMU stops supplying the autonomous device when the SoC drops to 50% in winter, while in summer, the battery is regularly overloaded. Despite this, the battery is often in full charge or heavily discharged when the primary energy source is solar. In this paper, a track to reduce worsening battery aging is proposed. In summer, the battery can thus age less while guaranteeing a greater reliability in winter, by adding a ventilation system, often forgotten in this kind of equipment and tolerated to maintain the mission for battery charges lower than that conventionally admitted in winter. In addition, to reduce sulphation caused by deeper discharges, it is then possible to perform full forced point charges in winter, while ensuring the autonomous device power supply for longer. Ultimately, operating under reduced temperatures but at lower charges increases sulphation, which must be controlled by pulsed currents, while reducing evaporation and without aggravating corrosion.

Author Contributions: Original draft: C.S.; Review and editing: C.S. and E.Y.

Funding: This research received no external funding.

Conflicts of Interest: The authors declare no conflict of interest.

References

1. Zairi, S.; Niel, E.; Zouari, B. Global generic model for formal validation of the wireless sensor networks properties. *IFAC Proc. Vol.* **2011**, *44*, 5395–5400. [CrossRef]
2. Rashvand, H.F.; Abedi, A.; Alcaraz-Calero, J.M.; Mitchell, P.D.; Mukhopadhyay, S.C. Wireless sensor systems for space and extreme environments: A review. *IEEE Sens. J.* **2014**, *14*, 3955–3970. [CrossRef]
3. Singh, K.; Moh, S. A comparative survey of energy harvesting techniques for wireless sensor networks. *Adv. Sci. Technol. Lett.* **2018**, *142*, 28–33.
4. Belsky, A.A.; Dobush, V.; Ivanchenko, D. Wind-PV-Diesel Hybrid System with flexible DC-bus voltage level. In Proceedings of the 2014 Electric Power Quality and Supply Reliability Conference (PQ), Rakvere, Estonia, 11–13 June 2014; pp. 181–184.

5. Yakovleva, E.V.; Skamin, A.N.; Belskiy, A.A. Configuration of a standalone hybrid wind-diesel photoelectric unit for guaranteed power supply for mineral resource industry facilities. *Int. J. Appl. Eng. Resour.* **2016**, *11*, 233–238.

6. Savard, C. *Le Stockage de l'Énergie Électrique*; Editions Universitaires Europeennes: Paris, France, 2017.

7. Buckley, D.N.; O'Dwyer, C.; Quil, N.; Lynch, R. Electrochemical Energy Storage. *Issues Environ. Sci. Technol.* **2019**, *45*, 115–149.

8. Stritih, U.; Mlakar, U. Technologies for seasonal solar energy storage in buildings. In *Advancements in Energy Storage Technologies*; IntechOpen: London, UK, 2018. [CrossRef]

9. Ruetschi, P. Aging mechanisms and service life of lead-acid batteries. *J. Power Sour.* **2004**, *127*, 33–44. [CrossRef]

10. Bullock, K.R.; Weeks, M.C.; Chalasanik, S.; Murugesamoorthi, K.A. A predictive model of the reliabilities and the distributions of the acid concentrations, open-circuit voltages and capacities of valve-regulated lead/acid batteries during storage. *J. Power Sources* **1997**, *61*, 139–145. [CrossRef]

11. Liaw, B.Y.; Jungst, R.G.; Urbina, A.; Paez, T.L. Modeling of lithium ion cells. A simple equivalent-circuit model approach. *Solid State Ionics* **2004**, *175*, 835–839.

12. El Mehdi, L.; El Filali, A.; Zazi, M. Impact of Pulse Voltage as Desulfator to Improve Automotive Lead Acid Battery Capacity. *Int. J. Adv. Comput. Sci. Appl.* **2017**, *8*, 522–526. [CrossRef]

13. Lam, L.T.; Ozgum, H.; Lim, O.V.; Hamilton, J.A.; Vu, L.H.; Vella, D.G.; Rand, D.A.J. Pulsed-current charging of lead-acid batteries—A possible means for overcoming premature capacity loss. *J. Power Sour.* **1995**, *53*, 215–228. [CrossRef]

14. Chen, L.R. A design of an optimal battery pulse charge system by frequency-varied technique. *IEEE Trans. Ind. Electron.* **2007**, *54*, 398–405. [CrossRef]

15. Mulyono, A.E.; Mustika, T.; Sulaikan, H.P.; Kartini, E. Applications battery performance data acquisition system for monitoring battery performance inside solar cell system. In Proceedings of the 1st Materials Research Society Indonesia Conference and Congress, Yogyakarta, Indonesia, 8–12 October 2018.

16. Sauer, D.U. *Encyclopedia of Electrochemical Power Sources—Secondary batteries: Lead-acid Batteries, Lifetime Determining Processes*; Garche, J., Ed.; Elsevier: Amsterdam, The Netherlands, 2009.

17. Rand, D.A.J.; Moseley, P.T. *Lead-acid Battery for Future Automobiles—Lead-acid Battery Fundamentals*; Elsevier: Amsterdam, The Netherlands, 2017.

18. Gauri; Bicht, M.S.; Pant, P.C.; Gairola, P.C. Effect of temperature on flooded lead-acid battery performance. *Int. J. Adv. Sci. Res.* **2018**, *3*, 27–29.

19. Haddad, R.; El Shabat, A.; Kalaani, T. Lead-acid battery modeling for photovoltaic applications. *J. Electr. Eng.* **2015**, *15*, 17–24.

20. Xiong, R.; He, H.; Guo, H.; Ding, Y. Modeling for Lithium-Ion Battery used in Electric Vehicles. *Procedia Eng.* **2011**, *15*, 2869–2874. [CrossRef]

21. Ganesan, A.; Sundaram, S.; Sinkaram, A. Heuristic algorithm for determining state of charge of a lead-acid battery for small engine applications. In Proceedings of the 2012 Small Engine Technology Conference & Exhibition, Madison, WI, USA, 16–18 October 2012.

22. Goodmin, A. Lead-acid Battery Working—Lifetime Study. Powerthru. 2014. Available online: https://www.scribd.com/doc/305243921/The-Truth-About-Batteries-POWERTHRU-White-Paper (accessed on 10 October 2018).

23. Savard, C. Amelioration de la Disponibilite Operationnelle des Systemes de Stockage de L'energie Electrique Multicellulaires. Ph.D. Thesis, Universite de Lyon, INSA de Lyon, French, 2017.

24. Boufaden, A.; Pietrac, L.; Gabouj, S. L'usage des réseaux de Petri dans la théorie de contrôle par supervision. Available online: https://www.researchgate.net/publication/228449393_L%27usage_des_Reseaux_de_Petri_dans_la_Theorie_de_Controle_par_Supervision (accessed on 10 October 2018).

25. Khyzhniak, T.; Kolesnyk, V. Modeling of power-supply subsystems of microgrid using Petri nets. In Proceedings of the Electronics and Nanotechnology (ELNANO), 2013 IEEE XXXIII International Scientific Conference, Kiev, Ukraine, 16–19 April 2013.

26. Lu, D.; Fakham, H.; Zhou, T.; Francois, B. Application of Petri nets for the energy management of a photovoltaic based power station including storage units. *Renew. Energy* **2010**, *35*, 1117–1124. [CrossRef]

27. European Commission's Photovoltaic Geographical Information System. Available online: https:// photovoltaic-software.com/pv-softwares-calculators/online-free-photovoltaic-software/pvgis (accessed on 10 October 2018).
28. *Standard Implementing Agreement on Photovoltaic Power Systems—Task 3: Use of Photovoltaic Power Systems in Stand-alone and Island Applications*; Report IEA PVPS T3-11; IEA PVPS International Energy Agency: Paris, France, 2002.

Article

Fast Electrical Characterizations of High-Energy Second Life Lithium-Ion Batteries for Embedded and Stationary Applications

Honorat Quinard [1,2,†], Eduardo Redondo-Iglesias [1,*], Serge Pelissier [1] and Pascal Venet [2]

1 Univ Lyon, IFSTTAR-AME-Eco7, 69500 Bron, France; honoratquinard@gmail.com (H.Q.);
   serge.pelissier@ifsttar.fr (S.P.)
2 Univ Lyon, Université Claude Bernard Lyon 1, Ecole Centrale de Lyon, INSA Lyon, CNRS, Ampère,
   F-69007 Lyon, France; pascal.venet@univ-lyon1.fr
* Correspondence: eduardo.redondo@ifsttar.fr; Tel.: +33-472-14-24-76
† Former Ph.D. Student Funded by Carwatt at IFSTTAR and Ampere laboratory, 69675 Bron Cedex, France.

Received: 31 January 2019; Accepted: 12 March 2019; Published: 14 March 2019

Abstract: This paper focuses on the fast characterization of automotive second life lithium-ion batteries that have been recently re-used in many projects to create battery storages for stationary applications and sporadically for embedded applications. Specific criteria dedicated to the second life are first discussed. After a short review of the available state of health indicators and their associated determination techniques, some electrical characterization tests are explored through an experimental campaign. This offline identification aims to estimate the remaining ability of the battery to store energy. Twenty-four modules from six different commercial electric vehicles are analyzed. Well-known methodologies like incremental capacity analysis (ICA) and constant voltage phase analysis during CC-CV charge highlight the difficulty—and sometimes the impossibility—to apply traditional tools on a battery pack or on individual modules, in the context of real second life applications. Indeed, the diversity of the available second life batteries induces a combination of aging mechanisms that leads to a complete heterogeneity from a cell to another. Moreover, due to the unknown first life of the battery, typical state of health determination methodologies are difficult to use. A new generic technique based on a partial coulometric counter is proposed and compared to other techniques. In the present case study, the partial coulometric counter allows a fast determination of the capacity aging. In conclusion, future improvements and working tracks are addressed.

Keywords: second life battery; lithium-ion; electrical characterization; state-of-health (SOH); partial coulometric counter

---

1. Introduction

1.1. Deployment of Electric Vehicles and Arrival of Second Life Batteries

The democratization of the electric vehicle in our modern society is nowadays an undeniable fact. Since 2010, the share of electric car holder scaled up significantly, reaching 3 million units in 2017 [1] and the Paris Declaration [2] in 2015 encouraged this phenomenon. According to the International Energy Agency [1] the global electric car stock for 2030 should raise up to 56 million units.

However, the decarbonisation of our transportation system has an insidious environmental drawback. That global trend is going to generate millions of used batteries and recycling processes are not ready to handle such quantities especially with heterogeneous chemistries. Before the question of recycling, the idea of a possible circular economy market around the second life of batteries is an upcoming question and many pioneer projects are now using automotive lithium-ion second-life batteries to store energy.

For example, research programs are using second life lithium-ion batteries like at University of California San Diego with a 160 kWh storage built from BMW batteries [3]. Nissan equipped its regional office in France with 100 vehicle-to-grid bi-directional chargers supplied by Enel, with an energy storage system combining 64 Nissan Leaf packs installed by Eaton [4]. EDF (Electricité de France), Forsee Power, Mitsubitshi Motors Corporation, Mitsubitshi Corporation and PSA Peugeot Citroën announced to jointly study the possibility of the energy storage business in Europe utilizing lithium-ion batteries from electric vehicles and launched a demonstration project in September 2015 in France at Forsee Power's new Headquarter near Paris [5]. Even domestic applications are considered like the power wall from Tesla [6] and the xStorage by Eaton [7]. More recently, start-ups like Carwatt [8] are trying to use second life lithium-ion batteries in different electric vehicle conversion projects.

1.2. Heterogeneity of State of Health in Second Life Batteries

An example of a first life lithium-ion battery pack, sold by the car manufacturer Renault, is illustrated in Figure 1. After few years of service, modules become heterogeneous and according to the origin and the life of the battery pack the phenomenon can increase. In order to build a brand new homogeneous second life battery pack, modules have to be dismantled and selected according to their SOH.

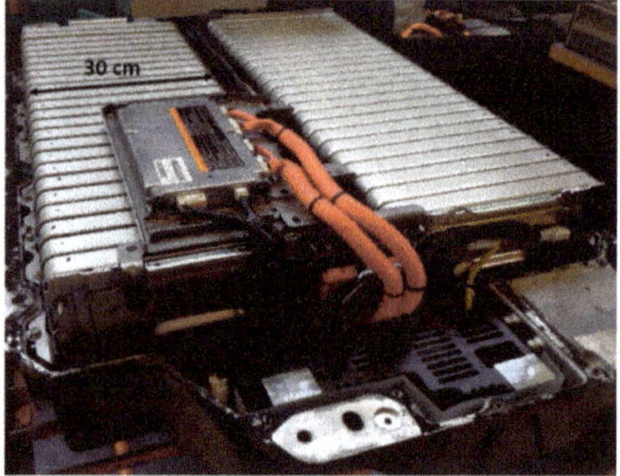

Figure 1. Renault KANGOO battery pack with AESC 65 Ah modules stored in CARWATT's workshop [8].

From stationary to embedded applications, scientific problems are addressed to the research community especially regarding the determination of the battery state of health (SOH). This identification has to be fast, cheap and robust in order to provide a competitive advantage compared to another substitute product like fresh battery, other technologies and to develop, by consequence, a real second life battery industry.

The main goal of a SOH measurement is to point out systemic limitations that could be brought by second life batteries to the future repurposed energy storage system (ESS).

Indeed, a lithium-ion battery degrades mainly by capacity and power fading but whether in the industry or in the research community, no clear consensus about a single definition of the state of health has been found. Most of the time, capacity and impedance are used to define the SOH. Frequently, car manufacturers consider that a loss of 20% of the initial capacity or a 200% impedance increase are good signatures to evaluate the end of the first life.

The SOH related to the capacity is generally defined by Equation (1):

$$SOH(Q)[\%] = 100\frac{Q_m}{Q_{nominal}} \tag{1}$$

where Q_m is the measured capacity during the test and $Q_{nominal}$ the nominal capacity of a fresh cell under the same experimental conditions.

The *SOH* related to the impedance is generally defined by Equation (2):

$$SOH(Z)[\%] = 100\left(2 - \frac{Z_m}{Z_{nominal}}\right) \tag{2}$$

where Z_m is the measured impedance during the test and $Z_{nominal}$ the nominal impedance of a fresh cell under the same experimental conditions.

Most of the time, second life battery modules are not produced anymore and electrochemical cell features are not available. Then, a classification based on a fresh reference is not possible. A way to solve this issue is to compare one module to each other with an average value provided by a specific methodology as presented in this paper. The characterization process has to be realized as fast as possible especially for economic reasons.

1.3. Need of Specific Methods for Determination of SOH

The basic method mainly used to measure the capacity is to fully charge or discharge a battery. This protocol is very time consuming since modules have to be tested one by one. Moreover, in practical case, second-life modules could be supplied by car manufacturers at various SOC (from 35% to 75%) adding preparation test time. These last observations lead to the impossibility to apply directly the aforementioned protocol.

New techniques have been explored by researchers through years. Berecibar et al. [9] reviewed the existing SOH estimation methods and classified in specific groups the different approaches. Adaptive methods like Kalman filter [10,11], neural networks [12,13], and many others are very accurate but computational resources and development are also complex [14]. On the opposite, empirical techniques mainly based on capacity and impedance measurements are simple and very effective methods but they are highly dependent on the operating conditions [15,16]. A third group using differential analysis can detect the degradation mechanisms with the advantages of both adaptive and empirical techniques, namely accuracy and low computational efforts [17].

Some approaches have already been tested and validated to evaluate the impedance of the battery within a short amount of time. Piłatowic et al. [18] define and describe various impedance estimation methodologies. Batteries' internal impedance is nonlinear and demonstrates different ohmic, capacitive and inductive behaviours according to the frequency. Then, it appears that the major concern is not really related to the measurement protocol itself but mostly how to interpret and how to give the right figure of merit to the internal impedance value.

On the opposite, there is an apparent lack of methodologies when it comes to a quick capacity fading estimation. Therefore, the present article is going to focus on fast offline approaches based on experimental measurements that can identify the capacity loss of a second life lithium-ion battery.

Incremental Capacity Analysis (ICA) is a well-known technique and this mathematical tool initially used by electrochemists, is gaining more and more credit in the battery community. Through peak area analysis (Riviere et al. [19]) or peak position analysis (Ansèan et al. [20]) a deep understanding of the aging mechanisms can be achieved. It requires static charge/discharge at very low C-rate (C/20) which generate very long periods of test. This initial drawback has been solved and different researches proved that higher C-rate (up to C/2) and truncated profiles (Li et al. [21]) could provide sufficient information about the capacity fading.

Eddahech et al. [22] analysed the charge current profile occurring throughout the constant voltage step. Indeed, the current time constant which in fact, represents the kinetic of the lithium-ions intercalation process into the negative electrode can give an accurate representation of the capacity fading phenomenon. Yang et al. [23] extended this methodology and proposed to study a partial CV charge reducing by consequence, the estimation time.

Results of the two previous techniques, standing on real aged second life lithium-ion batteries experimental campaign (realized in IFSTTAR and Ampere laboratories), demonstrate the impossibility to apply actual protocols with a good accuracy or within a short amount of time on a unique module. Consequently, the need of a specific approach to characterize an individual second life lithium-ion battery is corroborated and a new generic technique based on a partial coulometric counter is proposed.

2. Experimental Setup and Methods

2.1. Data Acquisition System

All the experiments were performed using a battery module tester Bitrode FTV2 (2 channels/0–250 A) with a resolution of 1 mV/10 mA and an electrochemical impedance spectrometer Biologic VSP+VMP (3 channels/0–20 A) with a resolution of 1 mV/20 mA test bench. In this experiment, data acquisition systems have a maximum sampling frequency of 10 Hz. The temperature is set and controlled at 25 °C into a climatic chamber.

2.2. Cell Specifications

The experimental part was realized on aged high energy modules made from pouch cells. These batteries sold by the manufacturer Automotive Energy Supply Corporation (AESC) are rated with a nominal capacity of 65 Ah.

Modules are extracted from six different aged battery packs which equipped the first generation of electric commercial vehicles, Renault KANGOO. Each pack is initially constituted by 48 modules in series as we can observe in Figure 1. The following study investigates the first four modules of each pack considering that this sample should give a good representativeness of the others.

A module is an association of two half modules in series. Each of them is composed of two unit cells in parallel. The module architecture is, therefore, 2P2S cells and voltage measurements are available for each half module as represented in Figure 2.

Half module 1   Half module 2

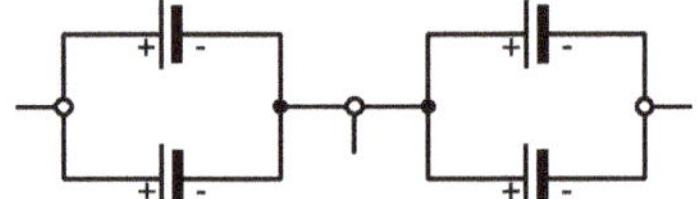

Figure 2. 2P2S unit cells.

By considering 24 samples represented in Figure 2, 48 half modules measurable separately are tested in coming paragraphs. Consequently, each tested sample is referenced as "Pack_i_m_j_halfmodule_k" where "i" is pack number (1 to 6), "j" is module number (1–4) and "k" is the half-module number (1 or 2). Each of these tested samples, made of two unit cells in parallel is considered as a unique cell for the rest of the work. Considering the single cell characteristics given in Table 1, the initial value of the capacity of the tested samples was 65 Ah.

Table 1. Initial specifications of tested cells according to the manufacturer's datasheet.

AESC E5-M	
Cell type	Laminate Type
Positive electrode material	LMO with LNO
Negative electrode material	Graphite
Nominal capacity (at 0,3C)	32.5 Ah
Nominal voltage	3.75 V
Energy density	157 Wh/kg
Voltage limit range	2 to 4.2 V
Voltage operating range	2.5 to 4.15 V

The manufacturer AESC and the car manufacturer Renault recommendations give a voltage operating range for an optimal life expectancy of the cell which is from $V_{min,op} = 2.5$ V to $V_{max,op} = 4.15$ V. These values will frame the next experimental work.

2.3. Experimental Methods for SOH Estimation

Three different methods are presented in the following section and results will be compared in Section 3. The nominal capacity of each second life cell used as reference has been measured during a full CC discharge at 1C forerun by a wake up cycle.

2.3.1. Constant Voltage Analysis during CC-CV Charge

One possible way to analyse the capacity loss of a battery is to study the CC-CV charge as in Figure 3. The standard and very common charging protocol can be divided into two distinct parts. First the constant current (CC) phase applies a constant current to the battery until the cut off voltage is reached then the battery remains at this voltage (CV) threshold in floating mode until the current reaches a minimal value determined most of the time by the battery management system, a typical value is C/20.

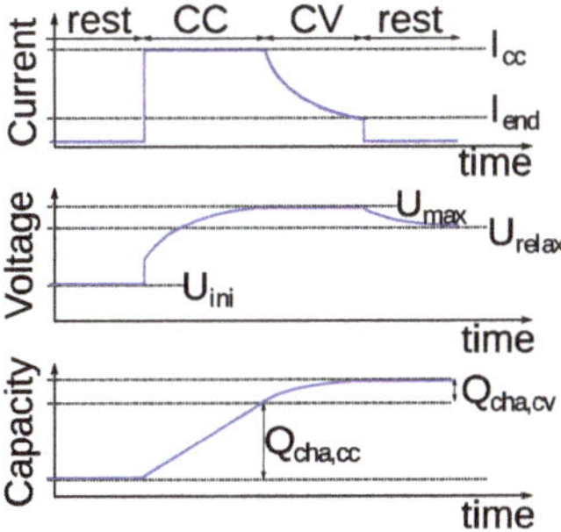

Figure 3. CC-CV charge cycle (Redondo-Iglesias et al. [24]).

The intercalation process of the lithium into the graphite negative electrode mainly takes place during the charge at constant current. For fresh cells only up to 5% of this chemical reaction occurs during the constant voltage step. Whereas for aged lithium-ion batteries an important part (from 32% to 45 %) of the available capacity is stored during the CV phase as shown in Figure 4.

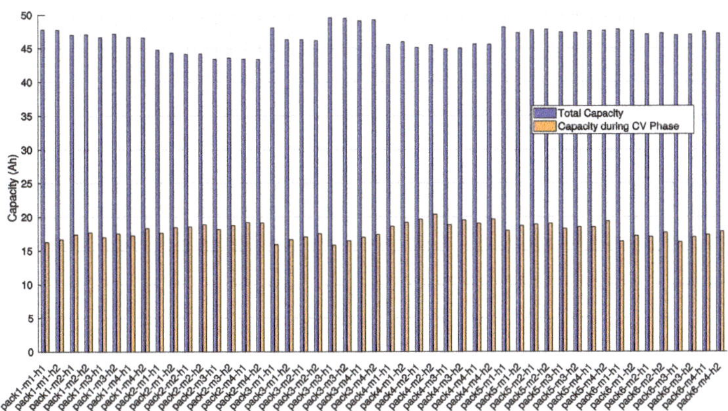

Figure 4. Capacity charged during the constant voltage phase and the full charge 1C for second life batteries.

This large amount of energy stored during the CV phase can be explained by the high value of the impedance. By consequence, the voltage upper limit during the CC phase is reached quicker than at its beginning of life increasing the role of the CV phase during the charge of the cell.

Eddahech et al. [22] state that the kinetic of the potentiostatic mode (CV) at 1C is a pertinent indicator of the battery state of health and can indicate the capacity loss. In their approach, the measured CV current is fitted by a non-linear regression expressed by Equation (3):

$$I(t) = Ae^{-Bt} + C \tag{3}$$

where B represents the kinetic of the battery during the CV charge and is used as an indicator of the state of health.

In Eddahech et al. [22], the coefficient B has a good linear relation with the capacity loss of a battery for a capacity fading greater than 4% and up to 30%.

The described methodology previously applied by Eddahech to follow the aging of one cell, was applied to our set of second life batteries composed of cells from different packs and at different *SOH*.

Figure 5 shows the correlation between the exponential regression and the experimental curve $I(t)$ during the CV phase voltage. The quality of the fitting identified with the least square method is satisfactory ($R^2 = 0.99$).

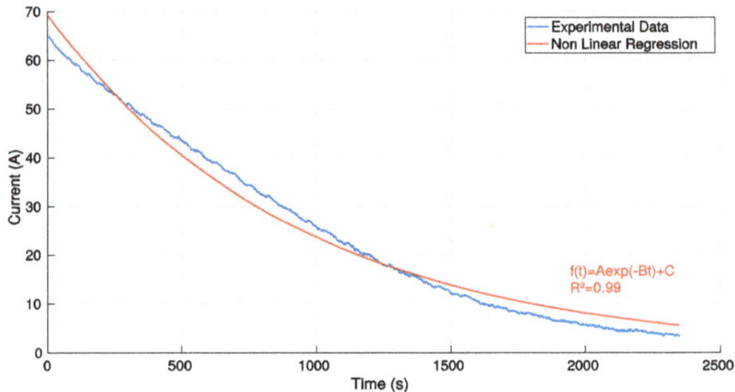

Figure 5. Comparison between measured and simulated current during CV step.

Experimental results from the aged battery test campaign are illustrated by Figure 6, where the quality of the correlation between the coefficient B and the capacity loss is identified using the least square method.

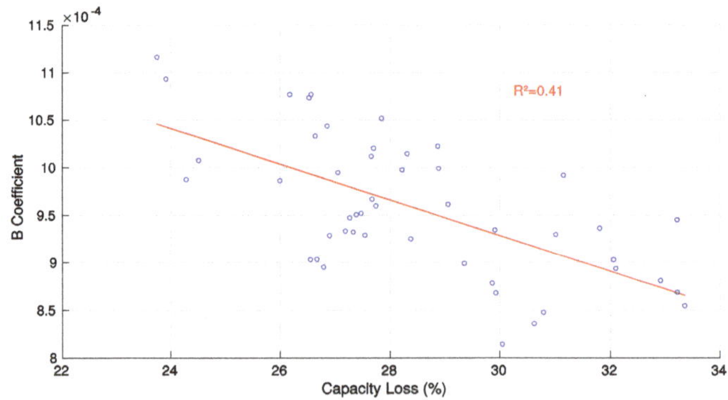

Figure 6. Evolution of the coefficient B according the capacity loss.

The R^2 coefficient value ($R^2 = 0.41$) of the correlation shows that the capacity fading cannot be correctly represented by any trend.

To sum up, this methodology seems difficult to use for a SOH determination in the context of the classification of second life lithium-ion batteries. At this point, the observation of the kinetic of the lithium insertion in the negative electrode is not sufficient anymore and advanced considerations will be presented in the final section of this paper.

2.3.2. Incremental Capacity Analysis

Incremental capacity analysis (ICA) is a mathematical tool based on the variation of the electric charge inside the battery considering the variation of the voltage across the electrodes. It can be described by Equation (4):

$$ICA\left[\frac{Ah}{V}\right] = \frac{dQ}{dV} = \frac{\Delta Q}{\Delta V} \tag{4}$$

where Q is the capacity and V the voltage.

The method is based on the differentiation of the battery capacity over the battery voltage emphasizing the flat regions and the quick variations of the open circuit voltage curve (OCV). Each resulting peak (corresponding to the flat region of the OCV) and valley (corresponding to the evolution of the OCV) (Figure 7a,b) is unique and reflects electrochemical processes taking place into the battery. Then, peak area, position and shape can be analysed and interpreted. Characteristic features of the incremental capacity analysis are extremely correlated to the battery aging mechanisms and very accurate estimations of the capacity loss can be obtained (Dubarry et al. [25]).

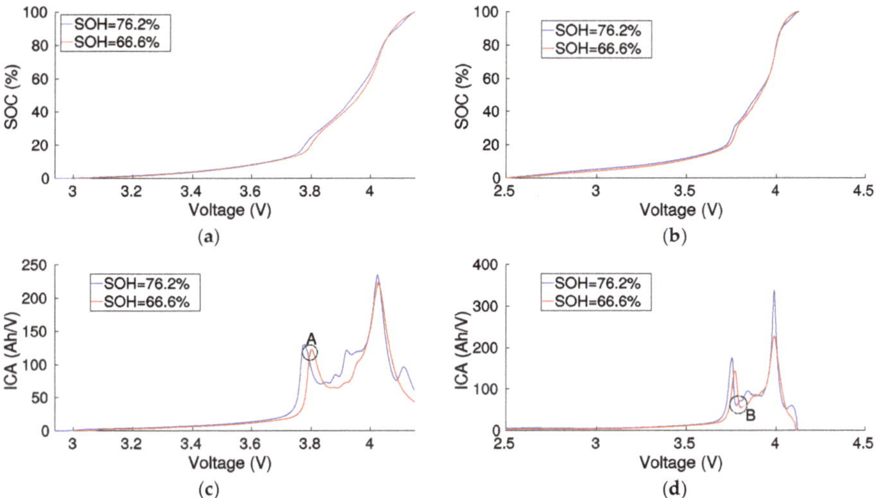

Figure 7. Comparison of results for the best sample (SOH = 76.2%) and the worst sample (SOH = 66.6%). (a) OCV charge C/20; (b) OCV discharge C/20; (c) ICA charge C/20; and (d) ICA discharge C/20.

Therefore, the initial step is to identify the rightful indicator for the given chemistry. C/20 charges and discharges have been realized using the 48 half modules. This low rate is synonymous of a quasi-thermodynamic equilibrium and gives a clear vision of the phenomena occurring inside the battery especially the transition phases. Then, an *ICA* curve comparison from different cells (Figure 7c,d) gives an overview of the changes in peaks and valleys. That observation allows the determination of the aging mechanisms occurring in the studied technology. Of course, these kinds of experiments are very slow and the final target is to be able to increase the charge or discharge rate to its maximum so as to accelerate the measurement protocol.

Most of the time, the raw experimental signal is not directly exploitable. Indeed, in *ICA* Equation (4), if $\Delta V = 0$, $ICA \rightarrow \infty$ moreover additive noises are very symptomatic and have to be filtered. Riviere et al. [19] used in their study a numerical Butterworth filter but Li et al. [21] recently highlighted the performance of a moving average filter (MA) represented by Equation (5):

$$y(i) = \sum_{j=1}^{M} MA(j){\cdot}x(i+j-1) \qquad with \ MA(j) = \frac{1}{M} \tag{5}$$

combined with a Gaussian filter (G) in Equation (6):

$$z(i) = \sum_{j=1}^{M} G(j){\cdot}y(i+j-1) \qquad with \ G(j) = \frac{1}{\sigma\sqrt{2\pi}}exp - \left(\frac{(j-\mu)^2}{2\sigma^2}\right) \tag{6}$$

In Equations (5) and (6), M is the number of points considered in the "average window", x is the input signal y an intermediary signal and z the output signal, μ is the mean value of the considered data and σ, called the standard deviation, is a parameter which controls the bandwidth of the filter.

This last technique allows more parametric choices for more flexibility of the experimental datasetting system rather than for example a typical Butterworth filter. Indeed, using Butterworth filter, the time step through the all experimentation has to be the same even during rest times leading sometimes to a useless increase of the dataset. Therefore, the combination of the moving average and the Gaussian filter has been chosen for its flexibility and its simplicity in the coming analysis.

As a result, the voltage variation of peak A and valley B, respectively during C/20 charge and discharge (Figure 7c,d), are linearly proportional to the battery aging state as shown in Figure 8a,b. The quality of the correlation is identified using the least square method.

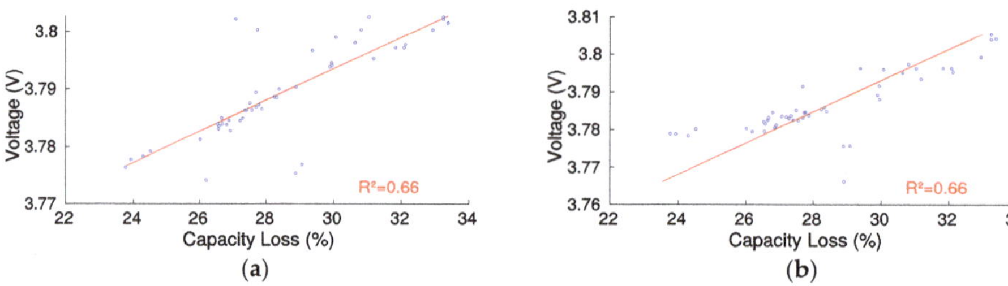

Figure 8. (**a**) Position of peak A variations according the capacity loss; (**b**) Position of valley B variations according the capacity loss.

A C/20 charge or discharge is a very slow process. In an attempt to reduce and optimize the test time protocol, C/10, C/4, C/3 and C/2 rates were employed and demonstrated that only the peak A, under such charging or discharging constraints, could provide sufficient information about the capacity fading as shown in Figure 9. It should be noted that a peak area analysis was also realized but did not provide any satisfying results.

A C-rate close to the nominal 1C leads to an important distortion of the ICA profile and by consequence to a significant loss of information. For the aforesaid AESC cells, a C/3 charge may afford sufficient information for an estimation of the capacity loss within an optimized processing time.

Even with a C/3 charge rate, the process is still very long. The peak A occurs around 20% of SOC. To save some time, an ICA on a partial charge from 0% up to 30 % of SOC could be considered in our study, instead of using the whole SOC range.

To conclude, this technique could provide an estimation of the SOH for the second life lithium-ion batteries. But the accuracy is directly linked to the C-rate used during the test. The lower is the C-rate the better will be the estimation. Then, a compromise between the experimental time and the desired accuracy will be necessary.

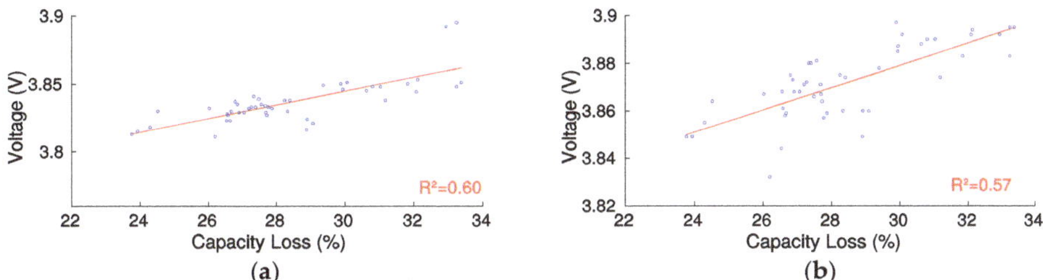

Figure 9. Position of peak A variations analysis during charge, (**a**) C/3, and (**b**) C/2.

2.3.3. Partial Coulometric Counter

The (full) coulometric counter is a straightforward tool used for SOC determination but in some cases it can be employed as a SOH estimation method like by Kong Sooun et al. [26]. Most of the time, this technique requires a complete charge or discharge cycle operated between $V_{max,op}$ and $V_{min,op}$ as defined in Table 1, leading unfortunately to an important test time.

SOH estimation using coulometric counter through the traditional technique is commonly calculated with the following Equations (7) (for discharge) and (8) (for charge):

$$SOH(Q)[\%] = 100 \frac{Q_{discharged}\ V_{min,op}\ V_{max,op}}{Q_{nominal\ during\ discharged}} \tag{7}$$

or

$$SOH(Q)[\%] = 100 \frac{Q_{charged}\ V_{min,op}\ V_{max,op}}{Q_{nominal\ during\ charged}} \tag{8}$$

$Q_{discharged}\ V_{min,op}\ V_{max,op}$ and $Q_{charged}\ V_{min,op}\ V_{max,op}$ are, respectively, the discharge capacity and charge capacity between $V_{min,op}$ and $V_{max,op}$.

We propose to adapt the previous methodology to a partial charge or discharge in order to reduce the test time. After different attempts, the charging phase seems to be the best way to apply directly the partial coulometric counter and can be expressed by Equation (9):

$$SOH(Q)[\%] = 100 \frac{Q_{charged}\ V_{pulse-start}\ V_{Pulse-stop}}{Q_{nominal\ charged}\ V_{pulse-start}\ V_{Pulse-stop}} \tag{9}$$

where $Q_{charged}\ V_{pulse\text{-}start}\ V_{pulse\text{-}stop}$ is the charge capacity between $V_{pulse\text{-}start}$ and $V_{pulse\text{-}stop}$. $V_{pulse\text{-}start}$ and $V_{pulse\text{-}stop}$ have to be optimized in order to achieve the best compromise between the operating time and the sensitivity of the estimator. These voltages are defined under operation and include the influence of the current.

Since a partial charge is going to be used in the following protocol, caution about the starting voltage and the length of the experiment has to be taken into account. In fact, Figure 10 highlights that the transient behaviour of a charge pulse is not the same and mainly depends on the initial voltage level of the half module. To reach the charge profile of the reference curve, few seconds are necessary for a pulse at an initial low voltage (like Pulse 1 in Figure 10) while a 400-second interval is mandatory for a pulse at an initial high voltage (like Pulse 2 in Figure 10). This transient phase is consistent with the transition from the OCV curve—when the battery was at rest—to the 1C charge curve. The characteristics of this transient response depend on the battery impedance which is itself SOC dependent. We noted from other tests that the C-rate can also influence the shape of the transient.

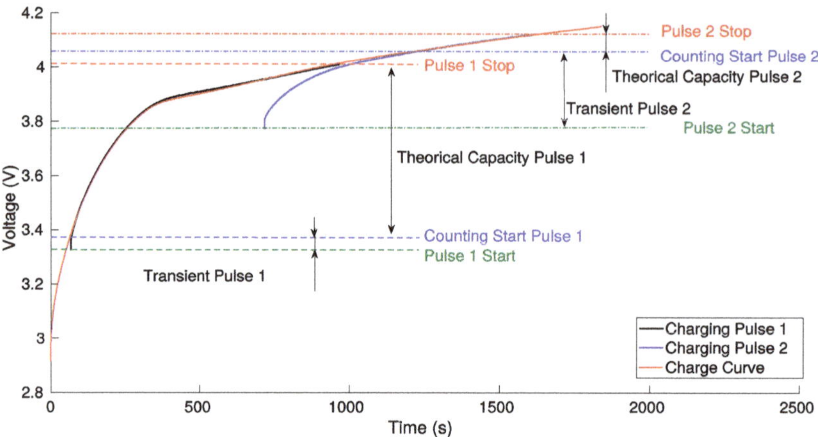

Figure 10. Voltage transient phenomenon according to the initial voltage level of a charge pulse 1C/CC.

According to the previous observation, the transient behaviour has to be taken into account for an accurate capacity measurement. Therefore, the capacity counting has to be delayed from the starting point of the pulse, as shown in Figure 11.

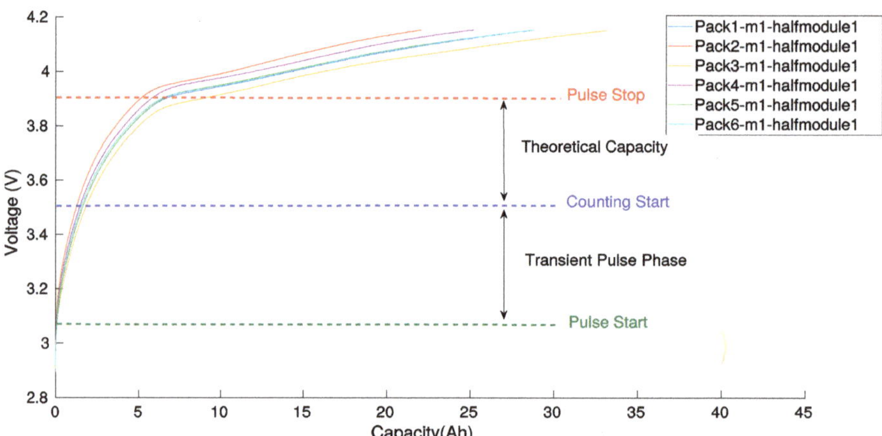

Figure 11. Example of pulse start, counting start and pulse stop limits represented on the reference curves.

Taking into account the previous observations, experimental results based on the second life battery test campaign are illustrated by Figure 12. The partial capacity of each cell is measured between two voltage thresholds (3.4 V and 3.8 V) that could correspond to a charge pulse of 300 s at 1C/CC with a counting start occurring 100 s after the beginning. Initially, the cell is considered as fully discharged. The quality of the correlation between the partial capacity and the capacity loss is identified using the least square method.

The R^2 coefficient value ($R^2 = 0.69$) of the correlation highlights a good trend between the partial capacity charged and the capacity loss. The rapidity of the test and its easy implementation should make this methodology a real asset for offline *SOH* estimations.

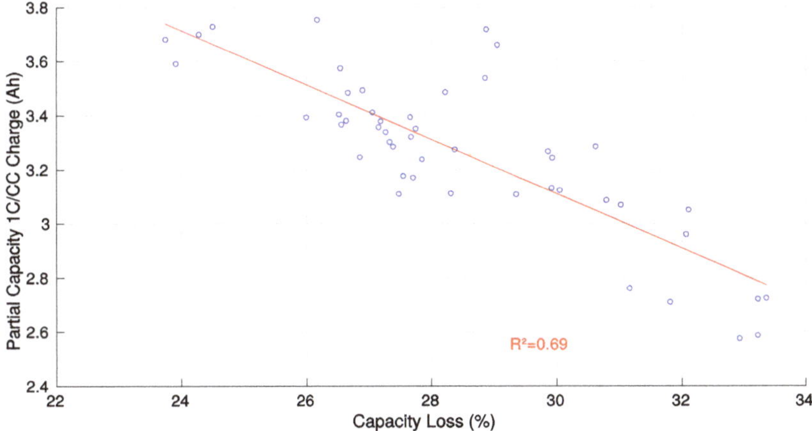

Figure 12. Evolution of the partial capacity charged between two voltage levels according the capacity loss.

3. Comparison between Methods for Fast Characterization of SOH

According to the foregoing section, the partial coulometric counter appears to be the most accurate method for a SOH estimation. Methods were previously based on a systematic measurement of all the modules under test. In order to shorten the duration of the SOH estimation campaign, we propose a protocol based on the full measurement of only 6 modules, considered as reference and a partial measurement of the others. Each of the 6 modules was selected randomly from six different vehicles. More "reference" modules could be used but it would lead to an increase of prerequisite data and test time. In the following analysis, the three SOH determination methods have been compared using the 6 same references. The quality of the correlation is again identified using the least square method.

3.1. CV Charge Phase Capacity Estimation Method

Through measurements during the CV charge phase according to Section 2.3.1, B coefficients for the six reference cells are determined and represented according to the total capacity of each cell in Figure 13.

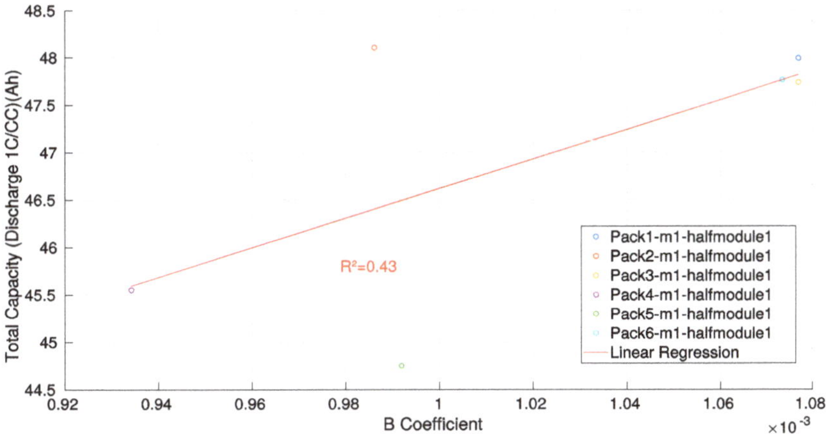

Figure 13. Linear relation between measured capacity and B coefficient for the six references.

From Figure 13, we can deduce in this specific case Equation (10):

$$TotalCapacity(Ah) = 15577 \times B + 31.042 \tag{10}$$

Then, from the experimental determination of the B coefficient of the different tested batteries, we can calculate with Equation (10), an estimation of the capacity illustrated in Figure 14. This representation includes 42 half modules (among the 48 half modules, six have been taken as reference). The average absolute error of the estimation is equal to 2.5% with a maximum of 5.7%.

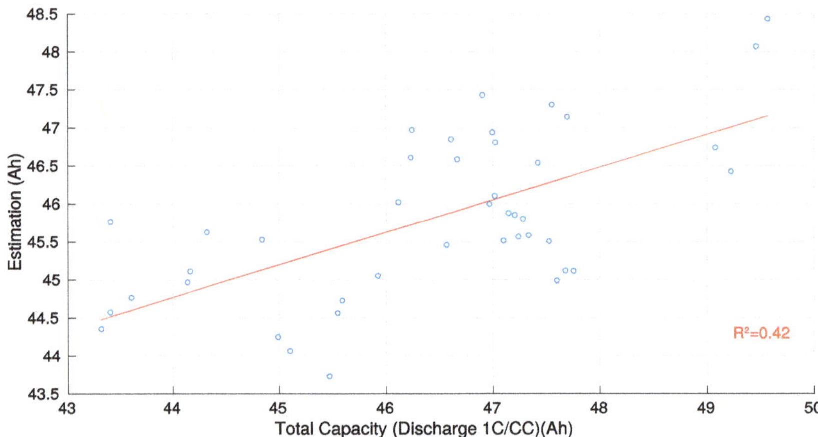

Figure 14. Capacity measurement compared with the capacity estimated from the CV phase technique.

3.2. ICA Capacity Estimation Methodology

Through the ICA analysis according to Section 2.3.2, peak A voltage position for the six reference cells, under a charge with a C/3 rate, is determined and represented according to the total capacity of each cell in Figure 15.

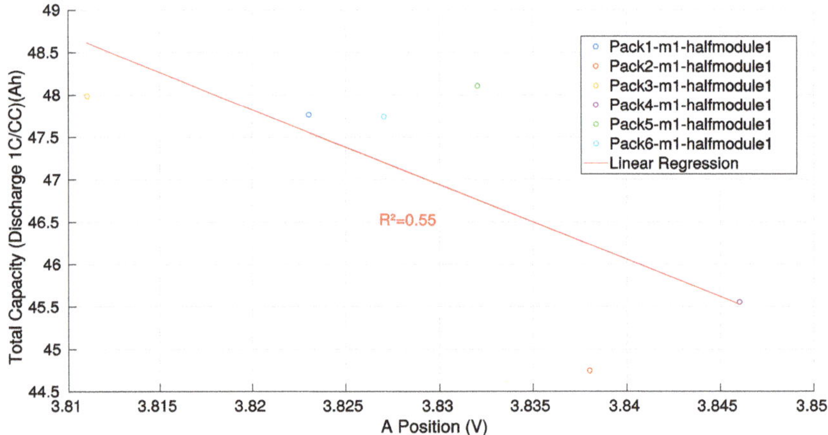

Figure 15. Relation between measured capacity and peak A position for the six references.

From Figure 15, we can deduce in this specific case Equation (11):

$$TotalCapacity(Ah) = -88.098 \times PeakAVoltage(V) + 384.35 \qquad (11)$$

Then, from the experimental determination of the Peak A voltage position of the different tested batteries, we can calculate with Equation (10), an estimation of the capacity illustrated in Figure 16. This representation includes 42 half modules (among the 48 half modules, six have been taken as reference). The average absolute error of the estimation is equal to 1.8% with a maximum of 5.1%.

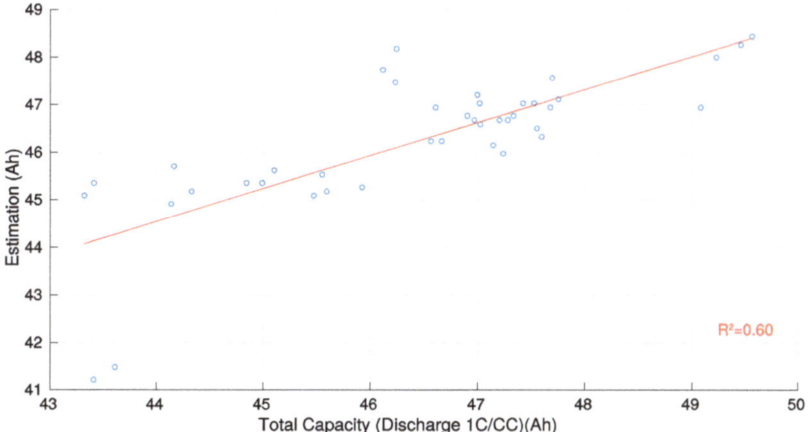

Figure 16. Capacity measurement compared with the capacity estimated from the ICA technique.

3.3. Partial Coulometric Counter

Through the measurement of the partial coulometric counter according to Section 2.3.3, partial capacities for the six reference cells are determined during a charge cycle and linked to the total capacity of each cell in Figure 17.

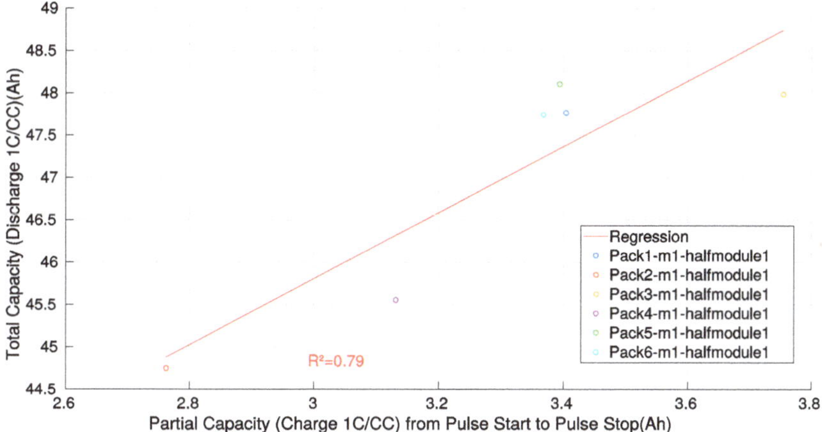

Figure 17. Capacity estimation using the regression curve based on the 6 references.

The linear relation in the case of this study can be modelled by Equation (12):

$$TotalCapacity(Ah) = 3.8967 \quad \times \partial Capacity Measured(Ah) + 34.1164 \tag{12}$$

As a result, Figure 18 shows the correlation between the nominal capacity measured during a full CC discharge 1C and the estimated capacity of the partial coulometric counter. The representation includes 42 half modules (among the 48 half modules, six have been taken as reference). The quality of the correlation is identified using the least square method (R^2 = 0.69). The average absolute error of the estimation is equal to 1.6% with a maximum of 5.1%.

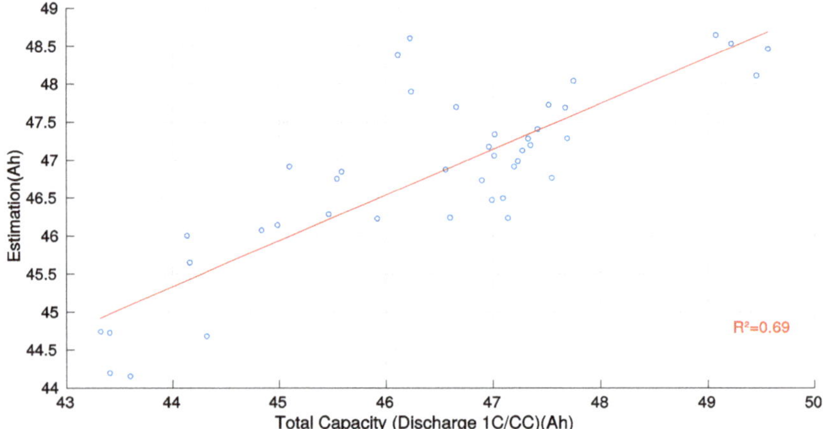

Figure 18. Capacity measurement compared with the capacity estimated from the coulometric counter technique.

3.4. Synthesis

The following Table 2 represents a synthesis of the different estimation methodologies.

Table 2. Comparison of the *SOH* methodologies applied to second life batteries.

Method	R^2	Average Absolute Error	Maximum Absolute Error	Estimated Test Time (s)	Pack Estimation Suitability
Phase CV	0.42	2.5	5.7	1050	-
ICA	0.60	1.8	5.1	3240	++
Partial counter	0.69	1.6	5.1	300	+

In Table 2, the estimated test time considers that the measurement is realized with a battery initially discharged both for the ICA and the partial counter techniques and partially charged (CC phase charge already realized) for the phase CV protocol. The practical duration would include the discharge/charge and rest phases which last approximately up to 90 min.

Simultaneity of the measurements could be time saving and protocols applicable directly to a pack should be promoted as much as possible. Therefore, the column pack estimation suitability indicates if the technique is compatible with a direct measurement on the battery pack.

Figure 19 highlights the cumulative distributions of the absolute error related to the different estimation methodologies. From this representation, one can see that if 90% of the modules are considered, the maximum error is lower for the coulometric counter than for the other methods. The phase CV technique has the worst accuracy. In addition, if we only focus on the R^2 of the different methodologies represented in Table 2, we can again eliminate the CV phase technique in such offline measurements.

From the previous comparison, we can also conclude that the coulometric counter offers the best accuracy regarding the average and the maximum absolute error.

It should be noted that the result of the ICA technique is quite similar. The difference between these two methodologies is the test time that is shorter for the coulometric counter taking into account that the estimation is performed module by module.

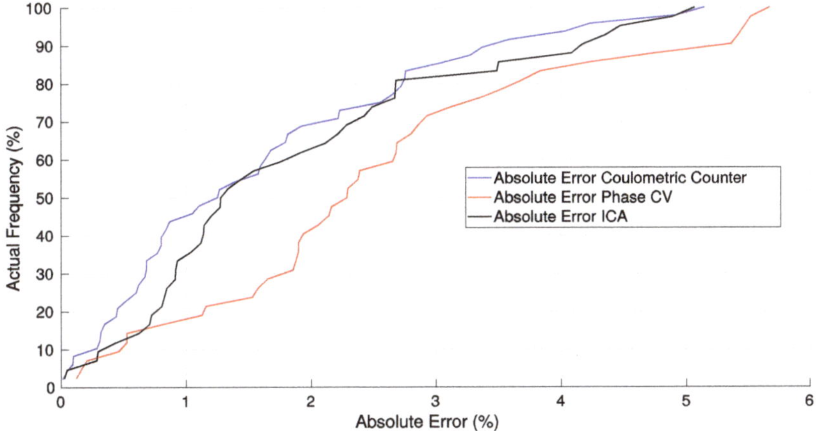

Figure 19. Cumulative distribution of the absolute error for the different estimations.

If the test of a full pack is considered, the performances of the previous methods change. Indeed, during the full discharge of a pack, the first cell reaching the low voltage limit will trigger the security of the BMS, leading naturally to the stop of the discharge protocol. By consequence, all the modules will not have the same voltage and some of them will not be totally discharged. The partial coulometric counter methodology was applied on cells with a SOC equals to 0% and it is very difficult to estimate without more investigations the impact of a measure starting from another SOC. Regarding the ICA technique, the peak A that gives the information necessary to the capacity estimation is situated around 20 % of SOC. The sensitivity of this methodology to the level of the initial SOC is expected lower than for the coulometric counter. By consequence, the ICA method could be more appropriate for the test of a full pack excluding the measurement of individual modules (based on condition of having sufficient test channels).

4. Conclusions

Three fast characterization methods to determine the full capacity fading of a battery have been tested in this paper. The CV phase protocol proved to be inefficient in the context of real second life lithium-ion batteries. While the ICA technique can only be applied to a whole battery pack mainly for time constraint reason, the generic methodology based on a partial coulometric counter has been proposed and experimentally validated. Each technique has an equivalent accuracy but can only be used on a pack level or on a module level, depending on the methodology chosen. In addition to giving an average absolute error of 1.6% and a maximum error equals to 5.1%, the new protocol has the advantage to determine the capacity fading without any fresh references. Indeed, in this case scenario only six second life reference batteries are necessary to build the needful trend employed to estimate the capacity of the entire experimental campaign.

Literature about second life batteries is quasi inexistent and comparisons of the results obtained during our experimental campaign with other studies are difficult. Despite that deficiency of information, several interesting points emerged from the previous analysis.

First, through the use of the ICA or the kinetic of the CV phase techniques, it clearly appears that real life aging does not match synthetic aging realized in laboratories, like in [19] or in [22]. Most of the time, cells are placed into a climatic chamber, then a predetermined cycle is chosen and applied for weeks. The atmosphere temperature around the tested cell is raised and precisely controlled in order to accelerate the aging phenomenon in accordance with Arrhenius' law as explained in Redondo-Iglesias et al. [27]. However, in our case, all the batteries originate from different vehicles, by consequence their aging behaviour cannot be the same through time. Indeed, the diversity of the cycle life of the vehicles induces a unique combination of aging mechanisms that leads to a complete heterogeneity of the modules from a vehicle to another. This observation can be extended to the battery pack, where a module could be different from another. This is mainly due to the existence of an unequal temperature gradient inside the casing and sometimes to unequal electrical constraints. These remarks could explain that an aging law which fits correctly for laboratory aging would not be suitable for real-life aging. Barré et al. [28] worked on statistical predictions that take into account future behavior but this approach implies to know all the cycle life of the studied battery.

Future improvements about the optimal counting start identification linked to the transient and the exact time length of the capacity counting have to be studied in order to consolidate the accuracy and the robustness of the proposed new methodology.

Further works would be necessary to complete the characterization method by classification of the second life batteries. This categorization would clearly lean on the measured SOH of the cells but should integrate other criteria. Indeed, the interaction of the multiple aging mechanisms during the first life of identical cells are even more complicated to predict during the second life of cells with different histories. Classification of second life cells could allow to put together cells that aged in a similar way (similar combination of aging mechanisms) and this point would facilitate a better SOH prediction for each group of cells.

Author Contributions: H.Q., S.P., P.V., and E.R.-I. conceived and designed the experiments; H.Q. performed the experiments; H.Q., S.P., P.V., and E.R.-I. analysed the data; H.Q. wrote the paper; and S.P., P.V., and E.R.-I. proofread the paper.

Funding: This research was funded by National French Association for Technological Research (ANRT) and the company CARWATT.

Conflicts of Interest: The authors declare no conflict of interest.

References

1. International Energy Agency. *Global EV Outlook 2018*; International Energy Agency: Paris, France, 2018. [CrossRef]
2. The Paris Declaration on Electro-Mobility and Climate Change and Call to Action. Available online: https://unfccc.int/news/the-paris-declaration-on-electro-mobility-and-climate-change-and-call-to-action (accessed on 29 November 2019).
3. New Life for Electric Vehicle Batteries—California Energy Storage Showcase. Available online: https://www.energy.ca.gov/research/energystorage/tour/ev_batteries/ (accessed on 29 November 2019).
4. Electric Vehicle Technology to Power New Nissan Office in Europe. Available online: https://europe.nissannews.com/en-GB/releases/release-143213-electric-vehicle-technology-to-power-new-nissan-office-in-europe/ (accessed on 29 November 2019).
5. Joint Japan-France Demonstration of Energy Storage System Project Utilization of used Lithium-Ion Batteries from Electric-Powered Vehicles. Available online: https://www.mitsubishi-motors.com/publish/pressrelease_en/corporate/2015/news/detailf710.html (accessed on 29 November 2019).
6. Powerwall—The Tesla Home Battery. Available online: https://www.tesla.com/powerwall (accessed on 29 November 2019).

7.   Eaton—Energy Storage. [En ligne]. Available online: https://www.eaton.com/gb/en-gb/products/energy-storage.html (accessed on 29 November 2019).

8.   Carwatt. Available online: http://www.carwatt.net/ (accessed on 29 November 2019).

9.   Berecibar, M.; Gandiaga, I.; Villarreal, I.; Omar, N.; Van Mierlo, J.; Van den Bossche, P. Critical review of state of health estimation methods of Li-ion batteries for real applications. *Renew. Sustain. Energy Rev.* **2016**, *56*, 572–587. [CrossRef]

10.   Plett, G.L. Extended Kalman filtering for battery management systems of LiPB- based HEV battery packs. *J. Power Sources* **2004**, *134*, 252–261. [CrossRef]

11.   Remmlinger, J.; Buchholz, M.; Soczka-Guth, T.; Dietmayer, K. On-board state-of- health monitoring of lithium-ion batteries using linear parameter-varying models. *J. Power Sources* **2013**, *239*, 689–695. [CrossRef]

12.   Ragsdale, M.; Brunet, J.; Fahimi, B. A novel battery identification method based on pattern recognition. In Proceedings of the 2008 IEEE Vehicle Power and Propulsion Conference (VPPC), Harbin, China, 3–5 September 2008.

13.   Zenati, A.; Desprez, P.; Razik, H.; Rael, S. Impedance measurements combined with the fuzzy logic methodology to assess the SOC and SOH of lithium-ion cells. In Proceedings of the 2010 IEEE Vehicle Power and Propulsion Conference (VPPC), Lille, France, 1–3 September 2010.

14.   Cho, S.; Jeong, H.; Han, C.; Jin, S.; Lim, J.H.; Oh, J. On-line monitoring of capacity fade in lithium-ion batteries. *J. Chem. Eng. Jpn.* **2012**, *45*, 983–994. [CrossRef]

15.   Dai, H.; Wei, X.; Sun, Z. A new SOH prediction concept for the power lithium-ion battery used on HEVs. In Proceedings of the 2009 IEEE Vehicle Power and Propulsion Conference (VPPC), Dearborn, MI, USA, 7–10 September 2009.

16.   Lu, L.; Han, X.; Li, J.; Hua, J.; Ouyang, M. A review on the key issues for lithium-ion battery management in electric vehicles. *J. Power Sources* **2013**, *226*, 272–288. [CrossRef]

17.   Bloom, I.; Jansen, A.N.; Abraham, D.P.; Knuth, J.; Jones, S.A.; Battaglia, V.S.; Henriksen, G.L. Differential voltage analyses of high-power lithium-ion cells. *J. Power Sources* **2005**, *139*, 295–303. [CrossRef]

18.   Piłatowicz, G.; Marongiu, A.; Drillkens, J.; Sinhuber, P.; Sauer, D.U. A critical overview of definitions and determination techniques of the internal resistance using lithium-ion, lead-acid, nickel metal-hydride batteries and electrochemical double-layer capacitors as examples. *J. Power Sources* **2015**, *296*, 365–376. [CrossRef]

19.   Riviere, E.; Venet, P.; Sari, A.; Meniere, F.; Bultel, Y. LiFePO$_4$ Battery State of Health Online Estimation Using Electric Vehicle Embedded Incremental Capacity Analysis. In Proceedings of the 2015 IEEE Vehicle Power and Propulsion Conference (VPPC), Montreal, QC, Canada, 19–22 October 2015.

20.   Anseán, D.; González, M.; Blanco, C.; Viera, J.C.; Fernández, Y.; García, V.M. Lithium-ion battery degradation indicators via incremental capacity analysis. In Proceedings of the 2017 IEEE International Conference on Environment and Electrical Engineering and 2017 IEEE Industrial and Commercial Power Systems Europe (EEEIC/I&CPS Europe), Milan, Italy, 6–9 June 2017.

21.   Li, Y.; Abdel-Monem, M.; Gopalakrishnan, R.; Berecibar, M.; Nanini-Maury, E.; Omar, N.; van den Bossche, P.; Van Mierlo, J. A quick on-line state of health estimation method for Li-ion battery with incremental capacity curves processed by Gaussian filter. *J. Power Sources* **2018**, *373*, 40–53. [CrossRef]

22.   Eddahech, A.; Briat, O.; Vinassa, J.-M. Determination of lithium-ion battery state-of-health based on constant-voltage charge phase. *J. Power Sources* **2014**, *258*, 218–227. [CrossRef]

23.   Yang, J.; Xia, B.; Huang, W.; Fu, Y.; Mi, C. Online state-of-health estimation for lithium-ion batteries using constant-voltage charging current analysis. *Appl. Energy* **2018**, *212*, 1589–1600. [CrossRef]

24.   Redondo-Iglesias, E.; Venet, P.; Pelissier, S. Global Model for Self-Discharge and Capacity Fade in Lithium-Ion Batteries Based on the Generalized Eyring Relationship. *IEEE Trans. Veh. Technol.* **2017**, *67*, 104–113. [CrossRef]

25.   Dubarry, M.; Truchot, C.; Liaw, B.Y. Synthesize battery degradation modes via a diagnostic and prognostic model. *J. Power Sources* **2012**, *219*, 204–216. [CrossRef]

26.   Ng, K.S.; Moo, C.-S.; Chen, Y.-P.; Hsieh, Y.-C. Enhanced coulomb counting method for estimating state-of-charge and state-of-health of lithium-ion batteries. *Appl. Energy* **2009**, *86*, 1506–1511. [CrossRef]

27. Redondo-Iglesias, E.; Venet, P.; Pelissier, S. Eyring acceleration model for predicting calendar ageing of lithium-ion batteries. *J. Energy Storage* **2017**, *13*, 176–183. [CrossRef]
28. Barré, A.; Suard, F.; Gérard, M.; Montaru, M.; Riu, D. Statistical analysis for understanding and predicting battery degradations in real-life electric vehicle use. *J. Power Sources* **2014**, *245*, 846–856. [CrossRef]

Article

Combining a Fatigue Model and an Incremental Capacity Analysis on a Commercial NMC/Graphite Cell under Constant Current Cycling with and without Calendar Aging

Tiphaine Plattard [1,2,*], Nathalie Barnel [1], Loïc Assaud [2], Sylvain Franger [2,*] and Jean-Marc Duffault [2]

[1] EDF R&D Les Renardières, 77250 Moret Loing et Orvanne, France; nathalie.barnel@edf.fr
[2] ICMMO-ERIEE, Université Paris-Sud, Université Paris-Saclay, CNRS UMR 8182, 91400 Orsay, France; loic.assaud@u-psud.fr (L.A.); jean-marc.duffault@u-psud.fr (J.-M.D.)
* Correspondence: tiphaine.plattard@edf.fr (T.P.); sylvain.franger@u-psud.fr (S.F.); Tel.: +33-160-736-059 (T.P.)

Received: 21 December 2018; Accepted: 13 March 2019; Published: 21 March 2019

Abstract: Reliable development of LIBs requires that they be correlated with accurate aging studies. The present project focuses on the implementation of a weighted ampere-hour throughput model, taking into account the operating parameters, and modulating the impact of an exchanged ampere-hour by the well-established three major stress factors: temperature, current intensity (rated), and state of charge (SoC). This model can drift with time due to repeated solicitation, so its parameters need to be updated by on-field measurements, in order to remain accurate. These on-field measurements are submitted to the so-called Incremental Capacity Analysis method (ICA), consisting in the analysis of dQ/dV as a function of V. It is a direct indicator of the state of health of the cell, as the experimental peaks are related to the active material chemical/structural evolution, such as phase transitions and recorded potential plateaus during charging/discharging. It is here applied to NMC/graphite based commercial cells. These peaks' evolution can be correlated with the here-defined Ah-kinetic and $\sqrt{t}$-kinetic aging, which are chemistry-dependent, and therefore, has to be adjusted to the different types of cells.

Keywords: lithium-ion; NMC; aging; ampere-hour throughput; incremental capacity analysis

---

1. Introduction

Rechargeable lithium-ion batteries (LIBs) appear to be the best system for energy storage in many applications, especially electric vehicles and stationary mass storage in the context of the energy transition. Any system employing Li-ion cells must be informed of the amount of energy that can be stored and the power that can be provided by the battery at any time. Therefore, reliable developments need to be correlated with accurate aging studies.

The aging of a battery leads mainly to loss of capacity, loss of power, and increase of internal resistance. The understanding of the underlying mechanisms is fundamental in order to provide an accurate and reliable aging prediction in the models. However, it is not always easy to build a strong model, since the battery cell consists of a complex system including interactions between several domains: mainly, physics, electrochemistry, and thermal sciences. Qualitative aging of the battery has been studied in all these fields, regarding electrolyte degradation, solid electrolyte interphase (SEI) formation, or mechanical deformations [1–4].

A LIB is expected to have a life span of more than 10 years. Therefore, the implemented model should predict the aging of a LIB in the long-term, and have a fast computation time.

In order to produce a quantitative aging estimation, cell degradation needs to be studied under various aging conditions, and models need to be strengthened with on-field measurements in order to generate predictions of energy storage capacity and power capability. Several generations of LIBs have been tested both at EDF Lab Les Renardières facility and Université Paris-Sud ICMMO Lab, using specific protocols and monitoring aging degradations.

Many different models have been developed so far to evaluate the electrochemical behavior of batteries:

- fundamental models or physical models;
- phenomenological models or empirical models; and
- mathematical models.

They are described in the following sections.

1.1. Fundamental Models

Physical models appreciate the aging of a battery by precisely describing the internal mechanisms, such as ion transport in the electrolyte or within the electrode active material, charge transfer at the electrode/electrolyte interface, thanks to partial differential equations, Fick's law, or the Butler–Volmer equation for example. Newman and Fuller have developed the porous electrode model [5], extended by Ramadass et al. to take into account the loss of capacity during each cycle [6]. An approximation of the lithium-ion concentration profile within the solid phase has been presented by Wang et al. [7] and by Subramanian et al. [8]. A simpler physical model has been introduced by Haran et al. that represents each electrode as a simple spherical particle [9], it has been extended to LIB by Ning et al. [10], and further studied by Delacourt and Safari [11].

The advantage of the porous electrode model's is that it takes into account almost all the physical processes taking place while battery is cycling, however solving it is time consuming and many required parameters cannot be verified. The simple spherical particle is easier to solve but does not encompass all the mechanisms. One has to remember that the computation time is crucial since the model has to be recomputed at every cycle. The influence of each separate aging parameter (temperature, state of charge, and rated current) on the model has, to our knowledge, never been developed.

1.2. Phenomenological Models

Despite physical models, the phenomenological models provide links between inputs and outputs of the system, without a comprehensive examination of the physical phenomena taking place. A simple structure is enough to solve the model. Instead of equations, this model consists of experimental and extrapolated curves. Experimental knowledge obtained with these models can circumvent an equational or theoretical problem.

Among these models designed for lithium-ion batteries, we find the fatigue models, like the one originally designed by Dudézert and Franger [12] and refined by Badey and Franger [13]. The model tends to modulate the impact of an exchanged ampere-hour by the conditions in which it is exchanged. Mathematical functions of the current, the state of charge, and the rated current are therefore established as weighting functions. Therefore, this model is natively able to decipher the contribution of each aging parameters: temperature, state of charge, and rated current. The second strength of this model is being able to distinguish the two contributions of aging to the cell: calendar aging proportional to the root square of time and cycling aging proportional to the total ampere-hours exchanged. Indeed, the aging of a battery occurs both when the battery is in use, corresponding to the cycling aging, and when it remains unused, corresponding to the calendar aging. The aging can be thus modeled by two different contributions: an active regime loss and a temporal one.

Previous attempts to do so include Baghdadi et al. [14], who suggested an exponentially decreasing capacity with time, or Bloom et al. [15], who proposed a model where the temperature follows an

Arrhenius law. Most of the authors use a root square dependency to describe the calendar loss of capacity [13–16]. Other authors suggest multiplying a degradation coefficient of the calendar aging to the cycling one. Badey's model includes the two most important features required in our case: decoupling between the calendar and cycling aging, thanks to a sum of two terms, and identification of each parameter's impact.

1.3. Mathematical Models

Mathematical models require a huge amount of data in order to find a link between inputs and outputs. They are based on numerical resolution, and use probabilistic or statistical methods. Generally speaking, they cannot extrapolate to a situation that was not given in the initial set of data. Given the ongoing increasing volume of data and their valorization, one can think these models will play an important role in the future. However, they do not embed any expertise. Furthermore, the error cannot be controlled.

The model chosen to establish our own aging model is based on the fatigue model of Badey. The functions have already been explained thermodynamically [17]. However, fatigue models drift with repeated solicitation, so that their parameters need to be regularly reevaluated. This is what the incremental capacity analysis (ICA) will allow.

In this contribution, we aim at investigating and establishing a link between the ICA and the loss of calendar and cycling capacity, as defined in our model. Our model would then be able to be reparametrized thanks to this metric computed on-field. In this paper, the evolution of the ICA peaks obtained for NMC/graphite cells are discussed to provide a comprehensive interpretation.

The first part of this paper focuses on the aging model implementation and experimental and model basics, whereas the second part deals with the integration of the incremental capacity analysis as a way to explain the degradation process in a quantitative manner.

2. Materials and Methods

In this piece of work, commercial NMC/graphite cells were considered. Due to confidentiality, the cell manufacturer cannot be named. The main technical characteristics of the cells are summarized in Table 1.

Table 1. NMC cell technical characteristics.

Cell	Capacity	Cut-Off Voltages	Charging Protocol
NMC/Graphite	64 Ah	4.2 V/3 V	CC-CV

The cells received were surveyed by weight and open-circuit voltage (OCV) measurements. Variations in OCV with temperature have also been measured.

NMC is very much considered nowadays because it enables very good cyclability as well as good capacity: 180–200 mAh/g. However, NMC suffers from noticeable manganese dissolution in the electrolyte.

All cells were tested with laboratory-made racks, enabling air ventilation. It is important to pay close attention to cell connections to the power channel, in order to minimize variability and achieve better consistency in tests results. All cells used in the tests were submitted to an initial check-up, also called the Reference Performance Test (RPT). Therefore, cells with close initial characteristics (capacity, weight) were selected to go through the aging tests, described by Delétang et al in [17] and in Table 2. The average OCV recorded at 30 °C is 3.60 V, which corresponds to a SOC of ~40%.

Table 2. Aging test matrix.

Cell Number	Temperature/°C	SOC	I Charge	I Discharge
1	T1 = 10	100%		
2	T1 = 23	100%		
3	T1 = 45	100%		
4	T1 = 55	100%		
5a	T1 = 45	15%	1C	1D
5b	T1 = 45	30%	1C	1D
5c	T1 = 45	50%	1C	1D
5d	T1 = 45	80%	1C	1D
6a	T1 = 45	50%	C/2	1D
6b	T1 = 45	50%	1C	D/2
7a	T1 = 23	50%	1C	1D
7b	T1 = 23	30%	1C	1D
8	T2 = 10	50%	1C	1D

The RPT procedure was performed at 23 °C and consists of the following steps; constant current phase charge at a regime of C/3, followed by a constant voltage phase at 4.2 V until the termination current C/20 was reached, then charge and discharge resistance pulses for 10 s at 1C current every 3.2 Ah discharged until the cutoff voltage is reached (3.0 V), the complete recharge at C/3 CC-CV, and then complete CC discharge at C/2 until the cutoff voltage is reached. This last discharge provides the remaining capacity. The detection of capacity fading therefore requires columbic counting of a full charge/discharge cycle. The cell is finally charged to 50% nominal SOC in order to achieve an electrochemical impedance spectroscopy measurement. This latter is done at SOC 50%. The measurements were performed using the battery testing bench VMP Biologic multichannel potentiostat.

Capacity remaining was thus measured at C/2 discharge regimes. The datasheet featured a 64 Ah capacity. The average value was 62.93 Ah, and the capacity measured was quite consistent among all cells (the median value being 62.92 Ah, the standard variation being 0.17 (0.27%), there is little cell-to-cell variation), demonstrating the cell quality in performance. This cell-to-cell variation of 0.27% is satisfying when compared to literature [18].

The resistance pulses allowed computing the polarization resistance of the cell. It was calculated from the difference in voltage drops using Ohm's law: $\Delta V = I\,R$.

A series of tests were launched so that cells undergo unitary aging tests.

The aging parameters were temperature (in °C), rated-current C-rate (abusively written I in A/Ah), and the state of charge of the battery (SOC in %). Cycling aging was performed at an approximation of the adequate value of SOC with a depth of discharge (DOD) of 10%, so that the aging of the cell can be considered as due to the defined SOC. The cycling process was the following: the cell was first charged to 100% SOC at C/3 in CC-CV mode, it is then discharged to the targeted SOC subtracted by 5% at a rate of C/3. Then a rest of 1 h is imposed before the ulterior cycles start.

The values of the parameters have been chosen in order to be aligned with the real applications in use. Especially, temperature does not exceed 55 °C and is never lower than 10 °C. The test facility is air conditioned to maintain a stable temperature. Due to difference in charge and discharge efficiency, the SOC, around which the cell is being cycled, tends to decrease with cycling. Therefore, every 100 cycles at 10% DOD, the re-SOC is imposed to the cell. It is a common test procedure, as described by Delacourt et al. [19]. Approximately every 3000 cycles at 10% DOD the cell was subjected to a RPT to assess changes in performance.

Impedance measurements associated with the RPT was performed in galvanostatic mode as recommended for LIB in which the voltage changes during testing. The current amplitude was set to 1 A, which shows a good signal-to-noise ratio and stays in the linear region. The frequency swept between 10 kHz to 10 mHz. All measurements were carried out at 23 °C.

3. Results

3.1. Cycling Aging

The usual test procedure, like the one mentioned in the European Standard CE NF 50,272 which specifies safety requirements for secondary batteries and battery installations, defines the end-of-life of a battery when it cannot deliver more than 80% of its initial capacity under a specific test protocol. At this time, the cells were at ~10/12%-capacity loss, constituting a huge volume of recorded data. Only some selected results are presented here.

Figure 1 presents the evolution of capacity over the 45 °C cycling for the different SOC. As explained in the above section, the discharge rate was C/2. It has to be noted that the higher the SOC, the higher the fade rate, except for SOC 15% that features a higher degradation rate than SOC 30%.

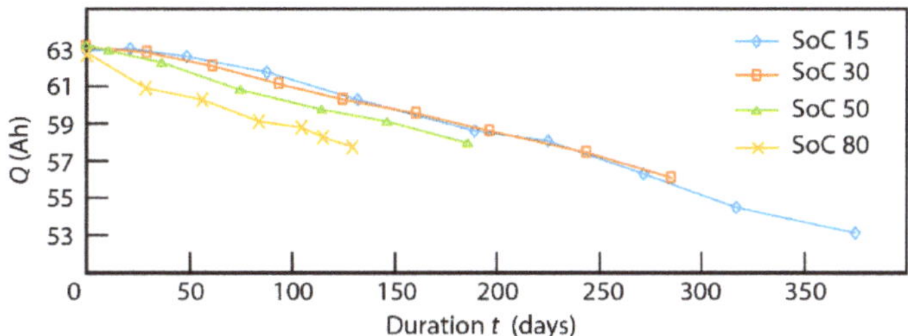

Figure 1. Evolution of the remaining cell capacity with respect of time, for aging at 45 °C at different state of charge (SOC).

The data were subsequently computed to apply the fatigue model [11,12], of which, the main capacity evolution equation is mentioned below.

$$\Delta Q = K^{cyc} f_1(T) f_2(I) f_3(SOC) \cdot Ah + K^{cal} g_1(T) g_2(I) g_3(SOC) \cdot \sqrt{t}$$

where K^{cyc} and K^{cal} are two constant coefficients specific to the cell considered. The loss of capacity is the sum of two terms: the first one is called the cycling aging and the second one is called the calendar aging. This separation is more and more specifically addressed in the literature [20,21].

For each data set, the linear regression of the function $\Delta Q/(\sqrt{t})$ with respect to $(CT\,(Ah))/(\sqrt{t})$ provides two coefficients, as seen on Figure 2, A,B defined as follows

$$A = K^{cyc} f_1(T) f_2(I) f_3(SOC),$$
$$B = K^{cal} g_1(T) g_2(I) g_3(SOC),$$

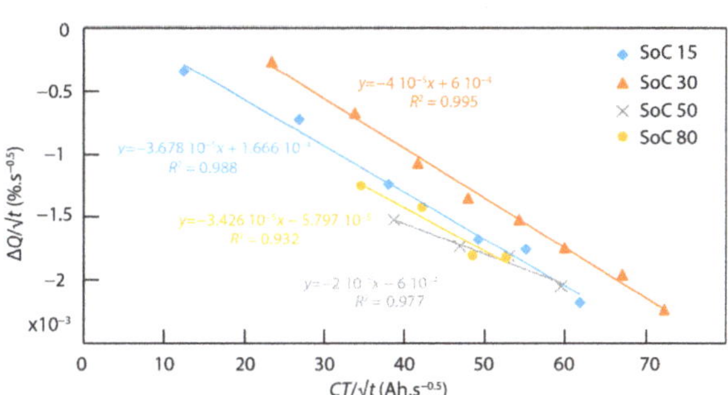

Figure 2. $\Delta Q\,(\%)/\sqrt{t}$ as a function of $CT\,(Ah)/\sqrt{t}$ (CT(Ah) stands for the total exchanged ampere-hours.

These two coefficients allow to extract the values of the weighted functions f_i and g_i, and to decipher the capacity fade due to calendar aging and the loss due to cycling aging [17]. The cycling aging of an aging test is equal to $A.Ah$ and the calendar aging is equal to $B.\sqrt{t}$. Therefore the cycling aging is proportional to the number of ampere-hours exchanged and the calendar aging is proportional to the square root of time. This latter square root of time dependency is, however, discussed by Dubarry [22], who compared it with a $t^{3/4}$ and a t dependency. The $\sqrt{t}$ obtained the best fit, so that the authors used it in the end.

3.2. Incremental Capacity Analysis

Figure 3a displays the evolution of the IC curves with time for cell aged at 45 °C at ~50% SOC at 1C/1D (denoted as (45°, SOC 50, 1C/1D)). In the same manner, Figure 3b stands for (45°, SOC 30, 1C/1D), Figure 3c stands for (45°, SOC 15, 1C/1D), and Figure 3d stands for (45°, SOC 80, 1C/1D). It exhibits two different peaks: the highest peak (maximum at ~3.6 V) and the second peak (maximum at ~3.4 V), which is coherent with what is observed by Berecibar et al. [23]. The highest peak has a symmetrical shape. Qualitatively, the most striking feature is that the intensity of the highest peak and of the second peak decreased for different states of health, which is coherent with the few NMC-based ICA study [21,23,24]. The average voltage of the maximum intensity of the highest peak is shown in Table 3. The average value was stable and the standard deviation was almost zero. This voltage remained stable with time (aging), and the highest peak is simply denoted as "3.5 V-peak" in the following sections. It seems that the curve remained in the same shape around the extrema of voltage values, indicating that the kinetics of the cell reaction has not been affected. As shown in Table 3, the average voltage of the maximum intensity was not as stable as the highest one. The standard deviation was also higher.

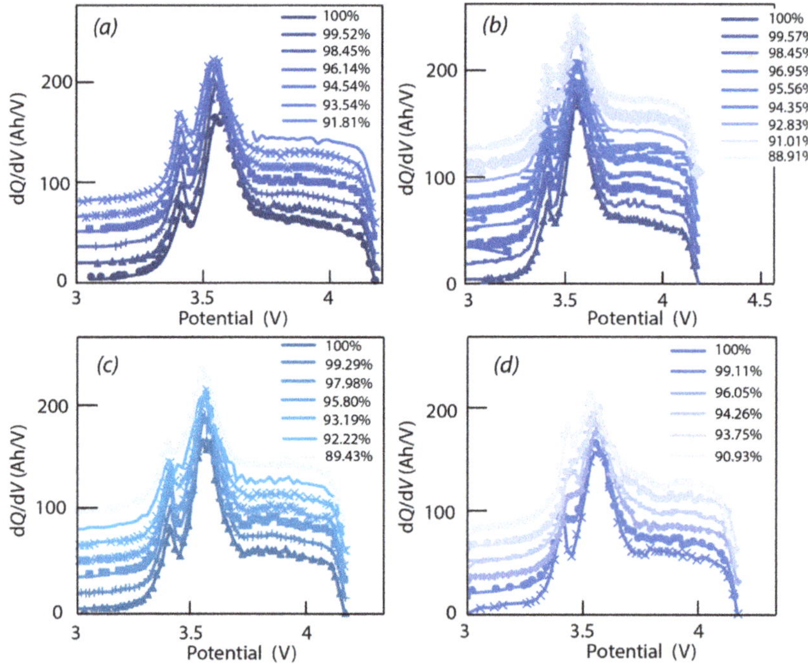

Figure 3. IC curves obtained for (**a**) (45°, SOC 50, 1C/1D), (**b**) (45°, SOC 30, 1C/1D), (**c**) (45°, SOC 15, 1C/1D), and (**d**) (45°, SOC 80, 1C/1D) at different states of health. A vertical offset was added for the sake of clarity: the curves have a 15 Ah/V-offset each.

Table 3. Average value of voltage and standard deviation for 3.5 V and 3.4 V peak maxima.

	Average Voltage of the Maximum of the 3.5 V Peak	Standard Deviation 3.5 V	Average Voltage of the Maximum of the 3.4 V Peak	Standard Deviation 3.4 V
(45 °C, 15%, 1C/1D)	3.57	0.01	3.42	0.03
(45 °C, 30%, 1C/1D)	3.56	0.01	3.41	0.01
(45 °C, 50%, 1C/1D)	3.55	0.01	3.43	0.04
(45 °C, 80%, 1C/1D)	3.56	0.01	3.43	0.02
(23 °C, 30%, 1C/1D)	3.57	0.00	3.41	0.00
(23 °C, 50%, 1C/1D)	3.55	0.01	3.42	0.03
(10 °C, 50%, 1C/1D)	3.56	0.01	3.43	0.01

Changing the temperature, the tests (23°, SOC 50, 1C/1D; 23°, SOC 30, 1C/1D) feature similar behaviors in IC curves, as shown in Figure 4.

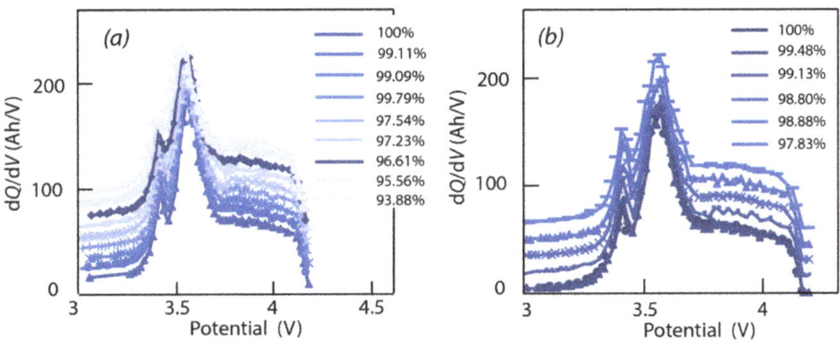

Figure 4. IC curves obtained for (**a**) 23°, SOC 50, 1C/1D and (**b**) 23°, SOC 30, 1C/1D. A vertical offset was added for the sake of clarity: the curves have a 10Ah/V-offset and 15 Ah/V-offset.

The peak area has been defined for the 3.5 V peak, as the area under the curve comprised between the peak's voltage value and is symmetrical, as described on Figure 5. It corresponds to the capacity involved in the related reaction happening in between the borders. For the 3.4 V peak, it is defined as the area under the curve comprised between the peak voltage value and the cutoff voltage of 3.0 V.

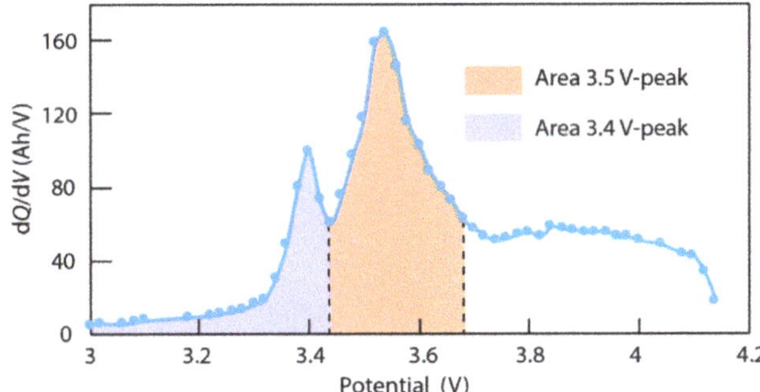

Figure 5. Incremental capacity analysis (ICA) versus voltage for one cell at the initial state, obtained for dV = 20 mV. AP-3.5 V is the area of the peak at 3.5 V and AP-3.4 V is the area of the peak at 3.4 V.

It can be observed that the 3.5 V peak area is linearly proportional to the total loss of capacity due to cell aging, which is coherent with [25,26].

The correlation coefficient of that left curve (Figure 6) is quite high (0.97), the other one is weaker. Others are also strongly correlated as shown in Table 4.

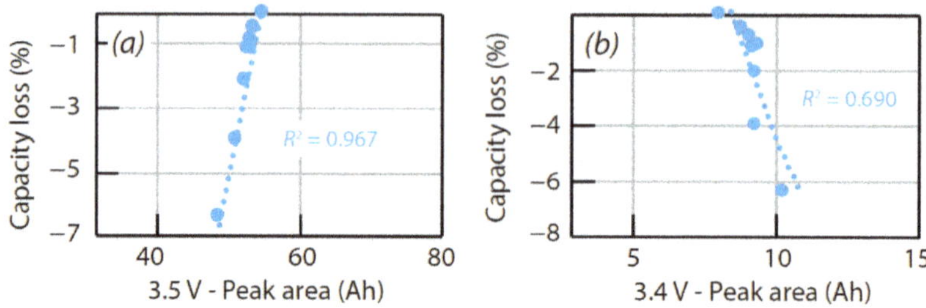

Figure 6. Capacity loss (%) as a function of peak area (3.5 V) (Ah) for cell (23°, SOC 30, 1C/1D) ((**a**) is for the 3.5 V peak and (**b**) is for the 3.4 V peak).

Table 4. Correlation coefficient of highest peak with the total loss of capacity for different tests.

Test Description	Correlation Coefficient HP-3.5 V//Loss Capacity	Correlation Coefficient HP-3.4 V//Loss Capacity
(45 °C, 15%, 1C/1D)	0.97	0.93
(45 °C, 30%, 1C/1D)	0.97	0.91
(45 °C, 50%, 1C/1D)	0.95	0.78
(45 °C, 80%, 1C/1D)	0.90	0.73
(23 °C, 30%, 1C/1D)	0.96	0.69
(23 °C, 50%, 1C/1D)	0.98	0.92
(10 °C, 50%, 1C/1D)	0.80	0.63

These first cycling tests show that the highest peak (3.5 V) seems to be strongly correlated with the total loss of capacity. In order to decipher between the calendar contribution ($B.\sqrt{t}$) and the cycling contribution ($A.Ah$) of our aging model, we were interested in determining if the area of the peak (labeled A) could be correlated to the Ah-exchanged or to the root square of t, as suggested by Cabelguen [18] and Riviere [27]. The correlation coefficient is shown in Table 5.

Table 5. Correlation coefficient of highest peak area with Ah and $\sqrt{t}$ for different tests.

Test	Correlation Coefficient Peak 3.5 V//Ah	Correlation Coefficient Peak 3.4 V//$\sqrt{t}$
(45 °C, 15%, 1C/1D)	0.97	0.98
(45 °C, 30%, 1C/1D)	0.98	0.71
(45 °C, 50%, 1C/1D)	0.94	0.76
(45 °C, 80%, 1C/1D)	0.82	0.32
(23 °C, 30%, 1C/1D)	0.81	0.97
(23 °C, 50%, 1C/1D)	0.98	0.93
(10 °C, 50%, 1C/1D)	0.88	0.53

With all the correlation coefficients above 0.81 for 3.5 V, in terms of modeling, it seems to be safe to consider the highest peak to be correlated to the Ah-exchanged. The second peak (3.4 V peak) seems to be correlated to the square root of time, although some quite low coefficients were determined.

The Ah-results summarized in Table 6 were not convincing enough as they do not show a clear trend about the 3.4 V peak, as discussed earlier. This is probably due to the difficulty in removing the cycling part in the calendar part with this criterion. Furthermore, considering only calendar aging tests at four different temperatures—10 °C, 23 °C, 45 °C, and 55 °C—the correlation between the area of the peak and the root square of the time was almost ideal, except for (55 °C, 100%, CAL). This confirms that cycling aging has an impact in the calendar aging contribution of the global capacity loss.

Table 6. Correlation coefficient of 3.4 V peak with Ah and $\sqrt{t}$ for different tests.

Test	Correlation Coefficient Peak 3.4 V//Ah	Correlation Coefficient Peak 3.4 V//$\sqrt{t}$
(45 °C, 15%, 1C/1D)	0.96	0.98
(45 °C, 30%, 1C/1D)	0.88	0.71
(45 °C, 50%, 1C/1D)	0.84	0.76
(45 °C, 80%, 1C/1D)	0.26	0.32
(23 °C, 30%, 1C/1D)	0.20	0.97
(23 °C, 50%, 1C/1D)	0.86	0.93
(10 °C, 50%, 1C/1D)	0.37	0.53
(55 °C, 100%, CAL)	–	0.56
(45 °C, 100%, CAL)	–	0.85
(10 °C, 100%, CAL)	–	0.99
(23 °C, 100%, CAL)	–	0.94

The last strategy was to seek if the area AP of that 3.4 V peak could follow the same kinetic as the loss of capacity, and then evaluate the predominant part of aging that affects that peak. We have thus computed the following equation, searching for alpha and beta coefficients.

$$\frac{\Delta \text{AP}}{\sqrt{t}} = \text{alpha} \cdot \frac{Ah}{\sqrt{t}} + \text{beta}$$

Table 7 shows the corresponding correlation coefficients.

Table 7. Correlation coefficient of 3.4 V peak with (alpha * Ah/$\sqrt{t}$ + beta) for different tests.

Test	Correlation Coefficient Peak 3.4 V//Alpha and Ah/$\sqrt{t}$ + Beta
(45 °C, 15%, 1C/1D)	0.70
(45 °C, 30%, 1C/1D)	0.92
(45 °C, 50%, 1C/1D)	0.97
(45 °C, 80%, 1C/1D)	0.18
(23 °C, 30%, 1C/1D)	0.78
(23 °C, 50%, 1C/1D)	0.66
(10 °C, 50%, 1C/1D)	0.26

The correlation coefficients were quite satisfying for three tests: (45 °C, 30%, 1C/1D), (45 °C, 50%, 1C/1D), and (23 °C, 30%, 1C/1D). The coefficients alpha and beta are not significant enough to be compared, as they are ~10^{-6}. This seems to be a promising path in order to better distinguish the different processes of aging taking place in the cell, to be further studied.

In terms of the model, we proposed assigning the Ah-kinetic of our aging model to the area of the 3.5 V peak and the Ah-kinetic and $\sqrt{t}$-kinetic aging to the area of the 3.4 V peak.

4. Discussion

4.1. Background on Aging Mechanisms

Aging can generally be due to three main causes:

- Loss of lithium inventory (LLI) is caused by the consumption of Li cations by parasitic reactions of which the major remains the well-known SEI growth at the negative electrode. Therefore, less and less lithium ions are available to shuffle back and forth in the electrodes, captured in the electrode bulk or inside the electrolyte aggregates. The reactions lead to by-products released in the electrolyte, which can further cause side reactions [28,29]. This aging cause can often explain the first stages of the aging. The impact on the potential of LLI, as explained by Bloom et al.

in [30], is that as the side reactions proceed, the lower potential regions of the NE would be removed, and thus, charging the cell would incompletely lithiate the NE. Compared to its initial state, at the end of charge, the NE is now at a higher potential. To maintain the upper limit value, there would be a shift to compensate at the positive electrode to higher potentials. However, in the graphite, the end of charge is a plateau and the potential of the PE will not very much.

- Loss of active material (LAM): insulation of active material with loss of contacts with the conductive matrix, dissolution of transition metal in the electrolyte, structural changes in the crystalline structure of anode, and cathode active materials due to repeated insertions/disinsertions of ions. LAM appearing at the negative electrode only (LAM NE) and LAM appearing at the positive electrode only (LAM PE) have to be differentiated. Dubarry et al. [28] even separate LAM on delithiated and LAM on lithiated electrodes. The impact of LAM on the potential of PE and NE is well explained in Figure 6 of ref. [31] and by Matadi [16] and Dubarry [32,33].

- Conductivity loss (CL): increasing of the faradic resistance due to mass transport slowdown or ionic conduction in the bulk of the electrode and increasing of the ohmic resistance due to contacts degradations in the electrodes or in the electrolyte conduction. CL only affects the cell voltage, not the capacity.

These mechanisms are schematically separated here, but they are coupled during the aging process.

Dubarry et al. used a very suitable analogy with a water clock [24], composed of two connected bulbs filled with a given amount of liquid. LAM affects the size of the bulb and LLI affects the amount of available liquid, if there was a leak in the water clock. They have also constructed a comprehensive toolbox "alawa" to simulate these phenomenas [31]; it is used in Section 4.4.

4.2. Background on ICA

The Incremental Capacity Analysis (ICA), meaning the analysis of dQ/dV as a function of V (dQ corresponds to the change in capacity that happens in the gap dV) is a direct indicator of the state of the cell. Differential Voltage Analysis (DVA), which analyses dV/dQ as a function of Q, is not appropriate for our study since the reference (x-axis) varies when capacity fades [26,31]. dV/dQ highlights single phase regions whereas dQ/dV curve represents the phase transformations [25,26,34,35]. The peaks represent the electrochemical mechanisms occurring in the cell. They are mainly related to the chemistry of the cell [23–33].

ICA is more beneficial when the cell is at equilibrium. Discharge regimes should therefore be small, approximately C/20 or C/25. However, these very low rates require too much time in real application (approximately 50 h to complete a charge and discharge cycle). As proven by Li [26], one can take benefit from this technique until 1C regime. In the above 1C regime, the results are erratic and not exploitable [32].

The evolution of the voltage of the cell during a discharge of 100% DOD shows the variations of voltage (that indicates what reaction is taking place) as a function of displaced charges (meaning capacity, which shows the extent of that reaction), as shown in Figure 7. Two clear voltage plateaus can be identified from that curve, approximately 3.4 V and 3.6V. A voltage plateau on a full cell means both electrodes are experiencing phase transformations (there is coexistence of two distinct phases on each electrode) [27]. As a phase depends on the quantity of lithium ions inserted in the electrode, the information about the battery aging mechanism is contained in the shift of these plateaus. The dQ/dV analysis estimates the capacity displaced in each incremental change of voltage in the reaction. However, it does not only provide pieces of information on electrochemical reactions occurring inside the cell (through voltage plateau for example) but its evolution with time reveal how these reactions are affected by aging.

The principle of the ICA is therefore clear: the peaks obtained at different aging states RPTs provide an aging signature and enable identification of the degradation modes (LAM PE, LAM NE, LLI, and CL). A peak is characterized by a voltage position and its area (linked to its height also called intensity, which is deeply studied by Berecibar et al. [23]).

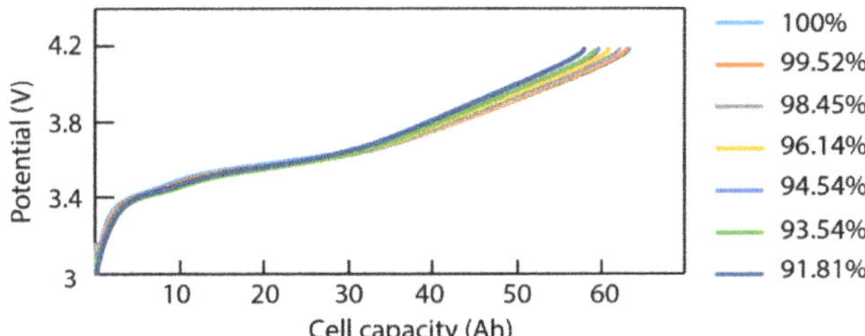

Figure 7. Evolution of the discharge curve of a cell with aging during a discharge at C/2 at 23 °C. Curves are shown according to their State of Health (SOH).

The final goal of using this metric is to be able to update the first fatigue model. Therefore, the correlation made previously between the peak and the calendar and cycling aging processes will be used in a numerical way. Having that correlation link, a discharge profile of the cell will lead to the updated real values of the weighted functions (and no more the predicted values), leading to the refinement of our model.

4.3. Data Analysis of ICA

The C/2 discharge data and very fast acquisition rate require mathematical filtering, to minimize the noise and enlighten the peaks. This filtering was done with Microsoft Excel, considering a fixed ΔV of 20 mV for discharge data (which is consistent with Bloom et al. [35,36] and Li et al. [37]). dQ then appears to be the exchanged charges during this given voltage step ΔV. For charge-related IC data, the ΔV depends on the rate of the charge so that the peaks can be observed and compared: 20 mV for 0.04 C, 10 mV for 0.2 C, and 2 mV for 0.33C. There is proportionality between these two values as shown in Figure 8.

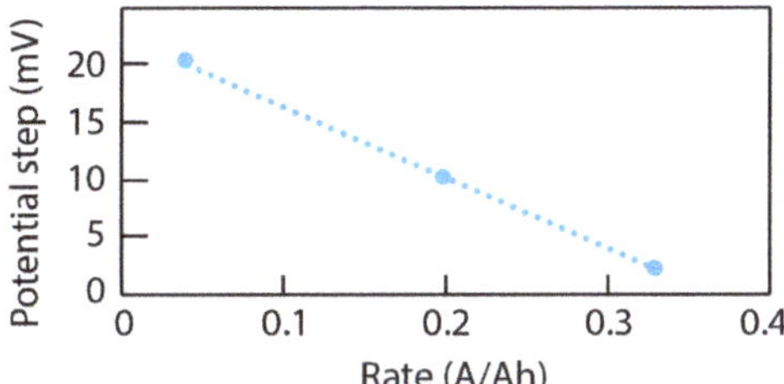

Figure 8. Voltage step chosen for charge ICA analysis as a function of the charge rate.

A plot of dQ/dV versus voltage for a cell at t = 0 during discharge is given in Figure 5. It contains primarily two peaks that can be assigned thanks to additional information, found in the literature. The first statement that can be made is that these two peaks correspond to two different electrochemical reactions taking place during discharge inside the whole cell (both anode and cathode). These peaks are strongly chemistry-dependent as mentioned in the literature [23,25,26].

Based on the literature, starting from the highest potential where the NMC electrode forms a solid solution, during discharge, the voltage of the cell decreases consistently from 4.2 V to 3.75 V (vs Li^+/Li), filling the NMC cathode of lithium ions [18]. In this stage, the graphite electrode content transfers from LiC_6 to LiC_{12} and then from LiC_{12} to LiC_{18} [18]. From the ripples in the IC curve of Figure 5, ~3.75 V can be due to the transition phase LiC_6 to LiC_{12}. This transition phase lasts much longer than the others in the graphite (it is kinetically slower) [34]. As no hysteresis between lithiation and delithiation in this phase was observed, this should be reversible and thus, not strongly affected by aging.

Then, the lithium intercalation inside the crystalline structure leads to a phase transformation of NMC at 3.75 V (vs. Li^+/Li) [18]. Convolution of this reaction vs. the graphite negative electrode implies a decrease of 0.122 V (vs. Li^+/Li) in the cell voltage, as the graphite transits from LiC_{12} to LiC_{18} (stage 2 transformation). This explains the observed peak at ~3.6V (3.75 V–0.122 V $\approx$ 3.6 V). The additional difference can be explained by additives in the electrolyte. This is the only NMC-based reaction, as confirmed by Li et al. [26].

The remaining reactions happening inside the cell are graphite-based. From LiC_{18} to LiC_{24}, and even to LiC_{36}, the graphite potential (vs. Li^+/Li) increases drastically up to 0.6 V. At this step, the PE is on a voltage plateau. Therefore, since PE potential is not changing, but the NE potential is, a new peak will appear. This could explain the second peak observed at 3.4 V.

LLI and LAM NE could be the aging cause for the 3.4 V-peak evolution, as this peak is graphite-related. It is well-known that LLI, leading mainly to SEI formation, occurs primarily at the negative electrode. Indeed, LLI induces a potential shift of the NE curve of V = f(SOC) towards higher SOC (see Figure 7 of [31]). The last graphite reaction (LiC_{18} to LiC_{24} and to LiC_{36}, stage 4–5 transformation), that is involved in the 3.5 V-peak, will be progressively convoluted with a flatter PE-curve. Thus the convolution will lead to a thinner and more intense peak.

The 3.5 V-peak corresponds both to the PE reaction and the NE stage 2 transformation. With LLI, the NE curve (V = f(SOC)) will be shifted towards higher SOC relative to the PE. Therefore, the stage 2 transformation of NE will be initiated at higher SOC and prior to the start of the PE-reaction. Thus, less of the capacity involved in the stage 2 will convolute with the PE and the peak area will go down. One could suppose that LAM PE could be responsible as well. Indeed, as SEI grows, there is production of species resulting from the electrolyte decomposition that might accumulate on the active grain surface, which may cause isolation of the grains inside the electrode matrix. LLI would therefore result in LAM PE. This would explain the difficulty to separate both modes of aging based on the 3.4 V peak's evolution. In any case, it is well-established that the cycling aging affects more the cathode electrode (NMC) [12]. Therefore, the fact that the area under the 3.5 V peak appears to be proportional to the number of exchanged-ampere hours seems legitimate. This hypothesis needs to be confirmed.

However, because the aging has reached less than 10% of capacity loss, the predominant mode of degradation seems to be LLI. This could be confirmed in the fact that in [31], the authors attributed to LAM PE an exponential evolution and to LLI a linear evolution (with capacity loss): at the beginning, the exponential is hardly differentiable from the straight linear.

4.4. Verification of Assumptions Based on 'Alawa' Tool

Different degradation of scenarios has been simulated on the 'alawa' toolbox developed by Dubarry et al. [31]. The data used to simulate our full cell come from the software (Graphite and NMC electrode data), so it was compulsory to adjust the Loading Ratio and the Offset to fit the IC curve obtained as close as possible to our own IC curve. Figure 9 shows the resulting IC curves obtained for a G//NMC cell with LR = 1.2, OFS = −5, resistance = 1, and a discharge rate of C/25.

LAM PE corresponds to LAMdePE (delithiated state), as LAM occurring on the lithiated state is equivalent to LAMdePE plus some LLI (same for LAM NE).

These simulations show the following tendencies.

- LLI increases the 3.4 V peak intensity, and reduces the 3.5 V peak intensity. It confirms that less and less lithium is involved in NE stage 2-reaction through cycling. The remainder of the capacity that is not engaged in that stage 2-reaction remains available for the other reactions, thus the growing and earlier start of the 3.4 V peak.
- LAM PE reduces the 3.4 V peak intensity, until it disappears, and moves the 3.5 V peak to lower voltages (slightly) while decreasing its intensity slightly.
- LAM NE decreases slightly the 3.4 V peak and increases slightly the 3.5 V peak intensity.

Therefore, what we observed on the 3.5 V peak of our own cell seems indeed to be mostly due to LLI. The intensity evolution of the 3.4 V peak seems indeed to be a mix of LLI (that increases its intensity) and LAM NE (that reduces its intensity). The 3.4 V peak area is shown to be slightly increasing (Figure 6b). It is however difficult to decipher between LLI and LAM NE which one is dominating, since the 'alawa' toolbox is not designed to take into account the temperature and precise SOC impact.

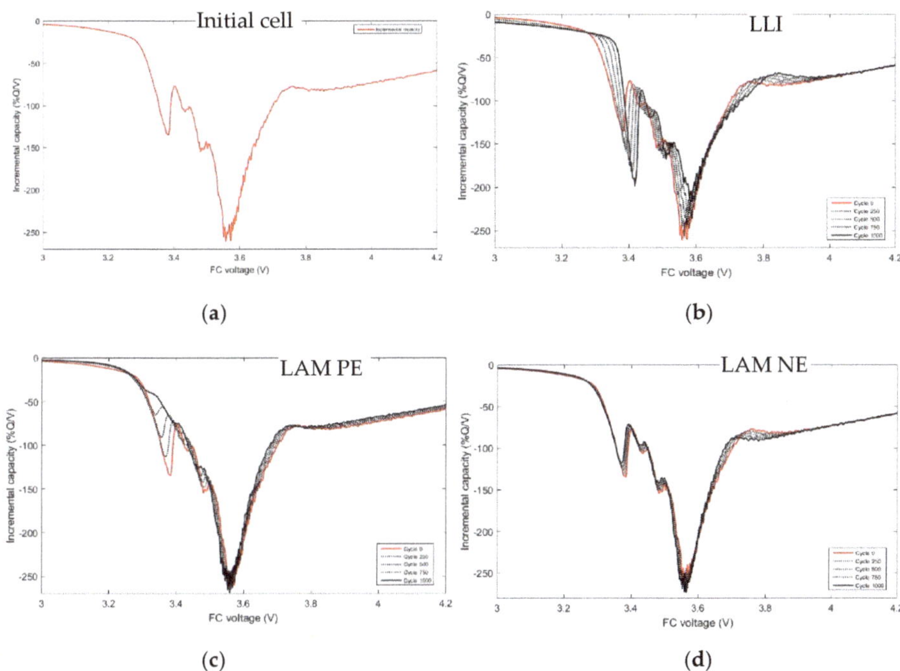

Figure 9. IC curves obtained from the 'alawa' toolbox: (**a**) Initial cell, (**b**) lithium inventory (LLI), (**c**) LAM PE, and (**d**) LAM NE. The red curve is the initial curve, and the other curves show the tendency of evolution every 250 cycles.

4.5. Connection between ICA and the Fatigue Model

The question arisen is how to combine the ICA observations and the fatigue model explained previously in Section 3.1.

The preliminary analysis of all the unitary aging tests leads to six weighting functions, and so six graphs (see the previously published graphs by Deletang et al. [17]). Being given a solicitation made of time steps, a rate, a temperature, and a SOC at each time step, the fatigue model will provide the infinitesimal (between two time steps) loss of cycling capacity ($A.Ah$), called *delta_f[i]*, and the infinitesimal loss of calendar capacity ($B.\sqrt{t}$), called *delta_g[i]*. With these two values, we can compute the value of the predicted remaining capacity at each time step. This is the direct pathway.

However, with on-field measurements, the value of the real capacity can be different from what has been predicted. So it is all about taking the reverse pathway: from the information of the on-field measurement, we extract the Incremental Capacity data and we update the *delta_f[i]* and *delta_g[i]* from which we then extract the weighted functions and, finally, we recompute the prediction with the updated weighting functions. To our knowledge, this has never been done before. Simulations from the developed software will be further published.

5. Conclusions

In this paper, we suggested a review to summarize the different aging models used for LIBs, and we presented the fatigue model used. Then we presented some results coming from an aging test campaign and suggested a link between an observable value (IC peak area), related to the discharge curve of a Li-ion cell, and the cycling and calendar capacity loss, defined in our fatigue model.

We observed two IC peaks for the studied NMC-based cell. One seems to be correlated to the number of exchanged ampere-hours, and thus could be correlated to what was defined as the "cycling aging" in our model. The other one is not clearly correlated to the root square of time, even if it is observed when considering cells aged only in calendar pattern (no exchanged ampere-hour). The difficulty in assigning a trend in the evolution of that peak would be due to the fact that calendar aging is also appearing during cycling. The alawa toolbox has been used to verify some assumptions.

The implementation of that result on another chemistry where the peaks' position will change will be further studied.

As the IC peak degradation shown in the data is related to the first 10% of aging, the correlation needs to be verified for further degradation, where LLI is not the main degradation mechanism any longer [38–40]. Continuation of aging will clarify the explanation regarding LLI and LAM in order to further explain the loss of capacity.

Author Contributions: T.P., N.B., L.A. and S.F. developed the theoretical formalism, performed the analytic calculations. T.P. and N.B. performed the numerical simulations. T.P., L.A. and S.F. designed and performed the experiments. J.D. assisted with electrochemical analysis. T.P. wrote the manuscript in consultation with L.A., S.F. and N.B.

Funding: This research was funded by EDF R&D, Université Paris Sud/Université Paris Saclay ICMMO, and ANRT Agence Nationale de la Recherche et de la Technologie of France.

Acknowledgments: The authors gratefully acknowledge facilities provided by the EDF Lab Les Renardières, and ICMMO lab. The authors gratefully thank the Hawaii Natura Energy Institute for the disposal of the alawa toolbox.

Conflicts of Interest: The authors declare no conflicts of interest.

References

1. Vetter, J.; Novak, P.; Wagner, M.R.; Veit, C.; Moller, K.C.; Besenhard, J.O.; Winter, M.; Wohlfahrt-Mehrens, M.; Vogler, C.; Hammouche, A. Ageing mechanisms in lithium-ion batteries. *J. Power Sources* **2005**, *147*, 269–281. [CrossRef]

2. Broussely, M.; Biensan, P.; Bonhomme, F.; Blanchard, P.; Herreyre, S.; Nechev, K.; Staniewicz, R.J. Main aging mechanisms in Li ion batteries. *J. Power Sources* **2005**, *146*, 90–96. [CrossRef]

3. Grolleau, S.; Delaille, A.; Gualous, H.; Gyan, P.; Revel, R.; Bernard, J.; Redondo-Iglesias, E.; Peter, J. Calendar aging of commercial graphite/LiFePO$_4$ cell—Predicting capacity fade under time dependent storage conditions. *J. Power Sources* **2014**, *255*, 450–458. [CrossRef]

4. Novak, P.; Joho, F.; Lanz, M.; Rykart, B.; Panitz, J.-C.; Alliata, D.; Kötz, R.; Haas, O. The complex electrochemistry of graphite electrodes in lithium-ion batteries. *J. Power Sources* **2001**, *97–98*, 39–46. [CrossRef]

5. Fuller, T.F.; Doyle, M.; Newman, J. Simulation and Optimization of the Dual Lithium Ion Insertion Cell. *J. Electrochem. Soc.* **1994**, *141*, 1–10. [CrossRef]

6. Ramadass, P.; Haran, B.; White, R.; Popov, B.N. Development of First Principles Capacity Fade Model for Li-Ion Cells. *J. Electrochem. Soc.* **2004**, *151*, A196–A203. [CrossRef]

7. Wang, C.Y.; Gu, W.B.; Liaw, B.Y. Micro-Macroscopic Coupled Modeling of Batteries and Fuel Cells I. Model Development. *J. Electrochem. Soc.* **1998**, *145*, A3407–A3417. [CrossRef]

8. Subramanian, V.R.; Tapriyal, D.; White, R. A Boundary Condition for Porous Electrodes. *Electrochem. Solid State Lett.* **2004**, *9*, A259–A263. [CrossRef]

9. Haran, B.S.; Popov, B.N.; White, R. Determination of a hydrogen diffusion coefficient in metal hybrids by impedance spectroscopy. *J. Power Sources* **1998**, *75*, A56–A63. [CrossRef]

10. Ning, G.; Popov, B.N. Cycle life modeling of lithium-ion batteries. *J. Electrochem. Soc.* **2004**, *151*, A1584–A1591. [CrossRef]

11. Delacourt, C.; Safari, M. Life Simulation of a Graphite/LiFePO$_4$ Cell under Cycling and Storage. *J. Electrochem. Soc.* **2012**, *159*, 1283–1291. [CrossRef]
12. Dudézert, C.; Reynier, Y.; Duffault, J.-M.; Franger, S. Fatigue damage approach applied to Li-ion batteries ageing characterization. *Mater. Sci. Eng. B* **2016**, *213*, 177–189. [CrossRef]
13. Badey, Q. Etude des Mécanismes et Modélisation du Vieillissement des Batteries Lithium-ion Dans le Cadre d'un Usage Automobile. Ph.D. Thesis, Université Paris Sud, Orsay, France, 2012.
14. Baghdadi, I.; Briat, O.; Delétage, J.; Gyan, P.; Vinassa, J. Chemical rate phenomenon approach applied to lithium battery capacity fade estimation. *Microelectron. Reliabil.* **2016**, *64*, 134–139. [CrossRef]
15. Cole, B.; Sohn, J.J.; Bloom, I.; Jones, S.A.; Polzin, E.G. An accelerated calendar and cycle life study of li-ion cells. *J. Power Sources* **2001**, *101*, 238–247. [CrossRef]
16. Matadi Pilipili, B. Etude des Mécanismes de Vieillissement des Batteries Li-ion en Cyclage à Basse Température et en Stockage à Haute Température: Compréhension des Origines et Modélisation du Vieillissement. Ph.D. Thesis, Université Grenoble Alpes, Grenoble, France, 2017.
17. Delétang, T.; Barnel, N.; Franger, S.; Assaud, L. Transposition of a weighted Ah-throughput model to another Li-ion technology: Is the model still valid? New insights on the mechanisms. In Proceedings of the Coupled Problems Conference, Rhodes, Greece, 12–14 June 2017.
18. Dubarry, M.; Truchot, C.; Cugnet, M.; Liaw, B.Y.; Gering, K.; Sazhin, S.; Jamison, D.; Michelbacher, C. Evaluation of commercial lithium-ion cells based on composite positive electrode for plug-in hybrid electric behicle applications. Part I: Initial Characterizations. *J. Power Source* **2011**, *196*, 10328–10335. [CrossRef]
19. Delacourt, C.; Kassem, M. Postmortem analysis of calendar-aged graphite LiFePO$_4$ cells. *J. Power Sources* **2013**, *235*, 159–171. [CrossRef]
20. De Hoog, J.; Timmermans, J.-M.; Ioan-Stroe, D.; Swierczynski, M.; Jaguemont, J.; Goutam, S.; Omar, N.; Van Mierlo, J.; Van den Boosche, P. Combined cycling and calendar capacity fade modelling of a NMC oxide cell with real-life profile validation. *Appl. Energy* **2017**, *200*, 47–61. [CrossRef]
21. Schmalstieg, J.; Käbitz, S.; Ecker, M.; Sauer, D.U. A holistic aging model for Li(NiMnCo)O$_2$ based 18650 lithium-ion batteries. *J. Power Sources* **2014**, *257*, 325–334. [CrossRef]
22. Dubarry, M.; Qin, N.; Brooker, P. Calendar aging of commercial Li-ion cells of different chemistries-A review. *Curr. Opin. Electrochem.* **2018**, *9*, 106–113. [CrossRef]
23. Berecibar, M.; Dubarry, M.; Omar, N.; Villarreal, I.; Van Mierlo, J. Degradation mechanism detection for NMC batteries based on Incremental Capacity curves. *World Electr. Veh. J.* **2016**, *8*, 350–361. [CrossRef]
24. Dubarry, M.; Devie, A.; Liaw, B.Y. The Value of Battery Diagnostics and Prognostics. *J. Energy Power Sources* **2014**, *1*, 242–249.
25. Weng, C.; Cui, Y. On-board state of health monitoring of lithium-ion batteries using incremental capacity analysis with support vector regression. *J. Power Sources* **2013**, *235*, 36–44. [CrossRef]
26. Li, Y.; Abdel-Monem, M.; Gopalakrishnan, R.; Berecibar, M.; Nanini-Maury, E.; Omar, N.; van den Bossche, P.; Van Mierlo, J. A quick on-line state of health estimation method for Li-ion battery with incremental capacity curves processed by Gaussian filter. *J. Power Sources* **2018**, *373*, 40–53. [CrossRef]
27. Riviere, E.; Venet, P.; Sari, A.; Meniere, F.; Bultel, Y. LiFePO$_4$ Battery State of Health Online Estimation Using Electric Vehicle Embedded Incremental Capacity Analysis. In Proceedings of the 2015 IEEE Vehicle Power and Propulsion Conference (VPPC), Montreal, QC, Canada, 19–22 October 2015.
28. Cabelguen, P.E. Analyse de la Microstructure des Matériaux Actifs D'électrode Positive de Batteries Lithium-Ion. Ph.D. Thesis, Université Grenoble Alpes, Grenoble, France, 2016.
29. Dubarry, M.; Truchot, C.; Liaw, B.Y. Cell degradation in commercial LiFePO$_4$ cells with high-power and high nergy designs. *J. Power Source* **2014**, *258*, 408–419. [CrossRef]
30. Bloom, I.; Walker, L.K.; Basco, J.; Abraham, D.; Christophersen, J.; Ho, C. Differential voltage analyses of high-power lithium cells 4. Cells containing NMC. *J. Power Sources* **2010**, *195*, 877–882. [CrossRef]
31. Dubarry, M.; Truchot, C.; Liaw, B.Y. Synthesize battery degradation modes via a diagnostic and prognostic model. *J. Power Sources* **2012**, *219*, 204–216. [CrossRef]
32. Dubarry, M.; Truchot, C.; Liaw, B.Y.; Gering, K.; Sazhin, S.; Jamison, D.; Michelbacher, C. Evaluation of commercial lithium-ion cells based on composite positive electrode for plug-in hybrid electric behicle applications. Part II: Degradation mechanism under 2C cycle aging. *J. Power Source* **2011**, *196*, 10336–10343. [CrossRef]

33. Dubarry, M.; Svoboda, V.; Hwu, R.; Liaw, B.Y. Incremental Capacity Analysis and Close-to-Equilibrium OCV measurements to quantify capacity fade incommercial rechargeable lithium batteries. *Electrochem. Solid State Lett.* **2006**, *9*, A454–A457. [CrossRef]

34. Fongy, C.; Jouanneau, S.; Guyomrd, D.; Badot, J.C.; Lestriez, B. Electronic and Ionic Wirings Versus the Insertion Reaction Contributions to the Polarization in LiFePO$_4$ Composite Electrodes. *J. Electrochem. Soc.* **2010**, *157*, A1347–A1353. [CrossRef]

35. Bloom, I.; Jansen, A.N.; Abraham, D.P.; Knuth, J.; Jones, S.A.; Battaglia, V.S.; Henriksen, G.L. Differential voltage analyses of high-power lithium cells 1. Technique and application. *J. Power Sources* **2005**, *139*, 295–303. [CrossRef]

36. Bloom, I.; Christophersen, J.; Gering, K. Differential voltage analyses of high-power lithium cells 2. Applications. *J. Power Sources* **2005**, *139*, 304–313. [CrossRef]

37. Li, X.; Wang, Z.; Zhang, L.; Zou, C.; Dorrell, D. State-of-health estimation for Li-ion batteries by combing the incremental capacity analysis with grey relational analysis. *J. Power Sources* **2019**, *410–411*, 106–114. [CrossRef]

38. Buchberger, I.; Seidlmayer, S.; Pokharel, A.; Piana, M.; Hattendorff, J.; Kudejova, P.; Gilles, R.; Gasteiger, H.A. Aging analysis of Graphite/LiNiMnCoO$_2$ cells using XRD, PGAA, and AC Impedance. *J. Electrochem. Soc.* **2015**, *162*, 2737–2746. [CrossRef]

39. Stiaszny, B.; Ziegler, J.C.; Krauß, E.E.; Zhang, M.; Schmidt, J.P.; Ivers-Tiffée, E. Electrochemical characterization and post-mortem analysis of aged LiMn$_2$O$_4$—NMC/graphite lithium ion batteries part ii: Calendar aging. *J. Power Source* **2014**, *195*, 61–75. [CrossRef]

40. Ansean, D.; Dubarry, M.; Devie, A.; Liaw, B.Y. Operando lithium plating quantification and early detection of a commercial LiFePO$_4$ cell cycled under dynamic driving schedule. *J. Power Sources* **2017**, *356*, 36–46. [CrossRef]

Article

Innovative Incremental Capacity Analysis Implementation for C/LiFePO$_4$ Cell State-of-Health Estimation in Electrical Vehicles

Elie Riviere [1,2,3], Ali Sari [2,*], Pascal Venet [2], Frédéric Meniere [1] and Yann Bultel [3]

[1] Battery Management System Department, EVE System (Electric Vehicles Engineering), 69440 Taluyers, France; elie.riviere@eve-system.com (E.R.); frederic.meniere@eve-system.com (F.M.)

[2] Univ Lyon, Université Claude Bernard Lyon 1, Ecole Centrale de Lyon, INSA Lyon, CNRS, Ampère, F-69000 Lyon, France; pascal.venet@univ-lyon1.fr

[3] Univ Grenoble Alpes, Univ Savoie Mont Blanc, CNRS, Grenoble INP, LEPMI, 38000 Grenoble, France; yann.bultel@grenoble-inp.fr

* Correspondence: ali.sari@univ-lyon1.fr

Received: 27 January 2019; Accepted: 15 March 2019; Published: 1 April 2019

Abstract: This paper presents a fully embedded state of health (SoH) estimator for widely used C/LiFePO$_4$ batteries. The SoH estimation study was intended for applications in electric vehicles (EV). C/LiFePO$_4$ cells were aged using pure electric vehicle cycles and were monitored with an automotive battery management system (BMS). An online capacity estimator based on incremental capacity analysis (ICA) is developed. The proposed estimator is robust to depth of discharge (DoD), charging current and temperature variations to satisfy real vehicle requirements. Finally, the SoH estimator tuned on C/LiFePO$_4$ cells from one manufacturer was tested on C/LiFePO$_4$ cells from another LFP (lithium iron phosphate) manufacturer.

Keywords: accelerated ageing; battery management system; battery management system (BMS); calendar ageing; cycling ageing; electric vehicle; embedded algorithm; incremental capacity analysis; incremental capacity analysis (ICA); lithium-ion battery; lithium iron phosphate; LFP; LiFePO$_4$; remaining capacity; state of health (SoH)

---

1. Introduction

The electrochemical storage system management remains challenging in electric vehicles (EV). Lithium-ion batteries appear as a promising alternative to address current energy needs in various applications such as electric vehicles (EV) and hybrid electric vehicles (HEV). A reliable State of health (SoH) estimation is mandatory to deal with the exact remaining energy available in the battery and thus to compute the vehicle remaining range. More broadly, a precise battery state knowledge is essential for EV development [1–8]. In this study focusing on pure electric vehicle, the remaining capacity is the most relevant parameter. State of health (SoH) is then defined as the ratio of the maximum capacity in the current state and the maximum capacity at the beginning of battery life (BoL) while, in case of HEV, the SoH can also be defined relative to the remaining available power instead of the capacity. Equation (1) presents the SoH percentage definition used in this work:

$$\text{SoH} = \frac{\text{Maximum capacity in current state (Ah)}}{\text{Maximum capacity at BoL (Ah)}} \tag{1}$$

The challenge of this work is to get an online and embedded SoH estimation compatible to the constraints of the EV. Indeed, the EV is rarely fully discharged and its DoD (depth of discharge) ranges usually from 40% to 80% during cycling and with charge regime ranging from C/6 to C/3. Namely, the estimator has to be able to measure the "maximum capacity in the current state" during the nominal use of the vehicle, i.e., without a full discharge of the battery (otherwise the vehicle is out of order). State of health has to be calculated by the battery management system (BMS) using sensors and electronic commonly used in vehicles embedded applications. This automotive electronic system should thus deal with strong cost constraints and wide environmental variations.

The instrumentation required to determine the SoH should be as simple as possible to ensure that the cost of the BMS is reasonable. The algorithms studied in this paper use only measurements of voltage, current and temperature, which are already present in the BMS. No active stimulation device of the cell is used (current injection for example).

The algorithms proposed use the cell response in real operating conditions of an electric vehicle. The classical methods usually employ the charge phase as the basis of analysis because this phase is much better controlled than the discharge. It is therefore necessary to obtain as much information as possible from the current, voltage and temperature measurements during the charging phases of the battery.

Approaches to estimate SoH may consist of using dynamic estimation of the parameters of an equivalent model. This requires adaptive algorithms such as estimators or observers ([9,10]), genetic algorithm [11] or neural networks [12]. Some authors suggest mixing several methods [12]. These algorithms often require a large amount of computing time and memory allocation, which can be annoying in some applications (e.g. EV).

An original method proposed by Guo et al. [13] is based on monitoring the evolution of the charging curves over the life of the battery. The algorithm for identifying the numerous parameters can be quite complex and the operation on partial charging curves (charging cycles starting from a state of charge different from one cycle to another) remains to be tested.

A method often used to estimate SoH is by using a coulometric counter. To take into account the constraints related to electric vehicles, the voltage thresholds reached during the cycle of operation of the vehicle are used [14]. Nonetheless, a steep charging curve is required which is not the case for $C/LiFePO_4$ cells. The proposed estimator is based on the well-known incremental capacity analysis (ICA) method [15]. If ICA is very powerful when applied in a controlled environment, it has hardly ever been used with uncontrolled environmental variations to propose an online capacity estimator considering EV constraints. Few authors proposed papers dealing with ICA implementation as an embedded SoH estimator. Weng et al. [16] proposed an embedded ICA implementation using a mathematical fit as data filter. They claim satisfying results. Nonetheless, they worked with small capacity cells and above all, they do not study the influence of environmental variation such as temperature or DoD on their estimator. Han et al. [17] also proposed a very interesting ICA implementation to directly obtain the dQ/dU curve through the "point counting method". This method is a way to implement the ICA with very few computing power, but does not provide robustness to environmental variations.

In this study, ICA is implemented on large capacity cells and with industrial constraints such as costs reduction, reliability as the main guidelines.

This work was divided in three main parts. First, large capacity cells were cycled using an EV driving profile to obtain aged cells. Secondly, on each aged cell, charging cycles were performed with temperature, charging current and depth-of-discharge changing. Thirdly, using recordings of these cycles, an ICA based SoH estimator was developed.

Next paragraph introduces the incremental capacity analysis method. The following one presents the experimental process and the last one deals with embedded and ICA implementation and results.

2. ICA Presentation

Incremental capacity analysis is an electrochemical technique, which provides information about the internal cell state using only the cell voltage and current measurements [18–21]. The incremental capacity (IC) depicts a capacity change associated with a voltage step and the IC curve equation is given by:

$$ICA(U_{Cell}) \, (\text{Ah/V}) = \frac{dQ(U_{Cell}) \, (\text{Ah})}{dU_{Cell} \, (\text{V})} \tag{2}$$

where Q is the charged capacity and U_{Cell} the cell voltage. Dubarry et al. [15] state that each peak in the increment capacity curve has a unique shape, intensity, and position, and it exhibits an electrochemical process taking place in the cell. Regarding C/LiFePO$_4$ cells, ICA allows to focus on graphite electrode phases when observing the charging curve (cell voltage function of capacity). ICA is particularly relevant with C/LiFePO$_4$ chemistry because the cell voltage (U_{Cell}) does only vary in a 150 mV range while charging state vary from 10% to 90% (Figure 1a). Thereby, a direct voltage analysis is inaccurate to determine any state of the battery for this battery technology. On the contrary, ICA allows to analyses the voltage curve shape (slopes and plateaus) instead of absolute value. To do so, Figure 1b,c present the algorithm applied to get the ICA curve from a voltage measurement and thus highlight the initial curve plateaus.

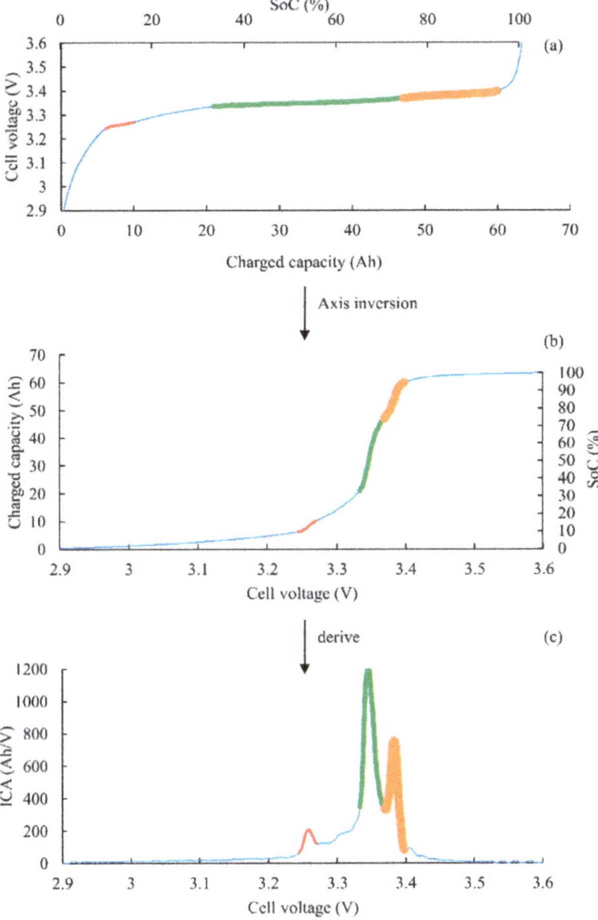

Figure 1. (a) LiFePO$_4$-C charging curve at C/20; (b) Incremental Capacity (IC) construction; (c) IC curve.

Three main peaks are observed on the IC curve, corresponding to the three voltage plateaus on the cell voltage versus state of charge $U_{Cell} = f(SoC)$ curve. The peaks observed in the incremental capacity curves (Figure 1c) correspond to the staging in the graphite negative electrode, convoluted with a single and very broad plateau of the LFP phase transformation on the positive electrode between 5% and 95% of SoC. Thus, all variations on the $U_{Cell} = f(SoC)$ curve in this SoC range are due to the graphite negative electrode.

Graphite is an insertion material with a layered structure [19,22,23]. During the insertion process (i.e., battery charging), lithium ions are inserted between graphene layers. Insertion occurs in several phases depending on the quantity of lithium ion inserted (i.e., the battery SoC). In other words, the active phase reflects the state of the negative electrode and Figure 2 shows the graphite structure corresponding to different phases. Each plateau in Figure 1a corresponds to the cohabitation of two graphite phases [20].

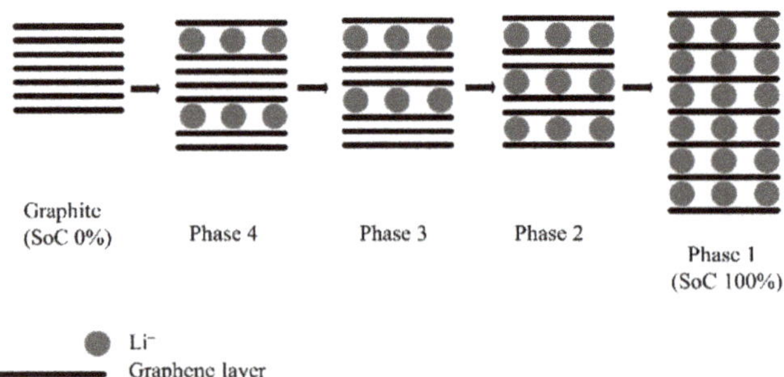

Figure 2. Graphite phases from discharged to fully charged graphite electrode (from [24]).

Numerous papers are using ICA as a powerful tool to analyze cell capacity degradation [15,18–20] or even degradation mechanisms [17,25]. Cell capacity is mainly deduced from peaks amplitude or area.

3. Experimentation

This section describes the experimental process of this study. Ageing process results and ICA implementation are presented in the next section.

3.1. Accelerated Ageing Process

The first step of this study is to produce aged cells using an accelerated ageing process. 60 Ah LiFePO$_4$-C prismatic cells are charged and discharged using an EV driving cycle. The New European Driving Cycle (NEDC), translated from speed profile to current profile using a small passenger car model, is considered for discharging the battery up to 2.5 V. Discharge average rate is C/3. CC-CV charge is then performed at a C/3 current, up to 3.7 V. In order to accelerate the capacity loss, cells are stored in a climatic chamber at 50 °C, still respecting the manufacturer recommendations. 700 charge-discharge cycles were done to reach 30% of capacity loss (i.e., SoH = 70%).

Charge and discharge cycles are done using a power test bench. Two cells are connected in series on each channel to increase the aged cell number.

An industrial battery management system (BMS) made by the manufacturer EVE System was adapted to the experiment and used to perform voltage, current and temperature continuous recording during cycling ageing. This point is very important: all dataset used to develop SoH algorithms are from an industrial BMS, with measures representatives of a real situation. The measurement accuracies are 1 mV for voltage, 0.1 A for current and less than 1 degree for temperature. The BMS also manages the balancing process between two series cells at the end of charge process. Figure 3 illustrates the BMS adapted with external 4-wires measures (2 for balancing circuit, 2 for voltage measurement).

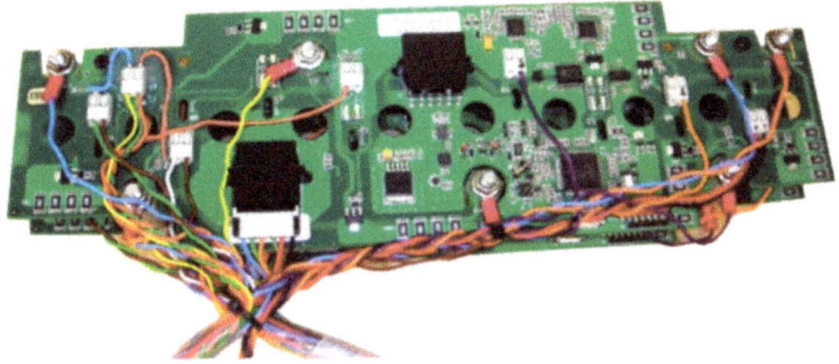

Figure 3. BMS with external measures.

In addition to cycling ageing, calendar ageing is also accomplished. Cells are kept at 50 °C, fully charged, without voltage floating.

At the end of the accelerated ageing process, eight cells are produced (four cells per ageing type, respectively at 70%, 80%, 90% and 100% SoH).

3.2. Electric Vehicle Charging Cycles

Once ageing cells are available, next step is to record different charging curves with environmental variations. Indeed, the main goal of this study is to develop a SoH estimator robust to temperature, current and DoD variations. Thus, several charging cycles are done and recorded with these 3 parameters separately varying. Only charging cycles are considered because, in an EV application, charging is the most controlled and repetitive phase of the battery use to deduce SoH.

3.2.1. Variable Depth-of-Discharge (DoD)

An electric vehicle is usually partially discharged and then charged up to 100% SoC. Depth-of-Discharge varies accordingly to the mission profile. DoD is rarely equal to 100% (i.e., SoC = 0% at the beginning of charge), in the practical use of EV. As a consequence, different DoD result in truncated charging curves and thus could significantly cause impact on the ICA algorithm.

To do so, 32 charging cycles are performed on different ageing cells with SoH ranging from 100% down to 70% and DoD varying from 30% to 100% with a 10% step. Charging occurs always after a partial discharge from 100% SoC. Charging profile is CC-CV at C/3. Charging temperature is 25 °C.

3.2.2. Variable Charging Current

Charging current can vary depending on the charge method using either vehicle internal charger or quick external charger for example, but also on the electric outlet used. Whereas quick charging can be considered as an exception in the vehicle life, the SoH estimator has to work for all charging current supplied by the internal charger regardless of the electric outlet (i.e., usually with a current from C/10 to C/4).

Sixteen charging cycles are performed on different ageing cells with a SoH ranging from 100% SoH down to 70% and with C/3, C/6, C/10 and C/20 currents. Temperature is set to 25 °C.

3.2.3. Variable Charging Temperature

The last parameter taken into account is the temperature. According to the manufacturer datasheet, charging temperature can only vary from 0 °C to 40 °C. Outside these limits, charging is not possible.

Sixteen charging cycles are also performed on different ageing cells varying from 100% SoH down to 70% SoH and at 5 °C, 10 °C, 20 °C and 40 °C, respectively. Charging current is set to C/3 in this case.

4. Embedded ICA Implementation and Results

Most authors practice an ICA with a very low current (C/20 or lower) with a DoD of 100% to record a fully defined curve and thus extract accurate information about ageing mechanisms. A C/20 charge is not acceptable for EV application. So, the aim of this study consists of proposing an IC measurement that reflects the cell remaining capacity during normal operation of the vehicle. Moreover, normal BMS measurements have to be precise for the chosen algorithm. ICA is then suitable for an online SoH estimator, as the method only needs voltage, current and temperature inputs.

The first section shows the sensitivity of the ICA to the battery ageing for EV application. The second part deals with the robustness of this estimator.

4.1. Capacity Estimation from ICA

Figure 4 shows several ICA curves from charging datasets of a C/LiFePO$_4$ cell during the cycling accelerated ageing process. The 100% DoD at 50 °C was chosen to accelerate ageing, even if it is not a representative EV type of use. The C/3 constant current charging cycle is commonly used and allows a full recharge in 3 hours. In Figure 4, the most obvious variation is a diminution of the last peak while capacity reduces. This peak-3 decrease highlights a capacity loss due to loss of lithium inventory (LLI) when the LiFePO$_4$-C prismatic cells considered are charged and discharged using an EV driving cycle (NEDC). The smaller decrease of the peaks 1 and 2 may indicate a lower loss of active material (LAM). So, the decrease of the peak-3 area measured using ICA may provide an estimation of the capacity fade and thus can be used as an efficient SoH estimator for the LiFePO$_4$-C cells considered. A slight right shift of the curves is also observed. It corresponds to an equivalent series resistance (ESR) rise. During charging, this induce a voltage increase and thus a right shift of the ICA.

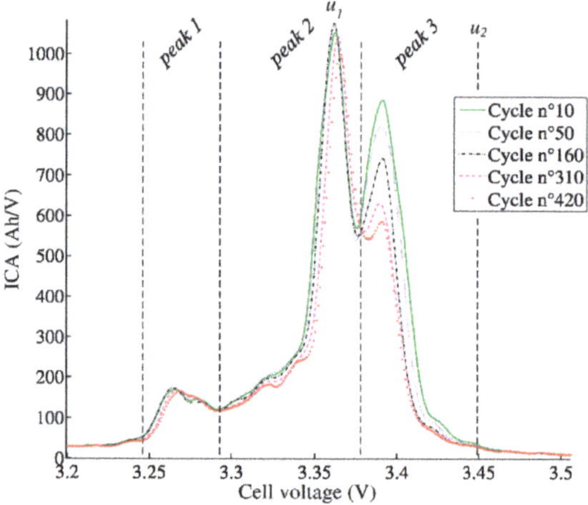

Figure 4. IC curves evolution during the cycling accelerated ageing (C/3 and 50 °C).

A first SoH estimator using ICA was proposed based on the peak 3 area. It can be easily demonstrated that the area under the IC curve is an image of the cell capacity between two voltage limits (u_1 and u_2) (Figure 5):

$$\int_{u_1}^{u_2} ICA(u)\, du = \int_{u_1}^{u_2} \frac{dQ(u)}{du}\, du = Q(u_2) - Q(u_1) \tag{3}$$

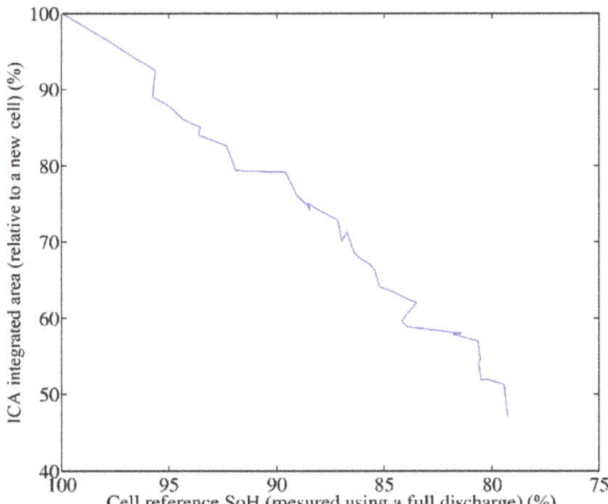

Figure 5. Integrated area function of the cell SoH.

A good estimate of SoH can be made with the area under peak 3. This latter phenomenon is fully described on a dedicated paper [26]. Since peak 3 is amortized and shifted with aging, it is not easy to calculate a representative area. In this paper, a robust protocol and method are proposed in order to estimate the SoH.

The aim is to ensure the operation of the algorithm in all conditions and throughout the battery state of health. To calculate this area, it is necessary to integrate the ICA curves between the two voltages u_1 and u_2. As shown in Figure 4, the beginning of peak 3 shifts to higher potentials as the battery ages. Moreover, since ICA is carried out for a positive current, this shift to the right corresponds to an increase of the ESR. Any increase in the ESR leads to a shift to the right of the integration terminals.

If the integration limits were fixed, the integrated area would not always correspond to peak 3's window, but would eventually take into account a part of peak 2 that shifts to the right too. An integration bounds detection algorithm has therefore been developed to overcome this issue. This includes two following steps:

- Take the absolute maximum of the IC curve that corresponds to the maximum of peak 2. The first voltage limit u_1 is taken at this peak 2's maximum instead of beginning of the peak 3 because it allows for better detection of any deformation of peak 3 with ageing. Indeed, the beginning of peak 3 may vanish with aging and therefore becomes difficult to detect. Contrarily, peak 2's maximum becomes more stable with aging.
- Define u_2 at a fixed distance from u_1, to ensure the area calculated under peak 3 is integrated over a fixed voltage interval. Thus, a decrease in the integrated area corresponds to a loss of capacity and not the integration width.

Regarding the operation conditions, this method leads to very good results and allows to compute the remaining cell capacity with an error smaller than 2%. Figure 5 shows the area evolution as a function of the remaining capacity. Note that the linearity is consistent with Equation (3). Once this relation is known, it is possible to measure the cell capacity with computing the ICA in order to measure the area and then, by using this law, obtain the capacity. The linearization of the measurements in Figure 5 gives Equation (4).

This estimator provides thus a SoH estimation in this quite ideal operation conditions with set temperature, current and DoD. Next paragraphs propose a sensitivity analysis of this latter estimator to these parameters to be implemented on an EV application.

$$\text{SoH (in\%)} = \frac{\text{ICA area (in\% of new cell area)} + 141.6}{2.43} \qquad (4)$$

4.2. Implementation Robust to Charging Cycles Variations

4.2.1. Depth of Discharge Effect on ICA

The depth of discharge is first investigated considering usual DoD for EV application ranging from 50% to 100%. Three ICA curves with different DoD are presented in Figure 6 for a SoH = 80%. Charging current is C/3 and temperature is set to 25 °C. Two main observations can be established. The DoD has a significant impact on the ICA curve shape and thus the peak 3 area. It is worth mentioning that the third peak is not well defined at C/3 when DoD becomes larger. Likewise, only a partial ICA is recorded and thus the low voltage part of the ICA is not available when the DoD is lower (i.e., SoC is high at the beginning of the charge). These observations highlight that the area under the ICA may depend on the DoD.

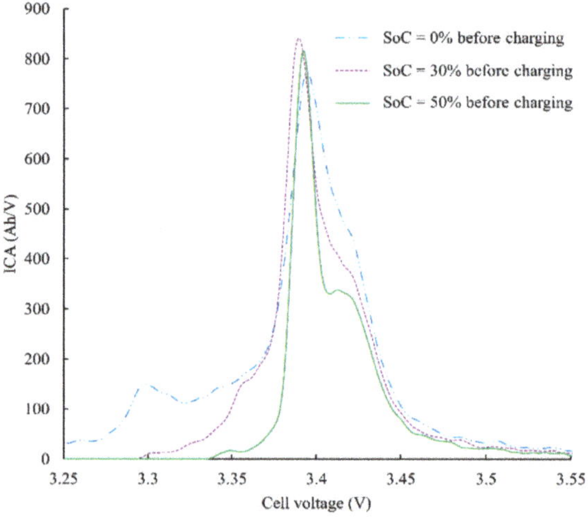

Figure 6. IC curves deformation function of the DoD (C/3-25 °C-SoH 80%).

This latter point is very critical for an EV application. To overcome this issue, the SoH estimator has to be the most robust possible to any variation of the DoD. As explained in Section 2, ICA peaks corresponds to voltage plateaus on the $U_{Cell} = f(\text{SoC})$ curve and may be related to the cohabitation of two graphite phases. Regarding the EV application, at relatively high charging current, the graphite electrode never reaches the thermodynamic stability during the charging at C/3 rate and thus, graphite phases are not well defined. In our application, the IC measurement cannot be performed under perfect thermodynamic condition leading to artefacts on the IC curves. Conversely, at low charging rate (e.g. C/20), the DoD does not have a significant impact on the ICA. Nevertheless, this latter effect of the DoD is beyond the scope of the paper.

In the case of an EV charging cycle, the charging current cannot be low enough to allow the DoD impact to be neglected. The developed solution is to add a pause during the charge, just prior to the ICA peak 2 at a SoC close to 50%. This allows the graphite electrode to reach the thermodynamic stability before the charge continues and ICA is computed. So, the graphite electrode is closer to the stability state than without the pause. Moreover, as the pause is always done at the same state (prior to peak 2), ICA is computed in similar conditions whatever the DoD. The optimal pause time to reach a satisfying stability in a reduce time was experimentally determined to be 30 minutes. The battery is placed in an open circuit during those 30 minutes. Nevertheless, a C/3 charge with a 30-minute pause remains much quicker than a C/20 charge without pause. Moreover, the SoH identification will not be done systematically with each recharge of the vehicle but occasionally.

Figure 7 shows different ICA with different DoD and with a pause of 30 minutes prior ICA peak 2. Curves are very close together and much better defined than in previous case (see Figure 6).

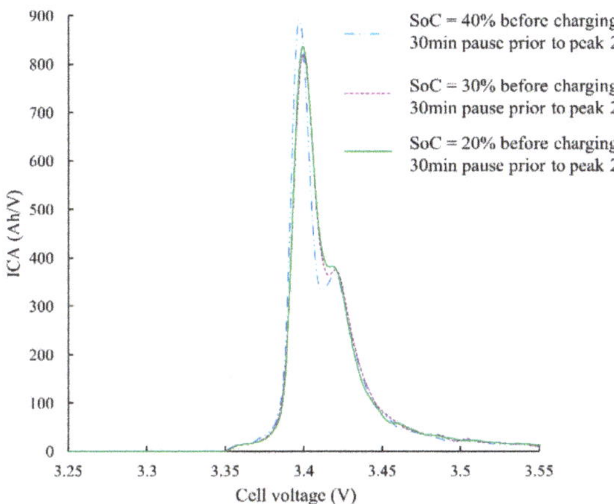

Figure 7. IC curves for different DoD and a pause of 30 minutes prior ICA peak 2.

Using this new ICA algorithm including a pause on aged cells with SoH ranging from 100% down to 70% at a C/3 charging current and a DoD between 80% and 60%, the remaining cell capacity is estimated within 4% error. Figure 8 illustrates the linear law (cf. Equation (5)) linking the measured area under the ICA and the cell capacity. It is important to note that for each SoH, several measurements are performed with different DoD. Nonetheless, the linear behavior observed in Figure 8 still remains valid with our new algorithm and the relevance of the pause is illustrated through the estimated area for a given capacity and for various DoD.

$$\text{Cell capacity (in Ah)} = \frac{\text{ICA area (in Ah)} + 5.96}{0.48} \tag{5}$$

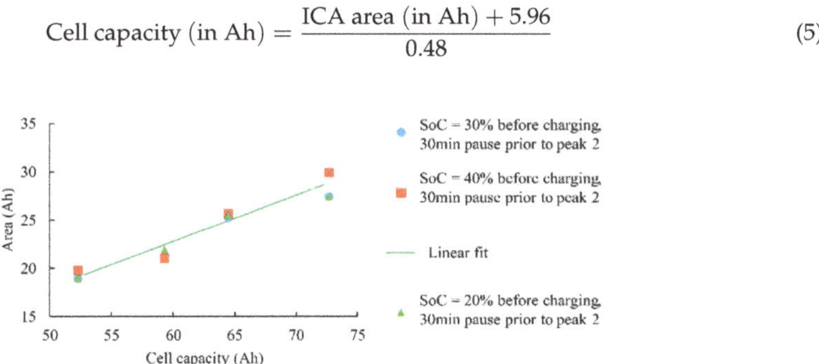

Figure 8. Measured area function of cell capacity for various DoD and with pause before peak 2 (C/3-25 °C).

4.2.2. Charging Current Effect on ICA

As previously introduced, the pause before the ICA peak 2 allows the graphite electrode to reach the thermodynamic stability prior to the ICA measurement. It is now proven that this pause is an efficient method to get similar ICA curves whatever the DoD, as it deletes the "history" of the cycle. This section investigates the charging C-rate impact because high charging current has a similar effect on the thermodynamic stability of graphite. Figure 9 presents IC curves of a SoH = 80% cell with different charging current and without the pause prior to peak 2. ICA deformation and peak shifting are distinctly noticeable. As shown in Figure 9, the charging current rate has a significant impact on the IC curves. Surprisingly, peak 3 starts to disappear when the C-rate increases. To overcome this issue, the same solution (pause before peak 2) is thus used to make the ICA from peak 2 maximum uniform, whatever the charging current.

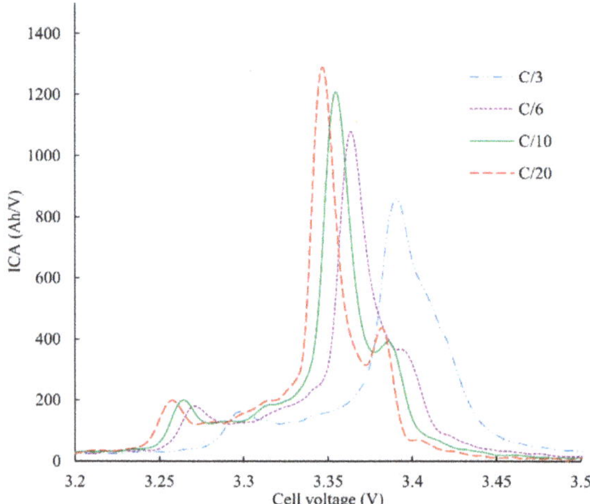

Figure 9. ICA deformation function of the charging current (100% DoD-25 °C-SoH 80%).

A second effect of the charging current is a right or left shift of ICA curves. Because of the cell ESR, in charge, the higher the current is, the higher the measured voltage. Nevertheless, as integration limits (from peak 2 maximum) are dynamically detected, this right shift does not impact the measured area. Figure 10 sums up final results with the pause before peak 2. As for DoD impact, the pause allows to measure very close areas whatever the charging current and lead to a 3% accurate SoH estimator. Equation (6) gives the linear fit of the equation, linking cell capacity to the ICA area.

$$\text{Cell capacity (in Ah)} = \frac{\text{ICA area (in Ah)} + 29}{0.94} \tag{6}$$

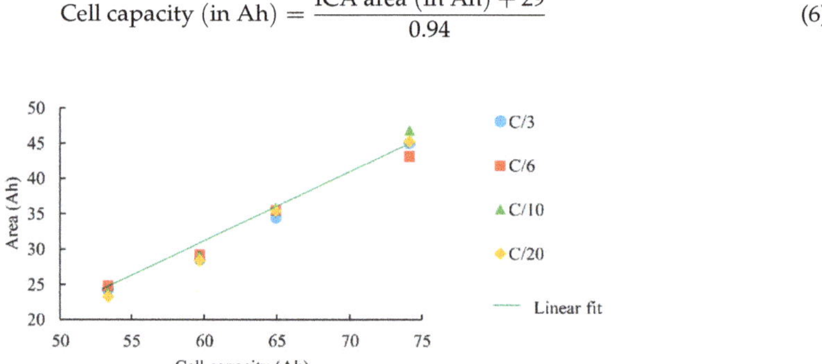

Figure 10. Measured area function of cell capacity for various charging currents and with pause before peak 2 (100% DoD-25 °C).

4.2.3. Temperature Effect on ICA

The last parameter studied in this work is the temperature. As a majority of chemical reactions, the graphite lithiation is activated by the temperature. In addition, the cell ESR reduces with temperature increase. Figure 11 shows different ICA with the pause before peak 2 and at different temperatures. The ESR variation is clearly visible. Another interesting point is the fact that the pause before peak 2 is not sufficient to cancel the effect of the temperature on the graphite lithiation: ICA at a high temperature is still better defined and the area is bigger. Figure 12 represents the area integrated as a function of the cell capacity and for different temperatures (Equation (7) gives the mean linear fit). This confirms the trend of the Figure 11 regarding the fact that the area is higher at high temperatures.

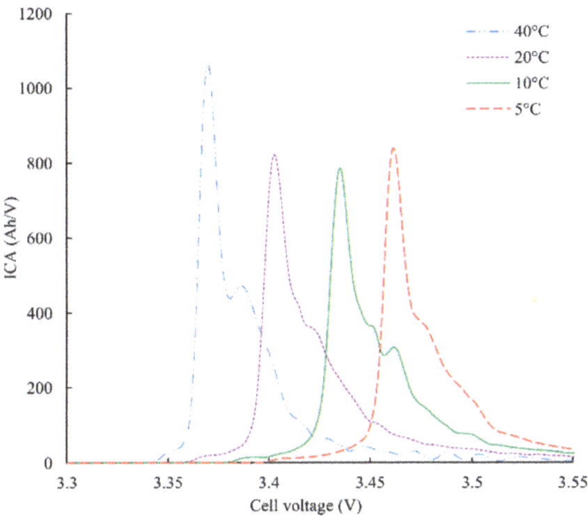

Figure 11. Temperature impact on ICA with pause prior to peak 2 (C/3-100% DoD-SoH 80%).

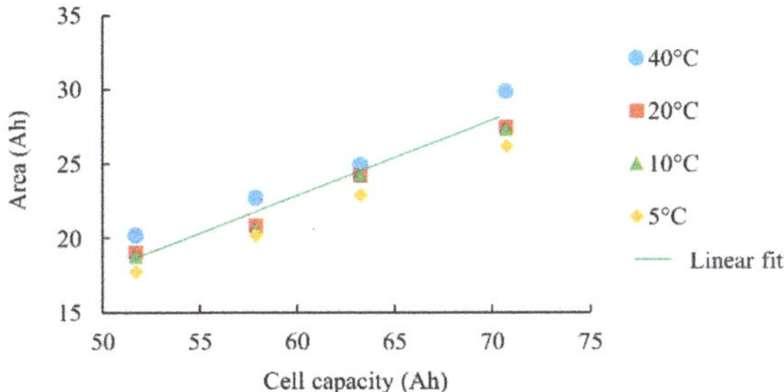

Figure 12. Measured area function of cell capacity for various temperatures and with pause before peak 2 (C/3-100% DoD).

It is thus not possible to have a single linear law giving the cell capacity from the measured area. Instead of this, a temperature compensation of the law is used. Offset and slope of the law are indexed to the temperature. Using this method, a 4% accurate SoH estimator is obtained with several different cells.

$$\text{Cell capacity (in Ah)} = \frac{\text{ICA area (in Ah)} + 7}{0.53} \tag{7}$$

5. Conclusions and Outlook

A powerful nonintrusive state-of-health estimator based on ICA measurement is proposed. This estimator is implementable on classic BMS for EV application and robust to DoD, current and temperature changes. In this work, a complete experimental part including accelerated ageing and many charge cycles with temperature, current and DoD variations was performed. A deep analysis of the impact of these latter parameters on the ICA was investigated to propose an innovative method to get the remaining capacity using embedded ICA. Most of previously papers using ICA as capacity estimator are not able to propose a method to obtain the cell capacity from partially discharged cells regardless the environmental conditions of the measure. Using the pause prior to peak 2 allows estimating the remaining capacity with a 4% accuracy even though the temperature ranges between 5 °C and 40 °C, with the current between C/20 and C/3, and the DoD between 80% and 60%.

The calibration of the capacity law function of the ICA area was done using a 60Ah C/LiFePO$_4$ cell. The same ageing process and same environmental variations measures were also done on another C/LiFePO$_4$ of 72 Ah from another manufacturer. It was demonstrated that the same law is efficient for both cells. This point is very interesting and means that it is not necessary to proceed to a time-consuming ageing process to calibrate the law in case of the use of a new cell. It also suggests an interesting robustness if a cell manufacturer can process the change. Finally, the same methodology (with some adaptations) was implemented on lithium manganese oxide batteries with a SoH estimation accurate within 4% [27].

Present work focused on separately variations only. Temperature, DoD and current effect were studied unitarily. Next part of this work will be to assemble all variations in a unique law and to check its efficiency in EV conditions. The final validation step will be to integrate the algorithm in an embedded BMS and to check its efficiency during a long period of use on a real EV. It should lead to a more accurate state-of-charge computation and thus should allow to reduce the safety margin, increasing the range of the vehicle.

Author Contributions: E.R., A.S., P.V., F.M. and Y.B. conceived and designed the experiments; E.R. performed the experiments; E.R., A.S., P.V., F.M. and Y.B. analysed the data; E.R., A.S., P.V., F.M. and Y.B. wrote the paper.

Funding: This research was funded by National French Association for Technological Research (ANRT) and the company EVE System.

Conflicts of Interest: The authors declare no conflicts of interest.

References

1. Khaligh, A.; Li, Z. Battery, Ultracapacitor, Fuel Cell, and Hybrid Energy Storage Systems for Electric, Hybrid Electric, Fuel Cell, and Plug-In Hybrid Electric Vehicles: State of the Art. *IEEE Trans. Veh. Technol.* **2010**, *59*, 2806–2814. [CrossRef]

2. Chan, C.C. The State of the Art of Electric, Hybrid, and Fuel Cell Vehicles. *Proc. IEEE* **2007**, *95*, 704–718. [CrossRef]

3. Cheng, K.W.E.; Divakar, B.P.; Wu, H.; Ding, K.; Ho, H.F. Battery-Management System (BMS) and SOC Development for Electrical Vehicles. *IEEE Trans. Veh. Technol.* **2011**, *60*, 76–88. [CrossRef]

4. Choi, M.E.; Lee, J.S.; Seo, S.W. Real-Time Optimization for Power Management Systems of a Battery/Supercapacitor Hybrid Energy Storage System in Electric Vehicles. *IEEE Trans. Veh. Technol.* **2014**, *63*, 3600–3611. [CrossRef]

5. Zandi, M.; Payman, A.; Martin, J.P.; Pierfederici, S.; Davat, B.; Meibody-Tabar, F. Energy Management of a Fuel Cell/Supercapacitor/Battery Power Source for Electric Vehicular Applications. *IEEE Trans. Veh. Technol.* **2011**, *60*, 433–443. [CrossRef]

6. Carter, R.; Cruden, A.; Hall, P.J. Optimizing for Efficiency or Battery Life in a Battery/Supercapacitor Electric Vehicle. *IEEE Trans. Veh. Technol.* **2012**, *61*, 1526–1533. [CrossRef]

7. Lièvre, A.; Sari, A.; Venet, P.; Hijazi, A.; Ouattara-Brigaudet, M.; Pelissier, S. Practical Online Estimation of Lithium-Ion Battery Apparent Series Resistance for Mild Hybrid Vehicles. *IEEE Trans. Veh. Technol.* **2016**, *65*, 4505–4511. [CrossRef]

8. Jaguemont, J.; Boulon, L.; Venet, P.; Sari, A. Lithium Ion Battery Aging Experiments at Sub- Zero Temperatures and Model Development for Capacity Fade Estimation. *IEEE Trans. Veh. Technol.* **2015**, *65*, 4328–4343. [CrossRef]

9. Haifeng, D.; Xuezhe, W.; Zechang, S. A new SOH prediction concept for the power lithium-ion battery used on HEVs. In Proceedings of the 2009 IEEE Vehicle Power and Propulsion Conference, Dearborn, MI, USA, 7–10 September 2009.

10. Remmlinger, J.; Buchholz, M.; Soczka-Guth, T.; Dietmayer, K. On-board state-of-health monitoring of lithium-ion batteries using linear parameter-varying models. *J. Power Sources* **2013**, *239*, 689–695. [CrossRef]

11. Chen, Z.; Mi, C.C.; Fu, Y.; Xu, J.; Gong, X. Online battery state of health estimation based on Genetic Algorithm for electric and hybrid vehicle applications. *J. Power Sources* **2013**, *240*, 184–192. [CrossRef]

12. Watrin, N.; Blunier, B.; Miraoui, A. Review of adaptive systems for lithium batteries State-of-Charge and State-of-Health estimation. In Proceedings of the IEEE Transportation Electrification Conference and Expo (ITEC), Dearborn, MI, USA, 18–20 June 2012.

13. Guo, Z.; Qiu, X.; Hou, G.; Liaw, B.Y.; Zhang, C. State of health estimation for lithium ion batteries based on charging curves. *J. Power Sources* **2014**, *249*, 457–462. [CrossRef]

14. Le, D.; Tang, X. Lithium-ion Battery State of Health Estimation Using Ah-V Characterization. In Proceedings of the Annual Conference of the Prognostics and Health Management Society, Montreal, QC, Canada, 25–29 September 2011.

15. Dubarry, M.; Liaw, B.Y. Identify capacity fading mechanism in a commercial LiFePO$_4$ cell. *J. Power Sources* **2009**, *194*, 541–549. [CrossRef]

16. Weng, C.; Cui, Y.; Sun, J.; Peng, H. On-board state of health monitoring of lithium-ion batteries using incremental capacity analysis with support vector regression. *J. Power Sources* **2013**, *235*, 36–44. [CrossRef]

17. Han, X.; Ouyang, M.; Lu, L.; Li, J.; Zheng, Y.; Li, Z. A comparative study of commercial lithium ion battery cycle life in electrical vehicle: Aging mechanism identification. *J. Power Sources* **2014**, *251*, 38–54. [CrossRef]

18. Dubarry, M.; Liaw, B.Y.; Chen, M.-S.; Chyan, S.-S.; Han, K.-C.; Sie, W.-T.; Wu, S.-H. Identifying battery aging mechanisms in large format Li ion cells. *J. Power Sources* **2011**, *196*, 3420–3425. [CrossRef]

19. Dubarry, M.; Truchot, C.; Liaw, B.Y. Cell degradation in commercial LiFePO$_4$ cells with high-power a18nd high-energy designs. *J. Power Sources* **2014**, *258*, 408–419. [CrossRef]

20. Groot, J. State-of-Health Estimation of Li-ion Batteries: Cycle Life Test Methods. Ph.D. Thesis, Division of Electric Power Engineering, Chalmers University of Technology, Göteborg, Sweden, 2012.

21. Kassem, M.; Bernard, J.; Revel, R.; Pelissier, S.; Duclaud, F.; Delacourt, C. Calendar aging of a graphite/LiFePO$_4$ cell. *J. Power Sources* **2012**, *208*, 296–305. [CrossRef]

22. Alzieu, J.; Robert, J. *Accumulateurs au Lithium*; Techniques de L'ingénieur: Saint-Denis, France, 2005.

23. Buchmann, I. *Batteries in a Portable World: A Handbook on Rechargeable Batteries for Non-Engineers*, 3rd ed.; Cadex Electronics Inc.: Richmond, BC, Canada, 2011.

24. Sole, C.; Drewett, N.; Hardwick, L.J. In situ Raman study of lithium-ion intercalation into microcrystalline graphite. *R. Soc. Chem. Faraday Discuss.* **2014**, *172*, 223–237. [CrossRef] [PubMed]

25. Dubarry, M.; Truchot, C.; Liaw, B. Synthesize battery degradation modes via a diagnostic and prognostic model. *J. Power Sources* **2012**, *219*, 204–216. [CrossRef]

26. Rivière, E.; Venet, P.; Bultel, Y.; Sari, A.; Ménière, F. LiFePO$_4$ Battery State of Health Online Estimation Using Electric Vehicle Embedeed Incremental Capacity Analysis. In Proceedings of the IEEE-Vehicular Power and Propulsion Conference 2015, Montreal, QC, Canada, 9–22 October 2015.

27. Rivière, E. Détermination In-Situ de L'état de Santé de Batteries Lithium-ion Pour un Véhicule Electrique. Ph.D. Thesis, University of Grenoble-Alpes, Grenoble, France, November 2016.

Article

Synthetic vs. Real Driving Cycles: A Comparison of Electric Vehicle Battery Degradation

George Baure and Matthieu Dubarry *

Hawai'i Natural Energy Institute, SOEST, University of Hawai'i at Mānoa, 1680 East-West Road, POST 109, Honolulu, HI 96822, USA; gbaure@hawaii.edu
* Correspondence: matthieu@hawaii.edu or Matthieu.Dubarry@gmail.com

Received: 4 April 2019; Accepted: 19 April 2019; Published: 1 May 2019

Abstract: Automobile dependency and the inexorable proliferation of electric vehicles (EVs) compels accurate predictions of cycle life across multiple usage conditions and for multiple lithium-ion battery systems. Synthetic driving cycles have been essential in accumulating data on EV battery lifetimes. However, since battery deterioration is path-dependent, the representability of synthetic cycles must be questioned. Hence, this work compared three different synthetic driving cycles to real driving data in terms of mimicking actual EV battery degradation. It was found that the average current and charge capacity during discharge were important parameters in determining the appropriate synthetic profile, and traffic conditions have a significant impact on cell lifetimes. In addition, a stage of accelerated capacity fade was observed and shown to be induced by an increased loss of lithium inventory (LLI) resulting from irreversible Li plating. New metrics, the ratio of the loss of active material at the negative electrode (LAM_{NE}) to the LLI and the plating threshold, were proposed as possible predictors for a stage of accelerated degradation. The results presented here demonstrated tracking properties, such as capacity loss and resistance increase, were insufficient in predicting cell lifetimes, supporting the adoption of metrics based on the analysis of degradation modes.

Keywords: incremental capacity analysis; lithium-ion; electric vehicles; driving cycles; cell degradation

---

1. Introduction

Because of its isolation, Hawai'i has been on the forefront of the sustainable energy movement, as evidenced by the Hawai'i Clean Energy Initiative, which endeavors to have 100% clean energy both on the grid and in all ground transportation by 2045. Reaching that goal will involve more renewable energy sources being incorporated into the grid and more electric vehicles (EVs) on the roads. Success will hinge on the integration and the combination of these technologies. What gets lost is how this integration, combination, and escalation in use will affect the durability of lithium-ion batteries that are essential to the efficacy of both.

What is more, battery degradation is path-dependent [1]. Different driving, storage, and charging conditions can all lead the battery down different paths of degradation. Hence, while testing for the effect of the evolving usage of EV batteries, it is crucial to evaluate correctly the impact of driving and not to just base prognoses on capacity loss or resistance increase. Several synthetic driving profiles were proposed in the literature to determine the range and durability of EV batteries, but a comparison with the real degradation upon driving was, to the best of our knowledge, never reported. Knowing the path dependency of degradation, it is essential to determine how well these profiles simulate typical real-world driving behavior.

The most common driving profiles are tested in this work: the Federal Urban Driving Schedule (FUDS), which a US standard, the dynamic stress test (DST), which is a simplified version of the FUDS cycle [2], and the New European Driving Cycle (NEDC), which is a European standard. Using properties such as capacity fade at C/35 as metrics, the measurements from these profiles are compared to the results from real driving data from the Hawai'i Natural Energy Institute's (HNEI) in-house 60,000 miles database [3] gathered in the early 2000s. In this work, a set of 30 commutes is used for two different tests that are compared to elucidate any effect of traffic. For the first test, a single commute with similar characteristics (power and average discharge current) to the synthetic cycles is chosen; while, for the second test, the entire set of commutes is used.

In addition, the mechanistic model detailed in previous work [4] and incremental capacity (IC) analyses were employed to determine if degradation of the cells subjected to simulated driving profiles differed from reality. IC analysis can enhance subtle changes in a cell's electrochemical behavior during aging [5], and it has been used to define degradation processes in different types of battery systems [6] and in different real-world scenarios [1,7–9]. This analysis is non-destructive [6] and can be used *in operando* [9,10] to provide real-time diagnoses. The novelty of this diagnostic and prognostic technique lies in its ability to anticipate breaks in initial trends and relate them with physical phenomena [1,9,11]. Other prognostic techniques have issues with dynamic conditions and sudden changes in cell properties [12].

This study aims to elucidate any significant discrepancies between simulated and real driving outcomes, expose the effect of traffic conditions, diagnose the cell as it ages throughout driving, and discover a technique to better predict cell lifetimes. The work presented here is part of ongoing research that delves into the effect of grid–vehicle interactions on the performance of Li-ion batteries and facilitates the determination of the causes of cell deterioration and the accurate prediction of capacity loss during real-world use. The knowledge gained is expected to provide suggestions that will improve property retention via highlighting best practices. The overall project, which spans several publications [1,13,14], is meant to investigate EV battery degradation when it is being driven, resting, charging, or interacting with the grid. Such complete understanding is essential to achieve the goal of 100% clean energy.

2. Materials and Methods

Panasonic cylindrical 3350 mAh NCR 18650B batteries were used in this work. They were consistent with the type of cells used by EVs such as the Tesla Model S [15,16]. A set of 15 batteries were selected from a batch of 100. Details on the entire batch of cylindrical 18650-size graphite (GIC)/$LiNi_xCo_{1-x-y}Al_yO_2$ (NCA) cells, as well as the cell-to-cell variation analysis, were presented in previous work [13,17]. The properties of the 15 cells chosen for this investigation were within the outlier boundaries of the cell-to-cell variation distribution [13].

As described in a previous publication [13], all cells were subjected to initial conditioning and characterization tests prior to the beginning of the experiment. Following these formation cycles and before the start of the duty cycle testing, a reference performance test (RPT) consisting of C/35, C/5, and C/3 full cycles was performed on all cells [13]. All RPTs and duty cycles were performed in an Amerex IC500-R chamber, which was set at 25 °C. During testing, the skin temperatures of each of the cells were recorded. The average skin temperature was 25.1 ± 0.95 °C. The voltage range tested was 2.5 to 4.2 V. The formation cycles were performed by a 40-channel Arbin BT-5HC tester, while the first RPTs were performed using a Biologic BCS-815 battery cycler. The following duty cycles and RPTs were carried out by a 40-channel Arbin BT-5HC tester.

The duty cycles consisted of a full charge at a constant C/5 rate followed by a constant voltage step with a 65 mA cut-off and a 2 h rest step. The cells were then discharged by applying the appropriate driving cycle as a power profile scaled to a maximum power of 400 W/kg (the U.S. Advanced Battery Consortium, USABC, goal). The synthetic power profiles for the DST and FUDS were taken from the USABC manual [2], while the NEDC power profile was extracted from literature [18]. Those three synthetic driving cycles are shown in Figure 1 with one of the real driving profiles (RD1). RD1 was the portion of the real-world data whose values for average discharge current (Table 1) most closely equaled synthetic cycles.

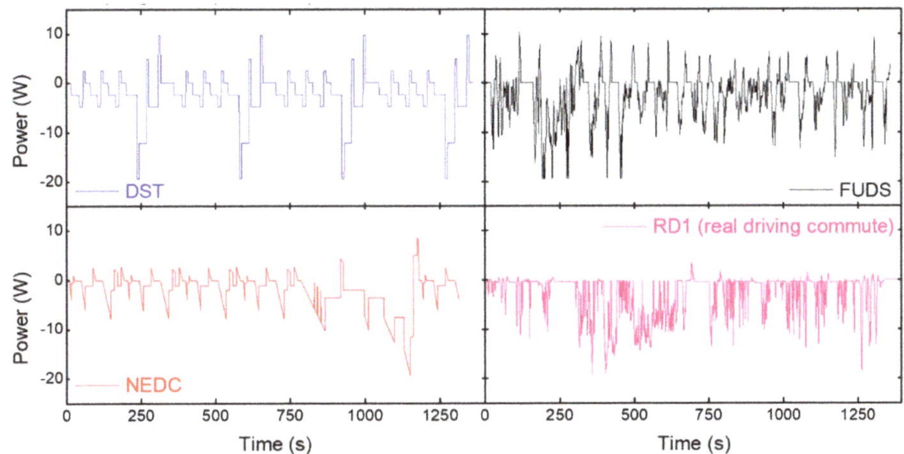

Figure 1. The power profiles for the dynamic stress test (DST), Federal Urban Driving Schedule (FUDS), New European Driving Cycle (NEDC), and real driving-1 (RD1) cycles.

Table 1. Properties of the driving profiles during discharge.

Driving Profiles	Avg. Current (A)	C-Rate Based on Avg. Current	Discharge Time (h)	Charge Capacity (Ah)	Discharge Capacity (Ah)
Dynamic Stress Test (DST)	−0.68	C/4.8	4.03	0.51	3.29
Federal Urban Driving Schedule (FUDS)	−0.70	C/4.6	3.74	0.56	3.20
New European Driving Cycle (NEDC)	−0.66	C/4.9	4.25	0.30	3.13
RD1: real driving (commute comparable to simulations)	−0.73	C/4.5	4.12	0.03	3.03
RD2: real driving (complete dataset)	−0.45	C/7.3	6.73	0.04	3.01

Figure 2 showcases the power profiles from several commutes similar to RD1 but with different traffic conditions. The real driving data were obtained from vehicles with limited regenerative braking. RD1 testing consisted of the single commute repeated over and over, while RD2 testing consisted of a set of 30 unique commutes in a loop.

All the driving cycles were applied continuously, end-to-end until the end-of-discharge (EOD) condition was met. For the cells subjected to the RD2 discharge, each of the commutes were applied for the same amount of time. The EOD condition was a measured cell voltage of less than 2.5 V. The cells rested for 4 h prior to a full charge, after which the cells were discharged with the same power profile again. Cycling was periodically interrupted to perform RPTs. Initially, cycling was interrupted every four weeks, but that decreased as the cell aged and exhibited significant capacity loss. Testing was stopped when the cell exhibited a capacity loss of at least 20% at C/35. The polarization resistance of the cells was estimated from the IR drop induced by the C/3 discharge using Ohm's law. The total test time of the cells ranged from 10 to 16 months. Three cells for each of the five driving profiles were tested to assess reproducibility, for a total of 15 cells were used in the experiment.

Computer simulations were performed using the proprietary '*alawa* toolbox that served as a user interface to facilitate application of the mechanistic degradation model [4]. Experimental validation supporting the simulation results based on the loss of lithium inventory (LLI) and the loss of active material (LAM) degradation modes has been reported by other groups [19–21]. More details in the emulation process and the analysis of the data can be found in previous work [1,13].

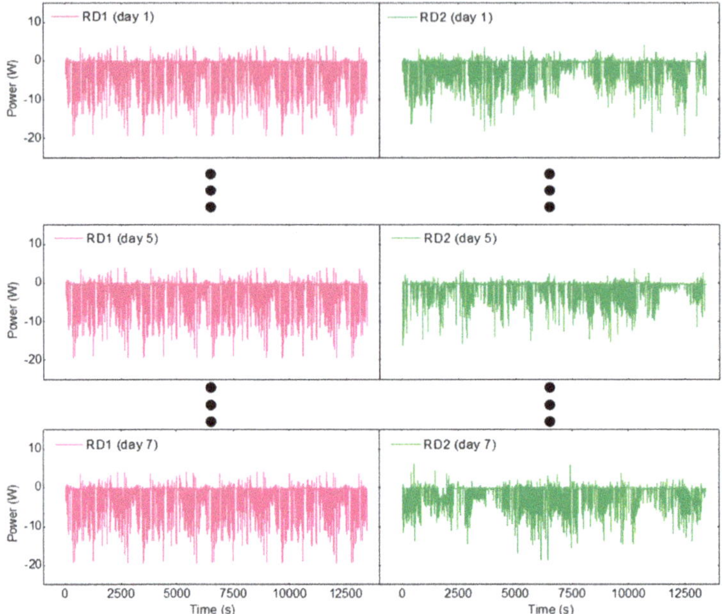

Figure 2. The power profiles from the real driving data (RD1 and RD2) selected from the Hawai'i Natural Energy Institute (HNEI) database.

3. Results and Discussion

Figure 3 plots the average normalized capacity and resistance increase for the driving conditions versus exchanged capacity. Initially, all cells lost capacity linearly ($R^2 > 0.98$) at about 0.008% ± 0.001% per Ah of exchanged capacity or 0.023% ± 0.003% per cycle. All 15 cells experienced an accelerated stage of degradation, 0.05% ± 0.02% per Ah of exchanged capacity, after a capacity loss of around 10%. This second stage of aging was predicted in a previous work using the '*alawa* toolbox [1]. The cell with the longest lifetime was subjected to the RD2 profile and proceeded to the accelerated aging stage after over 20% capacity loss, but most reached that stage by 11%. The cells subjected to the NEDC and RD1 driving cycles failed on average sooner than the others. Those cells reached the second stage of aging at an average exchanged capacity of 1191 ± 183 and 1216 ± 199 Ah, respectively. These values corresponded to around 350 cycles, in line with what was observed for C/1 charges and discharges [22]. Those profiles were also the ones that were charged the least during driving (see Table 1), indicating that regenerative braking could help capacity retention. This result corroborated what was found in literature [23]. The cells subjected to the DST and FUDS profiles experienced onset of the accelerated aging stage at similar values of capacity exchanged (1438 ± 38 and 1490 ± 78 Ah, respectively, and around 450 cycles). This outcome was unsurprising since DST was derived from FUDS [2]. The RD2 cells were cycled the longest before succumbing to the second stage (2146 ± 251 Ah, 650 cycles). These cells were subjected to a lower average discharge current (see Table 1), which might account for their comparatively long cell lifetimes.

In the initial stage, resistance increased more in the cells aged with the real driving profiles than the simulated ones (Figure 3b). It must be noted that the cells with the highest rates of resistance increase (RD2) were the ones with the longest lives. This highlighted that resistance increase and life expectancy were not correlated. Resistance increased linearly during the first stage of aging, then escalated nonlinearly in the second stage of degradation. Although there was an increase in rate of change of the resistance during the second stage of aging, in general, the disparity between stages was not as pronounced as the one seen in the plot of the capacity fade (Figure 3a).

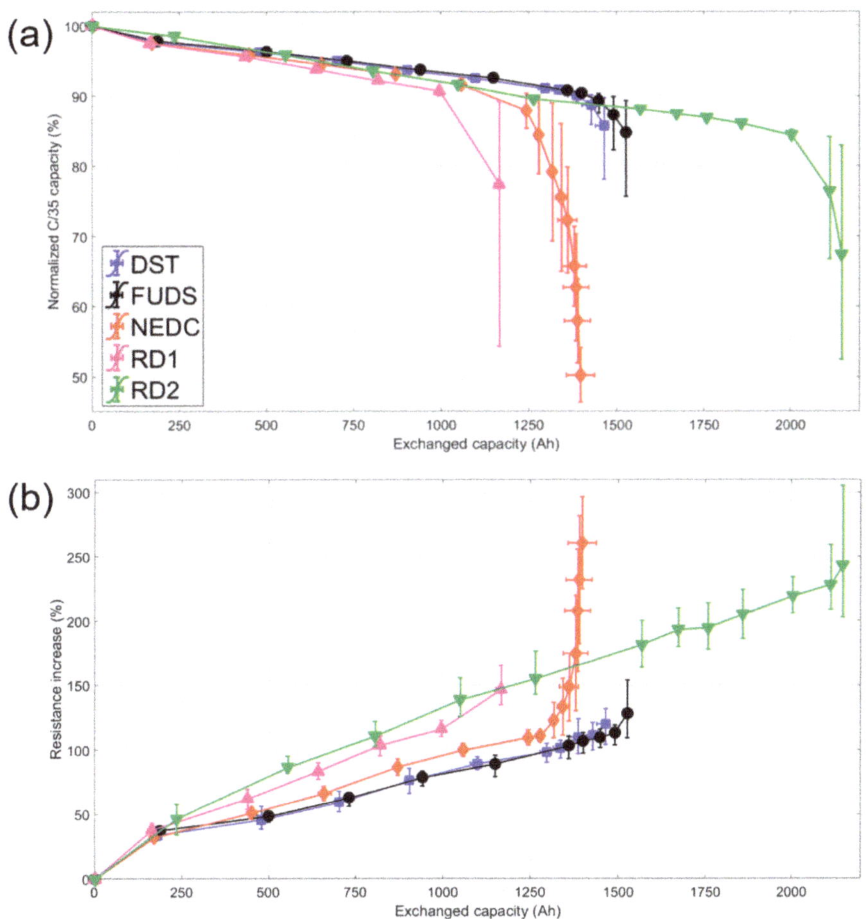

Figure 3. Plots of the average (**a**) normalized capacity and (**b**) resistance increase for each of the driving profiles vs. exchanged capacity.

Most interestingly, there was a considerable difference (>900 Ah or 275 cycles on average) between the final exchanged capacity of the cells subjected to the RD1 and RD2 profiles revealing a sizeable effect of traffic on property retention. This average difference between driving profiles was the largest in the experiment. From the C-rates of those two profiles (Table 1), time to complete discharge could vary by 3 h. As a consequence of these observations, it was reasonable to conclude that not one cycling profile with a fixed set of properties could account for the breadth of real driving outcomes. It may be possible to simulate the range of cell lifetimes by adjusting parameters, such as the maximum power, of a single driving cycle. Although, this range of outcomes may indicate unique degradation mechanisms leading to failure. Diagnosis of the cells was performed to illuminate disparities in the degradation induced by the five duty cycles.

To diagnose the cells, IC analysis was employed in conjunction with the features of interest approach [10,24] to quantify the different degradation modes and determine path dependency. This analysis was shown to identify metallic lithium deposition [7] as well as gas evolution in cells [11].

An extensive discussion on the IC analysis for these cells was previously published [14] and will not be repeated here.

Figure 4a shows the representative IC curve evolution of the C/35 charge. The most noticeable change to the IC signatures during the first few months of cycle aging was the disappearance of a local minimum, or arch, between 3.7 and 3.8 V. Figure 4b–e compares the IC curves at 5%, 10%, 15%, and 20% capacity loss for all the cells. Overall, the evolution of the curves was similar. At 5% capacity loss, the arch was barely visible (Figure 4b) and completely gone by 10% loss (Figure 4c). There was a sharp peak at low voltages that shrank and shifted to higher voltages during aging, but it did not quite disappear, even at 20% loss (Figure 4e). A slow shift to higher voltages of the beginning of charge was also observed, but there was no shift in the end-of-charge (EOC) voltage. It must be noted, however, that differences between duty cycles were visible in the IC curves at 15% and 20% capacity loss. The highest voltage peak between 4.1 and 4.2 V flattened with age more in the cells exposed to the real driving cycles (RD1 and RD2) than the synthetic ones. In addition, except for one of the RD1 cells, cycle aging with the real driving profiles caused a greater voltage shift and intensity decrease to the peak at 3.4 V. These disparities suggested that real-world battery degradation might not be the same as synthetically induced degradation.

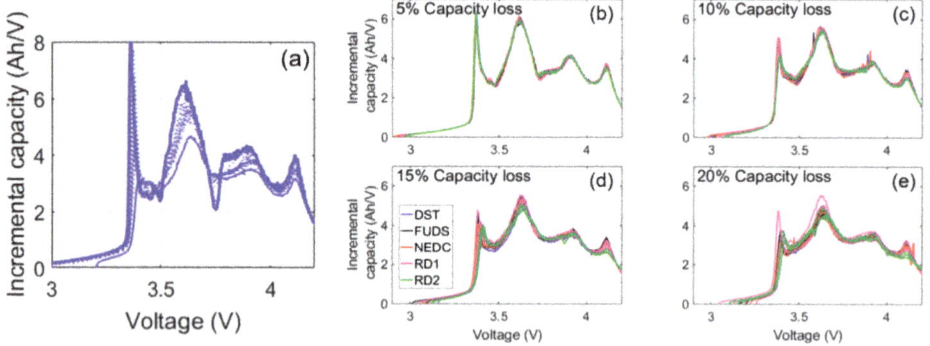

Figure 4. (**a**) Representative incremental capacity (IC) curve evolution up to 20% capacity loss. (**b–e**) Representative IC curves for each of the driving profiles at 5%, 10%, 15%, and 20% capacity loss.

Quantifying changes to the IC curves allowed determination of relative magnitudes of degradation modes occurring during aging, uncovering the differences in the effect of the five driving cycles. Similar to previous work [1], experimental changes were compared to the emulated IC evolution of a single mode using the *'alawa* toolbox and the degradation table (Figure 5) [14]. No single degradation matched the experimental variations; therefore, degradation was induced by a combination of separate modes. This complex combination included the loss of lithium inventory, loss of active material at both electrodes (LAM_{PE} and LAM_{NE} for the positive and negative electrodes, respectively), and some kinetic limitations (rate degradation factor, RDF). Scanning all the possible contributions would require an enormous amount of calculation [10], but a careful observation of Figures 4a and 5 reduced the number of cases to consider and allowed direct quantification of LLI, LAM_{PE}, and the rate degradation factor at the negative electrode (RDF_{NE}).

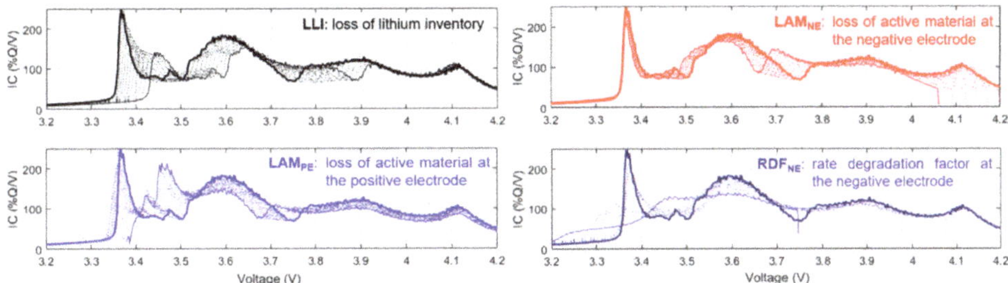

Figure 5. Simulated evolution of the IC curves for each of the prominent degradation modes generated by the 'alawa program. More details are available in a previous work [14].

First, it can be deduced that capacity fade was not caused by loss of active material. Capacity loss from LAM_{NE} would shift and change the features at the end-of-charge (the peak above 4 V), and capacity fade from LAM_{PE} would make the sharp peak between 3.3 and 3.4 V disappear—none of which were observed experimentally. Therefore, capacity loss was solely induced by LLI. As a result, LLI could be estimated directly from capacity loss. LLI was found to increase linearly during the initial aging stage to about 10% loss for most of the cells, then the value rose dramatically during the second stage.

Second, the decrease in intensity at 4 V could be used as a direct indicator of LAM_{PE} (Figure 5) since it was the only degradation that was shown to affect this intensity without receding the highest voltage peak. With exception of the RD2 cells, the LAM_{PE} remained below 10% during the initial stage. During the initial stage of aging, the real driving cycles induced LAM_{PE} at a faster rate than the synthetic cycles. At an exchanged capacity of 1000 Ah when all the cells were at the initial stage, the difference was about 2.5%. This degradation mode also accelerated during the secondary stage following the trend in capacity loss, though its value was always less than LLI regardless of the driving cycle.

Third, according to the 'alawa simulation and Figure 5, the disappearance of the arch between 3.7 and 3.8 V could only be attributed to decreasing reaction kinetics at the negative electrode, as all the other degradation modes only shifted this feature to lower or higher voltages. Thus, quantifying this change as the cell aged could allow direct estimation of the RDF_{NE}. It had to be noted that, according to simulations, the intensity of the arch increased then plateaued with increasing RDF_{NE} (inset of Figure 6); therefore, after reaching the plateau beyond a value of 10, the RDF_{NE} could not be estimated. Experimental data exhibited the plateau, but also a decrease in arch intensity after the second stage of aging started (Figure 6). This decrease in arch intensity during the second stage was unlikely to be induced by improving reaction kinetics at the negative electrode. Hence, the intensity change within this voltage range during the stage of accelerated capacity fade must be caused by some combination of all the degradation modes. The arch intensity decrease was not observed in previous work [1] since the cycle-aged cells experienced less than 10% capacity loss and did not undergo this level of degradation. The approach of the RDF_{NE} towards the plateau value was considered as an indicator of the changes to come, but the RD2 cells were among the first to reach the plateau value (Figure 6) yet remained at that level for a couple hundred cycles. All cells reached the RDF_{NE} plateau at different exchanged capacity values, and no distinctions between driving cycles were discerned using this metric.

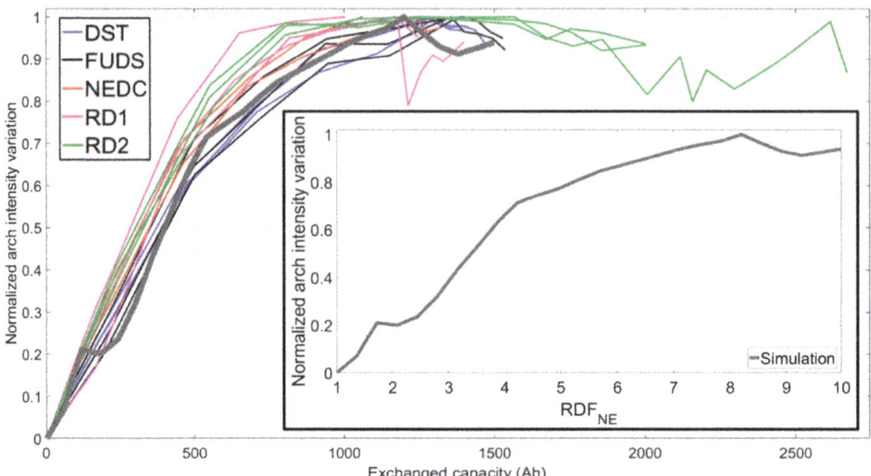

Figure 6. Evolution of the arch intensity during the initial stage of degradation for all the cells tested vs. the exchanged capacity. The inset is a plot of the simulated intensity variation vs. rate degradation factor at the negative electrode (RDF_{NE}).

The last parameter to decipher was LAM_{NE}. LAM_{NE} could not be estimated automatically from any single feature on the IC curve because of its combined effect with all other degradation modes. Based on the abovementioned assumptions, the relative values of three of the degradation modes, LLI, LAM_{PE}, and RDF_{NE}, were calculated and used to fit the IC curves for all the cells as they aged. Those simulations were performed with various amounts of LAM_{NE}, with 1% intervals, until the best fit to experimental data was obtained (Figure 7a). Since this was the only degradation mode that had to be obtained manually, the estimated values may not have been as reliable as the others, because it accounted for all (or a big part of) fitting errors from the other mechanisms. Initially, little to no LAM_{NE} was needed to fit the experimental curves. However, as the cell aged, the LAM_{NE} value became more significant. Figure 7b compares the experimental and the simulated IC curves with no LAM_{NE} (solid red curve) and 15% LAM_{NE} (solid blue curve) at the start of the second aging stage for one of the cells. Adding LAM_{NE} better simulated the shape and position of the peaks at low and high voltages (at 3.4 and 4.1 V, respectively) without significantly changing the fit of the features between those two peaks. The LAM_{NE} value shifted the peak around 3.4 V, and this shift was necessary to simulate the experimental IC curves approaching and during the second stage of aging. There were no clear differences in the LAM_{NE} induced by the five duty cycles during the initial aging stage. All LAM_{NE} values increased considerably at the onset of the second stage of aging for each cell.

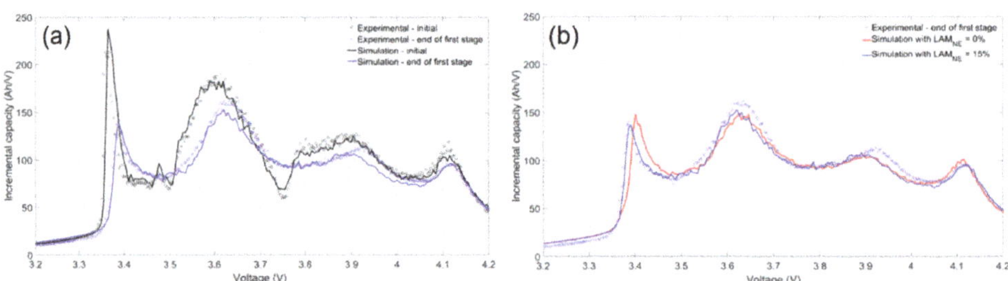

Figure 7. (**a**) Representative fits for the experimental IC curves at the start and end of the initial stage of aging. (**b**) Fits for the experimental IC curves at the end of the initial stage of aging with and without considering the loss of active material at the negative electrode (LAM_{NE}).

To highlight differences in degradation caused by the five driving profiles, the cells were analyzed at a similar point in their aging process: the transition from the initial to the advanced aging stage. Hence, the degradation modes for all the cells at the onset of the second stage were calculated and summarized in a spider plot, presented in Figure 8. The plot revealed that the real driving profiles degraded the battery differently than the synthetic ones, with the former on average inducing more LAM_{PE} and LAM_{NE}. The RD1 and RD2 profiles induced 6.2% ± 1.4% and 10.2% ± 2.6% LAM_{PE}, respectively, while the average LAM_{PE} caused by the synthetic driving cycles at the transition was 4.7% ± 0.5%. In the case of the negative electrode, the RD1 and RD2 profiles induced 15.5% ± 1.3% and 19.5% ± 2.0% LAM_{NE}, respectively, whereas the synthetic driving cycles induced an average LAM_{NE} of 14.5% ± 0.5% at the transition. Interestingly, the percent LLI caused by both real driving profiles was higher at the transition (11.9% ± 1.7% for RD1 and 17.8% ± 3.2% for RD2) than the loss of lithium inventory induced by the synthetic ones (10.6% ± 0.3% for DST, 10.5% ± 0.3% for FUDS, and 9.9% ± 1.6% for NEDC) despite, on average, the RD1 cells advancing to the second stage sooner than all the other cells, while the RD2 cells reached the transition the latest. Thus, the magnitude of LLI did not portend the coming of a secondary aging stage, as the average LLI value of the RD2 cells at the onset was much higher than the others. There was a clear difference between degradation induced by the two real driving cycles. Conversely, the distinction between degradation caused by DST and FUDS was negligible. The average LAM_{NE} caused by both driving cycles was equal (14.6% ± 0.2%), and the average RDF_{NE} values were very close (7.7% ± 1.9% for DST vs. 7.6% ± 1.7% for FUDS). This comparison proved similar aging paths led to similar degradation.

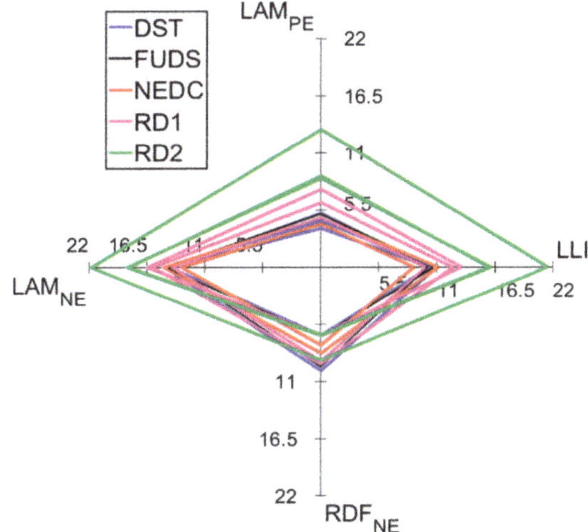

Figure 8. Spider plot of the calculated degradation modes at the onset of the accelerated stage of aging for all 15 cells.

Although the values varied, there were clear trends for each of the degradation modes irrespective of the driving cycle. Figure 9 presents the results of the IC curve analysis throughout aging for one of the cells. The plots revealed the representative trend that the degradation modes followed as all the cells aged. LLI, LAM_{PE}, and LAM_{NE} all accelerated, seemingly exponentially, during the secondary stage. As the cell aged, the LAM_{NE} value increased until it equaled or surpassed the LLI. The RDF_{NE} value tended to reach its plateau value before the acceleration of the other degradation modes. Beyond this point, RDF_{NE} could not be estimated correctly.

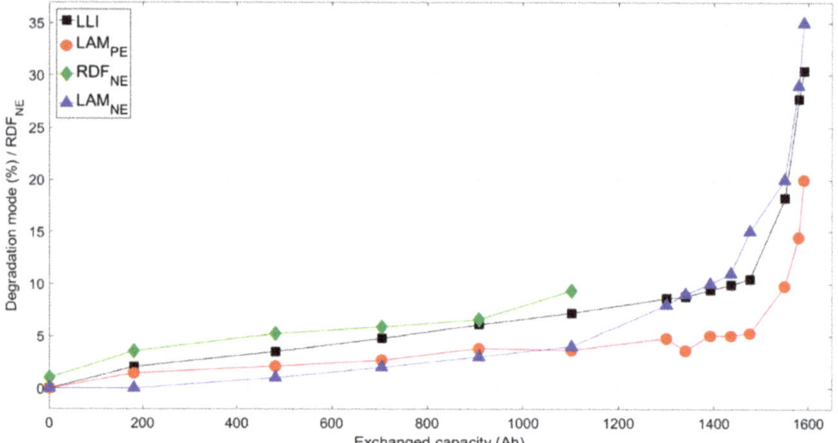

Figure 9. Plot of the calculated degradation modes vs. exchanged capacity for one cell subjected to the DST profile illustrating the representative trends of each.

In brief, none of the synthetic driving cycles adequately simulated the effect of either of the real driving profiles in terms of cell lifetime or degradation. There was a range of outcomes from the real driving data made evident by the difference in the capacity loss, resistance, and degradation mode calculations, elucidating the impact of traffic conditions and further complicating the representability of synthetic cycling. In contrast, those same metrics indicated the DST and FUDS cycles aged the cells almost identically. The cells aged with NEDC cycle exhibited similar lifetimes to the RD1 cells but with a different degradation profile. Despite the relative values being dissimilar, the degradation modes followed the same general trend for all cells, suggesting that the range in property retention and degradation can be mimicked by identifying and adjusting certain cycle parameters such as the maximum power.

The quantified degradation modes were then investigated further to determine the cause of the accelerated aging. This battery chemistry has been shown to be susceptible to fast capacity fade instigated by lithium deposition [25]. In a previous work [1], it was hypothesized that when the effect of the LAM_{NE} overcame the influence of the LLI, an accelerated stage of aging would transpire brought on by plating and its repercussions [26]. Hence, the LAM_{NE} value in relation to the LLI may be significant as it could, nondestructively, hint at the possible initiation of the secondary stage of degradation observed in the experiment. This hypothesis was developed from the states of charge (SOC) of the electrodes at the full cell's EOC voltage. The EOC condition was assumed to be attained when the state of charge of the positive electrode (SOC_{PE}) was 100%. As expected [27] and confirmed by half-cell testing, the GIC negative electrode (NE) exhibited a higher capacity than the positive electrode (PE). In other words, at full charge, the SOC_{PE} was 100%, but the state of charge of the NE (SOC_{NE}) was greater than 0%, and the NE was not fully lithiated [4]. If lithium plating was to occur during aging, provided that the resistance increase was limited (Figure 3b) and the cells were kept at ambient temperature, the NE would have to lose enough active material to be unable to store the totality of incoming lithium ions, Figure S1. As the cell approached full charge, SOC_{NE} would be 0% before 100% SOC was reached by the PE, necessitating the deposition of lithium onto the NE. Therefore, in battery chemistries with GIC as NE, for Li plating to occur, there must be a threshold value at which the PE attains 100% SOC while SOC_{NE} equals 0%. This threshold value was dubbed the plating threshold ($LAM_{NE,PT}$) and was influenced by the cell characteristics (loading ratio and initial offset), the capacity loss, and the other degradation modes, LLI and LAM_{PE}. Values above this threshold would lead to deposition of lithium on the NE [4].

The plating threshold, for $LAM_{PE}s < 10\%$, was estimated from EOC conditions of the graphite using Equation (1). The derivation of this equation is provided in Supplementary Materials. It was important to note that the lithium plating was completely irreversible, as evidenced by the absence of additional electrochemical features in the voltage response during discharge [9,28–30] and the shift in the voltage at the beginning of discharge to lower values (Figure 10). Lithium stripping was not observed in the C/35 IC curves during the accelerated stage or even at failure for all the cells tested. Hence, the plating only added to the rate of the LLI, which accelerated capacity fade. If only irreversible lithium plating were to occur in these cells, the transition to the accelerated aging stage would correspond with the point at which the plating threshold was reached. From this hypothesis, the percent LAM_{NE} during the second stage could be determined from Equation (1), where LR_{ini} is the initial ratio between the capacities of the negative and positive electrodes or loading ratio, and OFS_{ini} is the SOC_{PE} offset, relative to the negative electrode, due to the formation of the SEI layer [4]. The term in parentheses is constant and only depends on initial characteristics. The percentage of LLI, $\%LLI$, is the value that increases during aging. Using this equation, plating threshold values were estimated throughout aging and compared to the LAM_{NE} obtained manually from the best fit to the IC curves (Figure 11). During initial aging, the LAM_{NE} values estimated from the fit were well under the plating threshold. However, at the onset of the second stage, the fit LAM_{NE} percentages came within 2% of the equation values, indicating that irreversible plating could lead to accelerating aging. This conclusion will be validated further by postmortem analysis in a subsequent work.

$$LAM_{NE,PT} = \left(100 - \frac{100}{LR_{ini}} + \frac{OFS_{ini}}{LR_{ini}}\right) + \frac{\%LLI}{LR_{ini}} \tag{1}$$

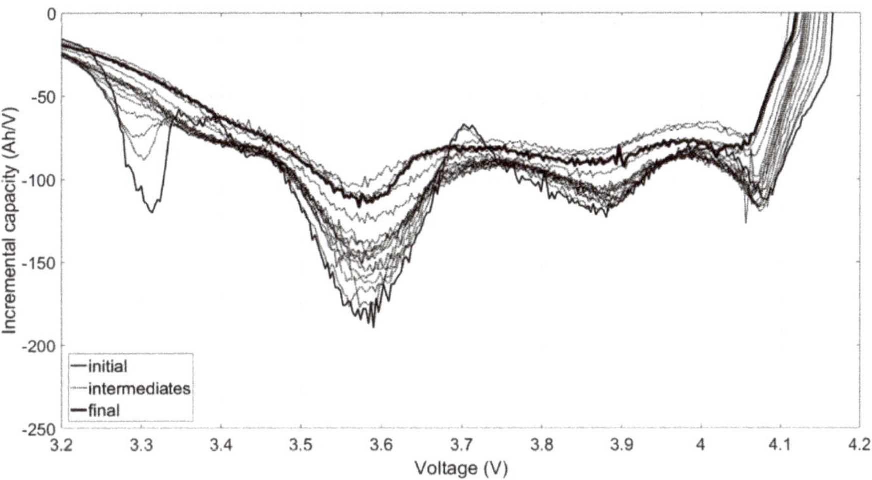

Figure 10. Representative IC curve evolution of a cell aged to failure (>20% capacity loss) during discharge.

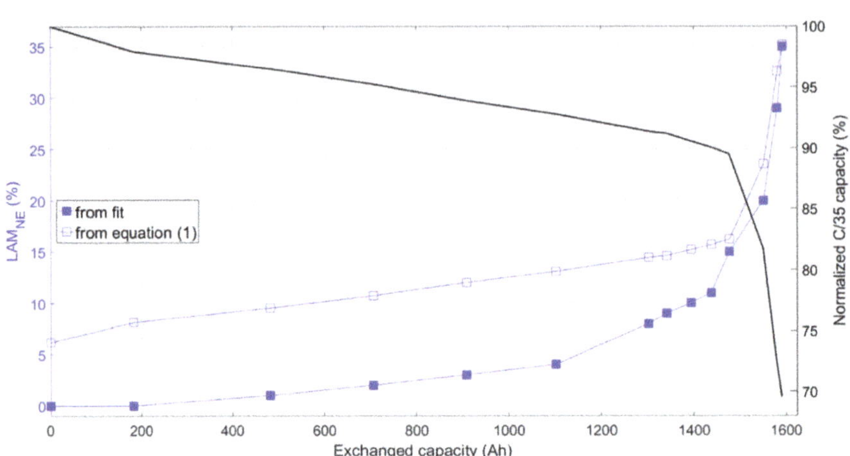

Figure 11. The percent LAM$_{NE}$, plating threshold, and normalized capacity of the corresponding cell as it aged.

Although diagnosis via IC analysis was achieved, accurate, reliable prognosis of lithium plating was challenging because of the complexity and path dependency involved in cell degradation. Tracking capacity and resistance were simply incapable of anticipating this phenomenon. From the evolution of the IC curve, the disappearance of the arch feature between 3.7 and 3.8 V always occurred before the transition in aging, but it could not be used to detect if the transition was forthcoming. Degradation mode quantification led to a metric, *LAM$_{NE,PT}$*, directly related to Li plating. Though promising, plating threshold analysis needed further investigation for confirmation. Moreover, it was logical to conclude a combination of metrics would be required for such a complicated prognosis. Hence, data at the transition from the initial to the secondary stage were scrutinized further. As suggested by a previous paper [1], the ratio of the LAM$_{NE}$ to the LLI may be a more decisive parameter in predicting the relative lifetimes of these cells. As shown by Figure 12, the second stage began sooner for the cells that exhibited a higher LAM$_{NE}$: LLI ratio at the transition. The cell with the lowest ratio (RD2, cell 2) exhibited a large decrease in capacity much later in life than the other cells. The LAM$_{NE}$: LLI ratio was significant phenomenologically because a value greater than one meant the plating threshold would be reached during aging and prompted a precipitous dip in capacity. Once the cell reached this point, the second stage of aging began, the loss of lithium inventory increased nonlinearly, and complete deterioration of the battery quickly followed. This ratio in conjunction with the RDF$_{NE}$ value and the plating threshold may lead to a non-heuristic approach to diagnosis and prognosis of batteries.

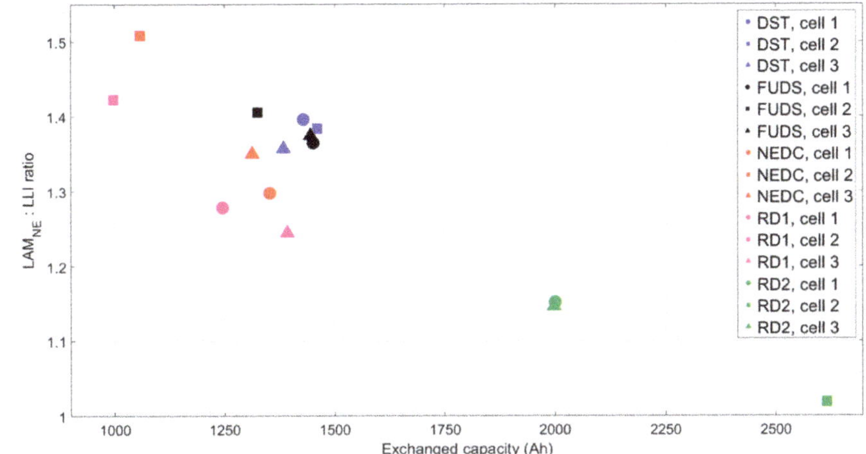

Figure 12. Plot of the LAM_{NE}: loss of lithium inventory (LLI) ratio versus the exchanged capacity at the start of the accelerated aging stage for all cells tested.

4. Conclusions

This study addressed three important points for the battery community. First, it determined whether synthetic driving cycles were representative of real driving. Second, it investigated the origin of the second stage of aging. Third, it proposed metrics to forecast the inception of this second stage.

Based on conventional metrics of capacity and resistance, NEDC was the synthetic driving profile that most closely replicated the trend in capacity fade caused by one of the real driving profiles used in this study (RD1). Since the NEDC driving cycle had the lowest charge capacity during discharge, it best simulated the capacity loss in EVs that traveled a regular daily commute. However, any charge-while-driving technology can change this evaluation. Because of the below average discharge current of real driving data from a varied commute (RD2), none of the synthetic driving profiles reproduced the relatively long lifetimes of those cells. Therefore, it was found that the average current and charge capacity during discharge were important parameters, and they should be considered in determining the appropriate synthetic profile for any EV battery lifetime test. More importantly, traffic greatly influenced cell degradation and cycle life. As a result, realistic EV battery testing might also require looping several iterations of the same synthetic cycle at different intensities since none of the synthetic cycles sufficiently imitated the range of lifetimes and degradation profiles caused by real driving.

All cells experienced this second stage of aging, in which the capacity fade accelerated by more than a factor of six. Changes to certain features in the charging IC curves during aging, such as the decrease in the intensity of the arch between 3.7 and 3.8 V and the shifting of the sharp peak at 3.4 V, were found to coincide with this steep increase in cell degradation. The accelerated aging stage was forecasted by the '*alawa* software, which was used to quantify the degradation mechanisms at play in previous studies. Using advanced IC analysis techniques, the cells were diagnosed with presenting symptoms of irreversible lithium plating leading to the accelerated deterioration of properties.

It was found that the LAM_{NE}: LLI ratio was an important parameter in diagnosing the cell. Ratios greater than one were prerequisites to a second stage of aging characterized by lithium plating and accelerated degradation if the plating was irreversible. Values close to or less than one can delay or prevent the appearance of an accelerated aging stage. In addition, the LAM_{NE} value, which is needed to instigate Li plating at any moment of aging, can be calculated using the analysis outlined in this paper.

These findings also expose a critical flaw in conventional state of health (SOH) estimations. SOH is usually defined from capacity loss and/or resistance increase, but this study showed that those measurements were ineffective in predicting the second stage of degradation, consequently leaving the devices vulnerable to unexpected cell failure. A different set of criteria incorporating new parameters (such as the LAM_{NE}: LLI ratio, potentially) might be crucial to accurately define SOH in the future.

The change in arch intensity, the plating threshold, and the ratio of the LAM_{NE} to LLI were identified as possible harbingers of lithium plating and imminent cell death. These metrics can be evaluated *in operando* without opening the cell. Furthermore, property retention strategies can be proposed based on these metrics. These are capabilities that the conventional examination of capacity loss and resistance increase lack. Parameters based on IC analysis should be monitored and evaluated collectively to determine if intervention, such as applying different rest conditions or preventing complete charging, is required to lengthen cell lifetimes. This promising diagnosis and prognosis technique will be used and refined in future investigations to predict Li-ion battery degradation induced by different grid–EV interactions.

Supplementary Materials: The derivation of Equation (1) and Figure S1 are available online at www.mdpi.com/2313-0105/5/2/42.

Author Contributions: All the authors designed and performed the experiments, analyzed the results, and wrote the manuscript.

Funding: This work was funded by the state of Hawai'i and ONR Asia Pacific Research Initiative for Sustainable Energy Systems (APRISES) award number N00014-17-1-2206.

Acknowledgments: The authors are grateful to the Hawai'i an Electric Company for their support to the operations of the Hawai'i Sustainable Energy Research Facility (HiSERF). Lastly, the authors would like to acknowledge the advice and cooperation of Keith Bethune.

Conflicts of Interest: The authors declare no conflict of interest.

References

1. Dubarry, M.; Baure, G.; Devie, A. Durability and Reliability of EV Batteries under Electric Utility Grid Operations: Path Dependence of Battery Degradation. *J. Electrochem. Soc.* **2018**, *165*, A773–A783. [CrossRef]
2. *Electric Vehicle Battery Test Procedures Manual, Rev. 2*; USABC/US DOE/INEL, 1996. Available online: http://avt.inl.gov/sites/default/files/pdf/battery/usabc_manual_rev2.pdf (accessed on 28 March 2019).
3. Liaw, B.Y.; Dubarry, M. From driving cycle analysis to understanding battery performance in real-life electric hybrid vehicle operation. *J. Power Sources* **2007**, *174*, 76–88. [CrossRef]
4. Dubarry, M.; Truchot, C.; Liaw, B.Y. Synthesize battery degradation modes via a diagnostic and prognostic model. *J. Power Sources* **2012**, *219*, 204–216. [CrossRef]
5. Dubarry, M.; Svoboda, V.; Hwu, R.; Liaw, B.Y. Incremental capacity analysis and close-to-equilibrium OCV measurements to quantify capacity fade in commercial rechargeable lithium batteries. *Electrochem. Solid State Lett.* **2006**, *9*, A454–A457. [CrossRef]
6. Han, X.B.; Ouyang, M.G.; Lu, L.G.; Li, J.Q.; Zheng, Y.J.; Li, Z. A comparative study of commercial lithium ion battery cycle life in electrical vehicle: Aging mechanism identification. *J. Power Sources* **2014**, *251*, 38–54. [CrossRef]
7. Ouyang, M.G.; Ren, D.S.; Lu, L.G.; Li, J.Q.; Feng, X.N.; Han, X.B.; Liu, G.M. Overcharge-induced capacity fading analysis for large format lithium-ion batteries with LiyNi1/3Co1/3Mn1/3O2 + LiyMn2O4 composite cathode. *J. Power Sources* **2015**, *279*, 626–635. [CrossRef]
8. Devie, A.; Dubarry, M.; Liaw, B.Y. Overcharge Study in Li4Ti5O12 Based Lithium-Ion Pouch Cell I. Quantitative Diagnosis of Degradation Modes. *J. Electrochem. Soc.* **2015**, *162*, A1033–A1040. [CrossRef]
9. Ansean, D.; Dubarry, M.; Devie, A.; Liaw, B.Y.; Garcia, V.M.; Viera, J.C.; Gonzalez, M. Operando lithium plating quantification and early detection of a commercial LiFePO4 cell cycled under dynamic driving schedule. *J. Power Sources* **2017**, *356*, 36–46. [CrossRef]
10. Dubarry, M.; Berecibar, M.; Devie, A.; Ansean, D.; Omar, N.; Villarreal, I. State of health battery estimator enabling degradation diagnosis: Model and algorithm description. *J. Power Sources* **2017**, *360*, 59–69. [CrossRef]

11. Devie, A.; Dubarry, M.; Wu, H.P.; Wu, T.H.; Liaw, B.Y. Overcharge Study in Li4Ti5O12 Based Lithium-Ion Pouch Cell II. Experimental Investigation of the Degradation Mechanism. *J. Electrochem. Soc.* **2016**, *163*, A2611–A2617. [CrossRef]
12. Rezvanizaniani, S.M.; Liu, Z.C.; Chen, Y.; Lee, J. Review and recent advances in battery health monitoring and prognostics technologies for electric vehicle (EV) safety and mobility. *J. Power Sources* **2014**, *256*, 110–124. [CrossRef]
13. Devie, A.; Dubarry, M. Durability and Reliability of Electric Vehicle Batteries under Electric Utility Grid Operations. Part 1: Cell-to-Cell Variations and Preliminary Testing. *Batteries* **2016**, *2*, 28. [CrossRef]
14. Dubarry, M.; Devie, A. Battery durability and reliability under electric utility grid operations: Representative usage aging and calendar aging. *J. Energy Storage* **2018**, *18*, 185–195. [CrossRef]
15. Berdichevsky, G.; Kelty, K.; Straubel, J.B.; Toomre, E. *The Tesla Roadster Battery System*; Tesla Motors: Palo Alto, CA, USA, 2007; pp. 1–5.
16. Uitz, M.; Sternad, M.; Breuer, S.; Täubert, C.; Traußnig, T.; Hennige, V.; Hanzu, I.; Wilkening, M. Aging of Tesla's 18650 Lithium-Ion Cells: Correlating Solid-Electrolyte-Interphase Evolution with Fading in Capacity and Power. *J. Electrochem. Soc.* **2017**, *164*, A3503–A3510. [CrossRef]
17. Dubarry, M.; Vuillaume, N.; Liaw, B.Y. Origins and accommodation of cell variations in Li-ion battery pack modeling. *Int. J. Energy Res.* **2010**, *34*, 216–231. [CrossRef]
18. Li, J.H.; Barillas, J.K.; Guenther, C.; Danzer, M.A. A comparative study of state of charge estimation algorithms for LiFePO4 batteries used in electric vehicles. *J. Power Sources* **2013**, *230*, 244–250. [CrossRef]
19. Kassem, M.; Delacourt, C. Postmortem analysis of calendar-aged graphite/LiFePO4 cells. *J. Power Sources* **2013**, *235*, 159–171. [CrossRef]
20. Schmidt, J.P.; Tran, H.Y.; Richter, J.; Ivers-Tiffee, E.; Wohlfahrt-Mehrens, M. Analysis and prediction of the open circuit potential of lithium-ion cells. *J. Power Sources* **2013**, *239*, 696–704. [CrossRef]
21. Birkl, C.R.; Roberts, M.R.; McTurk, E.; Bruce, P.G.; Howey, D.A. Degradation diagnostics for lithium ion cells. *J. Power Sources* **2017**, *341*, 373–386. [CrossRef]
22. Frisco, S.; Kumar, A.; Whitacre, J.F.; Litster, S. Understanding Li-Ion Battery Anode Degradation and Pore Morphological Changes through Nano-Resolution X-ray Computed Tomography. *J. Electrochem. Soc.* **2016**, *163*, A2636–A2640. [CrossRef]
23. Keil, P.; Jossen, A. Impact of Dynamic Driving Loads and Regenerative Braking on the Aging of Lithium-Ion Batteries in Electric Vehicles. *J. Electrochem. Soc.* **2017**, *164*, A3081–A3092. [CrossRef]
24. Berecibar, M.; Devriendt, F.; Dubarry, M.; Villarreal, I.; Omar, N.; Verbeke, W.; Van Mierlo, J. Online state of health estimation on NMC cells based on predictive analytics. *J. Power Sources* **2016**, *320*, 239–250. [CrossRef]
25. Waldmann, T.; Kasper, M.; Wohlfahrt-Mehrens, M. Optimization of Charging Strategy by Prevention of Lithium Deposition on Anodes in high-energy Lithium-ion Batteries—Electrochemical Experiments. *Electrochim. Acta* **2015**, *178*, 525–532. [CrossRef]
26. Zhang, S.S. The effect of the charging protocol on the cycle life of a Li-ion battery. *J. Power Sources* **2006**, *161*, 1385–1391. [CrossRef]
27. Nitta, N.; Wu, F.X.; Lee, J.T.; Yushin, G. Li-ion battery materials: Present and future. *Mater. Today* **2015**, *18*, 252–264. [CrossRef]
28. Kim, C.S.; Jeong, K.M.; Kim, K.; Yi, C.W. Effects of Capacity Ratios between Anode and Cathode on Electrochemical Properties for Lithium Polymer Batteries. *Electrochim. Acta* **2015**, *155*, 431–436. [CrossRef]
29. Fan, J.; Tan, S. Studies on charging lithium-ion cells at low temperatures. *J. Electrochem. Soc.* **2006**, *153*, A1081–A1092. [CrossRef]
30. Petzl, M.; Danzer, M.A. Nondestructive detection, characterization, and quantification of lithium plating in commercial lithium-ion batteries. *J. Power Sources* **2014**, *254*, 80–87. [CrossRef]

Article

A Post-Mortem Study of Stacked 16 Ah Graphite//LiFePO$_4$ Pouch Cells Cycled at 5 °C

Arianna Moretti [1,2,*], Diogo Vieira Carvalho [1,2], Niloofar Ehteshami [3], Elie Paillard [3], Willy Porcher [4], David Brun-Buisson [4], Jean-Baptiste Ducros [4], Iratxe de Meatza [5], Aitor Eguia-Barrio [5], Khiem Trad [6] and Stefano Passerini [1,2,*]

1 Helmholtz Institute Ulm (HIU), Helmholtzstrasse 11, 89081 Ulm, Germany; Diogo.carvalho@kit.edu
2 Karlsruhe Institute of Technology (KIT), P.O. Box 3640, 76021 Karlsruhe, Germany
3 Helmholtz-Institute Münster (IEK 12), Forschungszentrum Jülich GmbH, Corrensstraße 46, 48149 Münster, Germany; n.ehteshami@fz-juelich.de (N.E.); e.paillard@fz-juelich.de (E.P.)
4 Grenoble Université Alpes, CEA-LITEN, 17 avenue des Martyrs, 38000 Grenoble, France; willy.porcher@cea.fr (W.P.); david.brun-buisson@cea.fr (D.B.-B.); jean-baptiste.ducros@cea.fr (J.-B.D.)
5 CIDETEC Energy Storage, Parque Científico y Tecnológico de Gipuzkoa, Paseo Miramón 196, 20014 Donostia-San Sebastián, Spain; imeatza@cidetec.es (I.d.M.); aeguia@cidetec.es (A.E.-B.)
6 VITO/EnergyVille, Thor Park 8300, 3600 Genk, Belgium; khiem.trad@vito.be
* Correspondence: arianna.moretti@kit.edu (A.M.); stefano.passerini@kit.edu (S.P.); Tel.: +49-(0)-731-5034109 (A.M.); +49-(0)-731-5034101 (S.P.)

Received: 22 March 2019; Accepted: 24 April 2019; Published: 7 May 2019

Abstract: Herein, the post-mortem study on 16 Ah graphite//LiFePO$_4$ pouch cells is reported. Aiming to understand their failure mechanism, taking place when cycling at low temperature, the analysis of the cell components taken from different portions of the stacks and from different positions in the electrodes, is performed by scanning electron microscopy (SEM), X-ray diffraction (XRD) and X-ray photoemission spectroscopy (XPS). Also, the recovered electrodes are used to reassemble half-cells for further cycle tests. The combination of the several techniques detects an inhomogeneous ageing of the electrodes along the stack and from the center to the edge of the electrode, most probably due to differences in the pressure experienced by the electrodes. Interestingly, XPS reveals that more electrolyte decomposition took place at the edge of the electrodes and at the outer part of the cell stack independently of the ageing conditions. Finally, the use of high cycling currents buffers the low temperature detrimental effects, resulting in longer cycle life and less inhomogeneities.

Keywords: lithium-ion; batteries; ageing; post-mortem analysis

1. Introduction

Li-ion batteries are required to operate for thousands cycles in a wide temperature range to satisfy the market and end-user expectations. Usually accelerated ageing protocols are performed to speed-up the degradation processes in order to study, in a relatively short time, the influence of key parameters (such as temperature, charge/discharge rates, state of charge (SOC), cut-off voltages) on the battery performance and to predict battery lifetime [1]. Performance degradation is mainly attributed to complex chemical processes involving active and non-active electrode components and the electrolyte. In part, they occur at the electrode/electrolyte interface due to the electrochemical instability of the electrolyte at low (below 1 V) and high (above 4 V) voltages [2,3]. If the formation of a solid electrolyte interface (SEI) at the anode is pivotal for ensuring battery operation, the uncontrolled growth of this layer during cell lifetime contributes to the capacity decrease as the Li inventory (corresponding to the cathode capacity) is continuously depleted [4]. Loss of lithium inventory (LLI) can also be due to its deposition at the anode close to 0 V. This can happen in case of non-ideal cell's capacity balance (normally a negative to positive capacity ratio ≥1.1 is used), or upon fast charging at low temperature [5,6]. In addition, transition metals can be leached out of the cathode structure, especially at high voltage where the reactivity between the delithiated cathode surface and the electrolyte is the highest, inducing the active material leaching (LAM) resulting in the anode poisoning [7,8]. Furthermore, the mechanical failure of the electrode and the loss of electric contact generate electrochemically inactive areas [9] that adds to the LAM. Ageing tests can be followed by cell autopsy and post-mortem analysis. Cell disassembly is generally conducted on the discharged cells to limit the risk of thermal runaway. There is no standard procedure available and thus different methods, depending on the cell architecture, are used to open the cell in an inert atmosphere to avoid altering the physical-chemical state of the cell components [10]. The samples have to be carefully labelled to track their position in the stack (in stacked cells) or in the jelly-roll (for wound cells). A series of analytical techniques are then combined to find out the morphological, chemical and electrochemical modifications that occurred in the components upon ageing [11].

Herein, we focus on the post-mortem analysis of stacked 16 Ah graphite//LiFePO$_4$ pouch cells designed for automotive applications and produced within the European project SPICY (Silicon and polyanionic chemistries and architectures of Li-ion cell for high energy battery). The cells were subjected to cycle ageing (between 0% and 100% SOC) at different charge current rates and temperatures (Figure S1). The results obtained at 5 °C, which is a moderately low temperature easily experienced in electric vehicle (EV) applications, were surprising as a low C-rate led to lower cycle life when compared to ageing results at 25 °C. That it is why cells cycled at 5 °C, were investigated deeply. Portions of the cell components, especially the electrodes, were analysed before and after ageing using scanning electron microscopy (SEM), X-rays diffraction (XRD) and X-ray photoemission spectroscopy (XPS), and used to reassemble half-cells cells to determine the evolution of their electrochemical properties. The tests were conducted using electrode parts taken from different portion of the stacks and from different positions in the electrodes (edge vs. center) to detect non-homogeneous ageing effects.

2. Results and Discussion

2.1. Description of Cells, Ageing Conditions and Initial Visual Inspection of Components

The cells specifications are reported in Table 1. The cells were aged by cycling with a constant charge current (0.3C or 2C) between 0% and 100% SOC (i.e., between 2.5 V and 3.6 V) at 5 °C inside a ventilated climatic chamber. The discharge C-rate is kept the same (1C). The tests were interrupted regularly to run a performance test at room temperature (RT) to follow up on the evolution of the cell's characteristics. The cell state of health (SOH) was determined from a 1C discharge constant current/constant voltage (CC-CV) test performed during this reference test.

Table 1. Cells specifications.

Parameter	Description
Cathode active material Cathode composition (37 sheets) Cathode areal capacity	LiFePO$_4$ (LFP) LFP/C/PVDF = 90.5/5/4.5 2.3 mAh/cm^2/face
Anode active material Anode composition (38 sheets) Anode areal capacity	Graphite (Gr) Gr/C/CMC/SBR = 96/0/2/2 2.5 mAh/cm^2/face
Electrolyte composition	1 M LiPF$_6$ in EC:PC:DMC (1:1:3 vol.) + 2 wt. % VC
Separator (76 sheets)	Celgard® PE/PP 2325 (Celgard, LLC)
Format	Soft prismatic (pouch-cell)
Weight	394 g
Average capacity at RT @C @C/10	14.7 Ah 16.2 Ah
Specific energy at RT @C	115 Wh/kg
@C/10	134 Wh/kg

Acronyms: LFP = LiFePO$_4$; C = conductive carbon; PVDF = Polyvinylidendifluoride; Gr = graphite; CMC = Sodium carboxymethylcellulose; SBR = Styrene Butadiene Rubber; LiPF$_6$ = Lithium hexafluorophosphate EC = Ethylencarbonate; PC = Propylencarbonate; DMC = Dimethylcarbonate; VC = vinylcarbonate; PE/PP = Polyethylene/polypropylene.

The ageing protocol was interrupted when the cell loses at least 20% of its initial capacity. As the SOH is calculated during the performance test, it was not possible to interrupt the ageing at the exact same SOH value. Nevertheless, they have very close SOH values (79.7% and 72%).

Table 2 summarizes the results of the ageing phase. The total capacity throughput indicates the total capacity (charge and discharge) that the cells have cycled during ageing (including the performance test). The equivalent cycle number is obtained by dividing the total capacity throughput by the capacity of one complete cycle (about 32 Ah). Three cells were used for the study: a cell that performed only the formation protocol (named fresh cell) and two cells aged by cycling at 5 °C and 0.3C (named Cell A) or 2C (named Cell B).

Table 2. State of health (SOH), total capacity throughput, equivalent cycle number, time required for ageing and voltage of the cells before disassembly.

Ageing Parameters	Fresh Cell	Cell A	Cell B
Ageing conditions	/	5 °C; 0.3C	5 °C; 2C
SOH (%)	100	72	79.7
Total capacity throughput (Ah)	/	16,982	20,753
Ageing time (Days)	/	118	86
Equivalent cycle number	/	533	660
Voltage (V)	2.462	3.02	2.63

After cycle ageing, the cells did not present any clear sign of degradation, such as swelling or electrolyte leakage. Although the cells were completely discharged to 2.5 V, a slightly higher value of OCV was detected for Cell A shortly prior to cell disassembly.

Figure 1a,b shows the pouch cell as made and after opening. Each cell contains a stack (Figure 1c) of double-side coated electrodes and separators as schematically depicted in Figure 1d. The electrodes were harvested at the beginning (B), middle (M) and end (E) of the stack. The samples, taken from the edge (e) and the center (c) of the electrode tape (Figure 1d), were dipped in dimethylcarbonate (DMC) (except some cathode samples from the fresh cell for SEM analysis which were dipped in acetonitrile (ACN) for 30 s and then dried under dynamic vacuum at room temperature for 30 min. Table S1 in the Supplementary Materials reports a detailed description of the analysis performed on each sample and its position in the stack.

Figure 2 compares the pictures of cells components harvested from the middle of the stack. The separator of Cell A does not show a significant difference respect to the one from fresh cell, while that taken from Cell B shows dark areas, probably due to detachment of active material from anode tape. Compared to the fresh cell, the cathode from cell A and cell B show a grey/brown coloration after cycling at 5 °C. Noteworthy, the cathode tape in different portions of the stack does not show any visual difference (not shown). The anode tape, on the other hand, was rather shiny at the edges with a darker area in the center. The silvery color may be an indication of plated lithium [5], in both cell A and cell B.

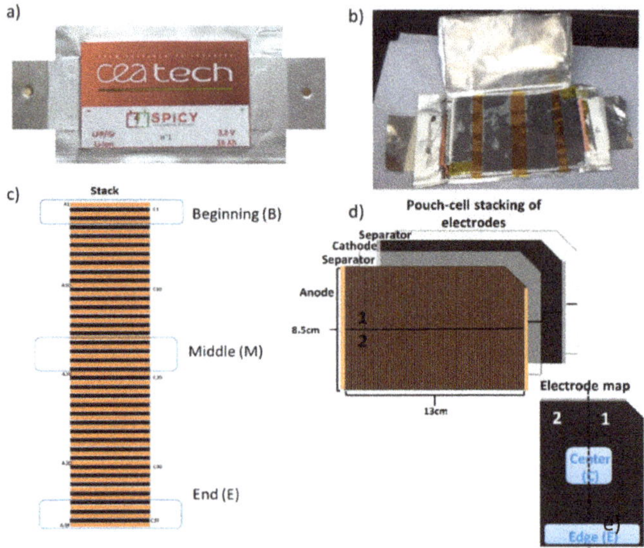

Figure 1. (**a**) Closed pouch cell; (**b**) view of the opened cell; (**c**) scheme of the stack defining the sampling portions; (**d**) simplified view of double coated electrode indicating the two areas sampled; and (**e**) definition of "edge" and "center" positions within a single electrode.

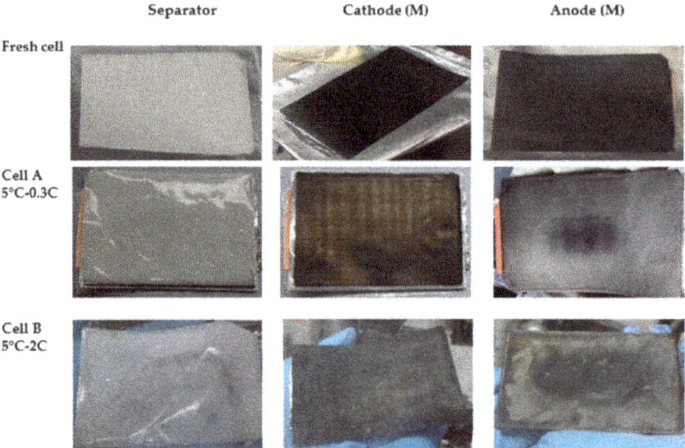

Figure 2. Visual inspection of cells components (separator, cathode and anode from the middle (M) of the stack).

2.2. Adhesion Test

Table 3 summarizes the thickness and adhesion test results of electrodes portions sampled from different position in the stack. The anode thickness increases already after cell formation while the cathode thickness only increases upon ageing.

Table 3. Thickness and adhesion strength of pristine and aged electrode tapes. Results are provided as average of at least two tests per electrode sample with the standard deviation in brackets.

Cell	Electrode (Position within the Stack B = Beginning, M = Middle, E = End)	Thickness (μm)	Peel Force (N/m)
Pristine	Anode	117 (3)	5.4 (2)
	Cathode	183 (5)	504 (17)
Fresh Cell	Anode	127 (2)	2.2 (0.4)
	Cathode	188 (2)	372 (56)
Cell A 5 °C–0.3C	Anode 3 (B)	147 (2)	5 (1)
	Anode 16 (M)	157 (2)	3.3 (9)
	Anode 34 (E)	150 (7)	3.2 (8)
	Cathode 3 (B)	195 (3)	336 (85)
	Cathode 16 (M)	202 (5)	358 (41)
	Cathode 34 (E)	192 (4)	534 (73)
Cell B 5 °C–2C	Anode 3 (B)	155 (6)	4.7 (2)
	Anode 16 (M)	154 (7)	8 (2)
	Anode 34 (E)	146 (1)	3.5 (3)
	Cathode 2 (B)	199 (1)	579 (53)
	Cathode 16 (M)	198 (1)	#
	Cathode 34 (E)	197 (2)	342 (8)

above equipment measurement range (>550 N/m).

Overall, the adhesion of the anode is quite poor compared to that of the cathode. Additionally, the anode adhesion significantly drops after formation (see fresh cell results), although it recovers nearly the values for the pristine electrode over cycle ageing (see Cell B results). The cathode adhesion appears also to be lower in the fresh cell than in the pristine electrode (never in contact with the electrolyte) but still retains high values (>330 N/m) on aged cells. For both electrodes adhesion values vary along the stack, which may be correlated with a different pressure distribution throughout the cell stack, also lead to variations of the electrode capacity.

2.3. Morphology of Cells' Components

The SEM images in Figure 3 show the appearance of the LiFePO$_4$ (LFP) electrodes at the beginning of the stack in the center position of the Fresh Cell, Cell A and Cell B. The electrodes are composed of a mixture of spherical particles and small amount of carbon fibers (used as conductive agent) of micrometric length. The larger spherical particles show a diameter comprised between 0.5 and 1 μm while the smaller particles have a diameter lower than 100 nm. Comparing the SEM images, no significant differences are observed. Additionally, no specific damage or deposit is visible.

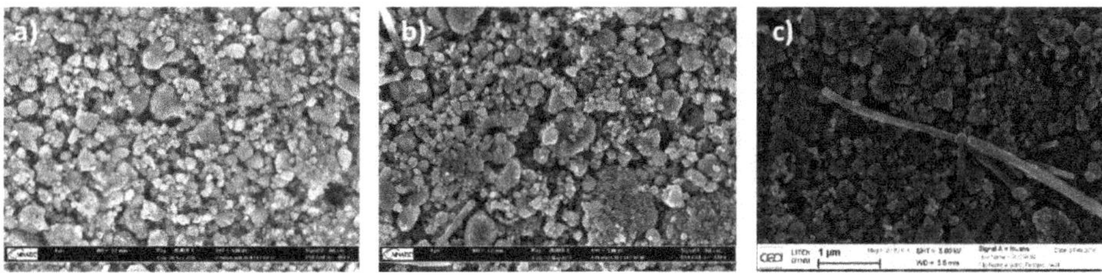

Figure 3. SEM images comparing the morphology of the LFP electrodes taken from (**a**) the fresh cell (after formation), (**b**) Cell A and (**c**) Cell B. All the samples were harvested from the beginning of the stack.

Also, the cathode morphology did not show any relevant difference when sampled from different position in the stack and in the tape (edge vs. center) (see, respectively, Figures S2 and S3 in the

Supplementary Materials). Furthermore, the energy dispersive X-ray (EDX) analysis (Figure S4 in the Supplementary Materials) does not reveal differences between Cell A and Cell B, nor the presence of contaminants. The elemental composition (Figure S4) is also similar to that of the fresh cell (only Fe, P, O and C are detected).

Figure 4 shows a few SEM images of the graphite electrode at the beginning of the stack in the center position, extracted from Fresh Cell (EDX spectra reported in Figure S5 in the Supplementary Materials), Cell A and Cell B. While no major difference between the electrodes from Fresh Cell and Cell A is observed, the anode of Cell B (cycled at 2C) shows the occurrence of graphite exfoliation (notice that several SEM images were taken using different samples to confirm the observed differences). Noteworthy, such exfoliation is not observed in the edge of the electrode (see Figure 5). Thus, a major difference between the center and the edge of the anode electrode is confirmed using both visual (see Figure 2) and SEM observations.

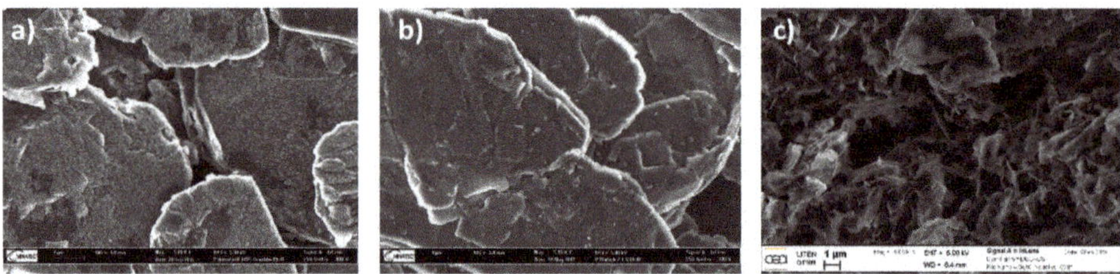

Figure 4. SEM images comparing the graphite electrodes of the (**a**) Fresh Cell (i.e., after formation), (**b**) Cell A and (**c**) Cell B (cycled at 0.3C and 2C, respectively). All the samples were taken from the electrode at the beginning of the cell stack.

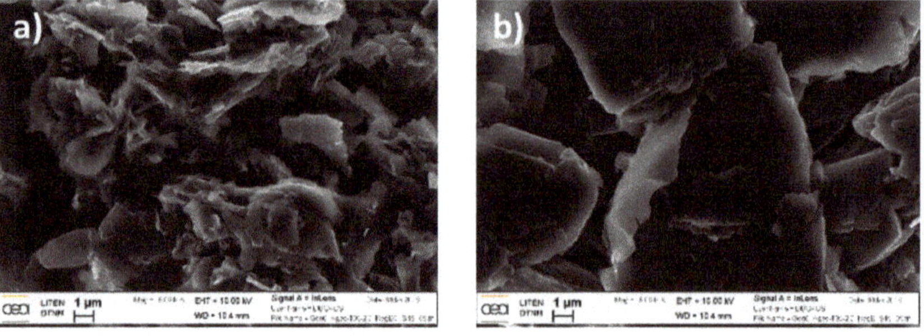

Figure 5. SEM images of the graphite anode of Cell B sampled at the (**a**) center and (**b**) edge of the tape. Both samples were taken from the middle (M) of the stack.

EDX analysis was performed on the samples extracted from both the center and the edge of the anode of Cell B (Figure S6 in Supplementary Materials). Surprisingly, no significant difference in terms of elemental composition was observed. EDX indicates the presence of O, P and F in both samples, due to electrolyte salt residue and the SEI. However, Fe traces were detected in the center (but not in the edge of the electrode) resulting from the cathode (LFP) decomposition. The Fe traces could be one of the reasons for the graphite exfoliation detected in the center of the electrode, even if Fe is not expected to be intercalated inside graphite.

The SEM pictures of the separator (Figure S7 in Supplementary Materials) do not show any damage upon ageing. The separator retained its porous structure upon cell ageing.

2.4. Residual Capacity Measurement

All samples used for the residual capacity determination were harvested from the electrode tapes taken from Area 2 in Figure 1d. The samples are named using the abbreviation given in Table 4, which describes their position in the stack and the electrode. Due to the poor adhesion of the anodic tape, the half-cell assembly with aged graphite anodes was not always possible therefore the residual capacity is analyzed using only the cathode.

Table 4. Description of sample positioning in the stack and within the electrode, sample coding and residual capacity.

Cell	Position in the Stack-Electrode Number	Position in the Electrode	Code	Residual Capacity [1] (%)
A 5 °C-0.3C	Beginning (B)-2	Edge	A_BE	70.7
		Centre	A_BC	62.8
	Middle (M)-15	Edge	A_ME	65.6
		Centre	A_MC	88.9
B 5 °C-2C	Beginning (B)-2	Edge	B_BE	79
		Centre	B_BC	68.4
	Middle (M)-15	Edge	B_ME	66.4
		Centre	B_MC	68.4

[1] The residual capacity is calculated from the first charge of the reassembled half-cells as a percentage of the capacity of the pristine electrode (2.53 mAh) in Figure S7.

The voltage profile obtained at 0.15 mA·cm^{-2} (corresponding to 0.065C) for a pristine cathode tape is reported in Figure S8 in the Supplementary Materials. The voltage profiles of the first cycle (charge comes first) of the aged cathodes performed in the same conditions are shown in Figure 6. It can be noticed that the residual charge capacity, i.e., the amount of Li$^+$ extracted from the electrode, is not homogeneous along the stack and across the electrode tape.

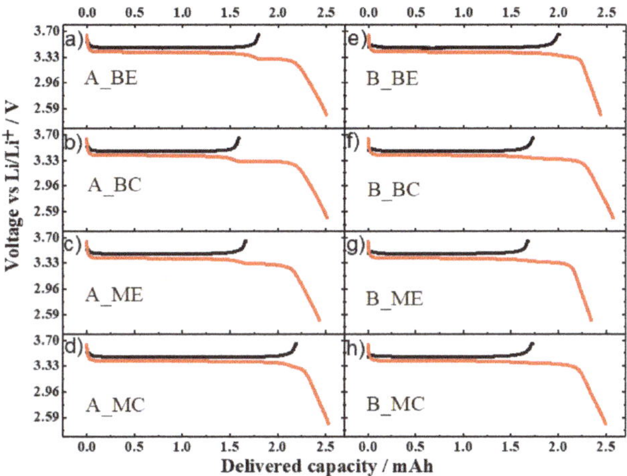

Figure 6. First cycle voltage profile of the aged cathodes in half-cells. (**a**–**d**) samples from cell A. (**e**–**h**) samples from cell B. Electrode area ≈ 1.13 cm^2; C-rate 0.065C.

For the cell aged at 0.3C, the sample A_MC delivered a high charge capacity of ca. 2.25 mAh, a value close to that of the pristine electrode. On the other hand, for the sample A_ME only 1.66 mAh were obtained. The situation is inverted for the samples taken at the beginning of the stack as the sample from the edge (A_BE) delivered a higher capacity (1.79 mAh) than that from the center (A_BC) (ca. 1.59 mAh). It is worth noting that, upon the subsequent lithiation, a "step" in the voltage profile appears. Its position matches well with the capacity value obtained during the previous de-lithiation. Therefore, we attributed this feature to the insertion of lithium into the LFP cathode, which did not occur during the ageing test. This means that, for all the samples, part of the cathode material was inactive during the ageing test, but not damaged. A similar situation is found for the voltage profiles of the aged cathodes from cell B (see Figure 6e–h). Here, the sample from the beginning of the stack and edge of the electrode (B _BE) delivered a capacity (ca. 2.0 mAh) higher than all the other samples (ca. 1.7 mAh). It is also noted that the resistance of Cell A cathode is higher than that of Cell B cathode as a marked overvoltage is present at the very beginning of the charge process (Figure S9 in the Supplementary Materials).

After the first discharge (lithiation of LFP), all the electrodes perform equally in the rate capability test shown in Figure 7 (cycling protocol described in Table S2 in the Supplementary Materials), indicating that the cathode material structure did not undergo significant damages independent on the position in the stack and across the tape. Therefore, the loss of active material (LAM) is negligible compared to the loss of lithium inventory (LLI).

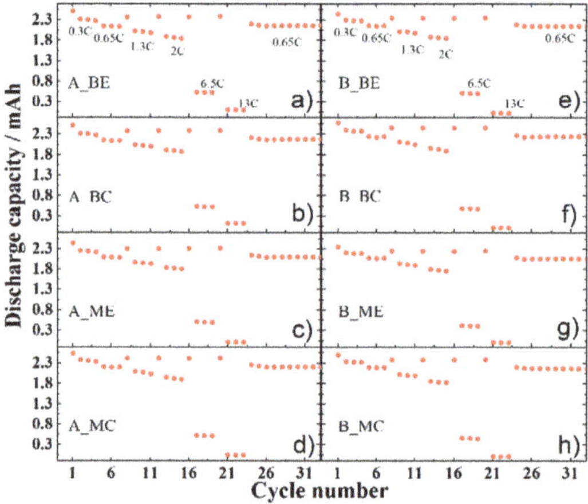

Figure 7. Rate capability test of cathodes harvested from aged cell A (**a–d**) and cell B (**e–h**) (1C = 2.3 mA·cm^{-2}). The detailed test protocol is reported in Table S2 in the Supplementary Materials.

Overall, the same trend of the residual capacity is observed for both aged cells, i.e., MC > ME and BE > BC, but the non-uniformity is more marked for Cell A than Cell B. The observed inhomogeneous LLI can be attributed to a different extent of side reactions (e.g., SEI formation at the facing anode) caused by variations of the electronic contact (due to differences in the internal pressure) and/or ionic conduction through the electrode (inhomogeneous electrode wetting).

It can be inferred that, in the middle of the stack and in the center of the electrode (MC), more LFP is reversibly cycled than at the edge, probably due to the better contact induced by the higher internal pressure. This would also result in a better anode SEI, with a lower consumption of the Li$^+$ inventory to repair it upon cycling. On the contrary, at the beginning of the stack, the edge part of the electrode (BE) is more electrochemically active. Recalling the SEM results for Cell B, graphite exfoliation was observed in the center of the electrode, but not at the edges.

2.5. Structural Analysis

The inhomogeneity revealed by the electrochemical tests is confirmed by phase quantification via Rietveld refinement of the XRD patterns of aged cathode electrodes (Figure S10 in Supplementary Materials). The results are summarized in Table 5. Considering that the cells were opened in the fully discharged state, the cathode material should mostly consist of LiFePO$_4$ but FePO$_4$ could also be present if the discharge process is not complete due to lack of available lithium. For Cell A, the trend that a higher fraction of LiFePO$_4$ is present in samples BE than BC is confirmed (and double checked with the samples taken at the end of the stack EE). On the other hand, discrepancies are found for electrodes in the middle of the stack for which the XRD investigation detects more $LiFePO_4$ at the edge than in the center, revealing that the inhomogeneity is more pronounced in this part of the stack.

Table 5. Phase quantification obtained by XRD of fresh and aged electrode. For comparison also the value obtained for a fresh cathode (after cell formation) is reported.

Sample Name	Electrode Number	$LiFePO_4$ (%)	$FePO_4$ (%)
Fresh	n/a	91	9
A_BC	3	70	30
A_BE	3	80	20
A_MC	16	75	25
A_ME	16	85	15
A_EC	34	74	26
A_EE	34	82	18
B_BC	3	75	25
B_MC	16	74	26
B_EC	34	97	3

The diffraction patterns of the anodes have been analysed for the shift of the first graphite peak at 26.8° (Figure S11). The shift towards lower angle of the *001* reflection is linked with graphite interlayer expansion upon lithium intercalation [12] The shift is larger in the Fresh Cell, while among the aged samples, Cell B showed the smallest shift, indicative of a lower amount of Li^+ remaining trapped into the graphite layers (the cells are opened in the fully discharged state, i.e., Li^+ ions should be fully removed). This translates into a higher lithiation of the cathode. Indeed, a higher $LiFePO_4$ content is observed in the cathode sample B_BC than A_BC (75% and 70%, respectively), which matches well with the slightly higher residual capacity (Figure 6) obtained for the B_BC cathode than the A_BC cathode (1.73 and 1.59 mAh, respectively).

2.6. Surface Analysis

The XPS spectra of graphite samples, harvested from the middle of the stack and at the center of the electrodes (MC), from fresh and aged cells are compared in Figure 8. The average atomic compositions are reported in Table 6. Sputter depth profiling was conducted to obtain information on the SEI composition and thickness.

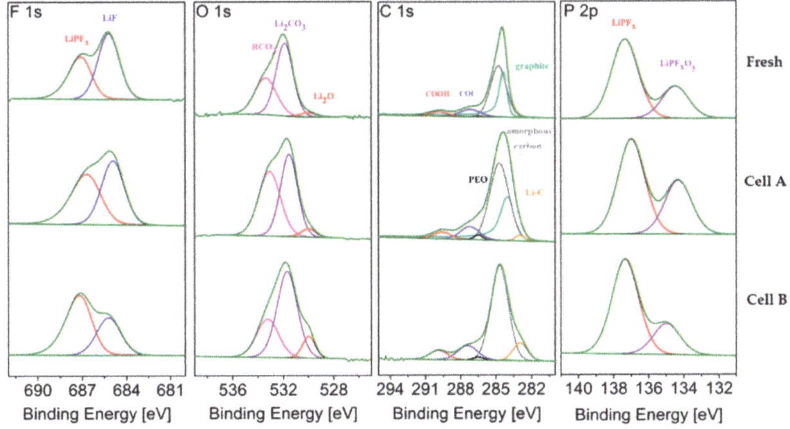

Figure 8. F 1s, O 1s, C 1s and P 2p core peaks XPS spectra of graphite electrode from fresh and aged cells (MC samples). Non-sputtered samples.

Table 6. Average atomic percentages (% at.) determined from three measurement spots on fresh and aged graphite electrodes.

Region	Components	Fresh Cell MC				Cell A MC				Cell B MC			
		Sputtering Time (s)											
		0	60	120	600	0	60	120	600	0	60	120	600
		Average Atomic Percentages (% at.)											
F 1s	$LiPF_x$	5.62	3.12	4.58	3.67	9.60	12.30	7.17	7.49	8.58	6.39	6.19	4.78
	LiF	6.17	9.81	7.32	2.88	8.99	9.81	13.84	4.65	3.66	3.98	4.37	4.36
O 1s	RCO_3	4.78	5.45	4.64	2.60	7.34	6.68	5.27	3.25	5.10	4.29	4.07	4.06
	Li_2CO_3	7.24	4.47	4.03	1.85	7.44	7.12	8.24	4.98	8.68	9.01	10.21	9.78
	Li_2O	0.09	0.11	0.16	0.04	0.19	0.29	0.84	1.46	1.32	1.31	1.49	1.63
C 1s	C1s	56.69	56.16	60.15	74.50	46.59	41.37	34.00	45.31	38.26	38.78	44.99	44.66
	CMC-COC	6.05	3.02	4.42	4.10	4.18	3.94	3.66	1.92	4.16	2.99	2.74	2.93
	CMC-COOH	3.02	2.83	2.85	2.30	2.69	2.66	2.84	1.35	2.91	2.40	2.44	1.90
	Li-C	0.00	0.00	0.00	0.00	1.57	3.00	10.35	19.29	2.96	2.15	2.62	2.02
	PEO	0.33	4.59	1.92	2.46	1.18	1.28	0.87	2.83	2.66	2.82	2.00	1.70
P 2p	$LiPF_x$	0.90	0.36	0.28	0.21	0.64	0.42	0.31	0.14	1.41	1.19	1.09	0.79
	$LiPO_xF_y$	0.31	0.41	0.39	0.21	0.36	0.49	0.41	0.21	0.43	0.45	0.54	0.46

Upon ageing, the SEI on graphite is expected to grow, consuming lithium and thus contributing to the LLI. Figure 8 shows the typical $LiPF_6$ (electrolyte salt) decomposition products, i.e., LiF, $LiPF_x$ and $LiPO_xF_y$, that are detected in the F 1s and P 2p regions [13]. Their relative ratio changes upon ageing. The sum of the atomic fractions derived from these three peaks after 600 s of sputtering (Table 6) represents 3.3, 5.0 and 5.6 % at. for the Fresh Cell, Cell A and Cell B, respectively, showing that a slightly higher salt decomposition occurred in Cell B upon cycling, as also highlighted by the higher signal detected for $LiPF_x$ (P 2p region).

The O 1s spectra of the graphite electrodes (Figure 8) consists of three peaks. The first peak, at the lowest energy, is assigned to lithium oxide, which is a well-known component of the inner part of the SEI [14], but it also corresponds to the degradation product of Li_2CO_3 induced by the sputtering process itself [15]. After ageing, the Li_2O feature is more pronounced in the cell cycled at 2C (cell B). The sputtering results in Table 6 indicate more Li_2O in the inner layer of the SEI of aged graphite electrodes. The second peak, assigned to Li_2CO_3, also grows slightly upon ageing. The peak at the highest binding energy, assigned to organic carbonates, increases more on the surface of the electrode during ageing at 0.3C (cell A), while after 600 s sputtering more organic carbonates are found in Cell B. Before sputtering, the oxygen containing components found in the O 1s region of the graphite from the fresh cell represent 12.11 % at., which increases to 14.98 % at. upon cycling at 0.3 C and 15 % at. 2C.

To eliminate the error caused by the overlapping of graphite and amorphous carbon signals, the sum of their contributions is labeled C 1s in Table 6. This contribution decreases upon cycling indicating that the SEI becomes thicker, thus hiding the graphite contribution. The Li-C signal at 600 s increases in the cycled electrodes, being more pronounced for Cell A (0.3C). This further confirms that the SEI is thicker in the cell cycled at higher C-rate (Cell B).

To investigate the effect of electrode position on the SEI build-up, Figure 9 compares the atomic fraction observed at the center and edge of the electrode tape from the beginning and the middle of the stack for the aged cells. The $LiPF_x$, LiF and $LiPO_xF_y$ % at. compositions are overall higher at the edge of the graphite tape, which means that $LiPF_6$ degrades more extensively there. Additionally, more $LiPF_6$ degradation products are detected in the outer layers of the stack. Overall, the amount of salt decomposition products is higher, whatever the position, in the cell cycled at 0.3C.

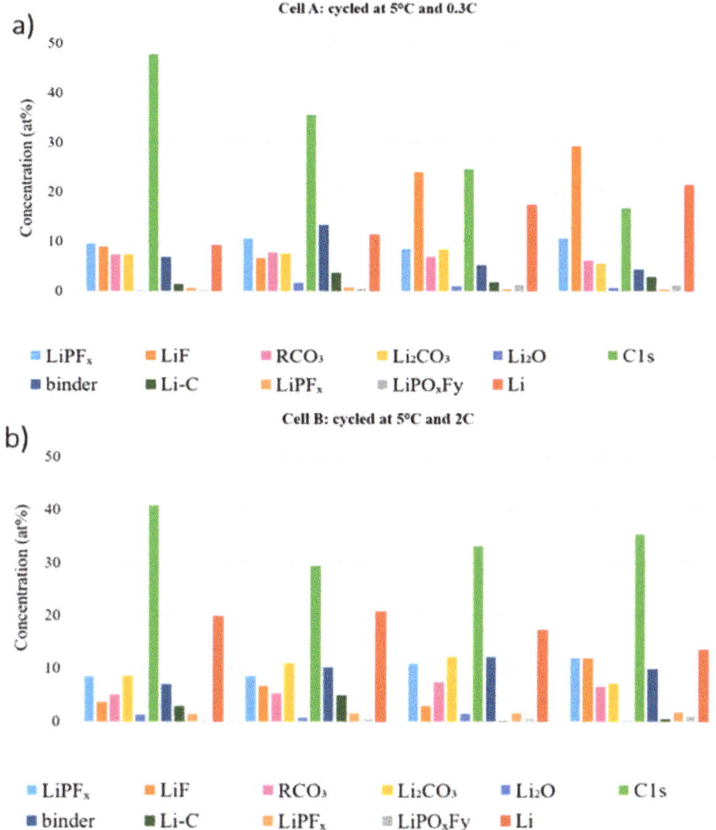

Figure 9. Atomic concentration at the center and edge of the electrode sheet in different position of the stack of (**a**) Cell A and (**b**) Cell B.

In Cell A, the C 1s and binder features decrease dramatically in the beginning of the stack, which indicates that the SEI is thicker than in the middle of the stack. On the other hand, the carbon features in the beginning of the stack in Cell B is higher than Cell A, which is an opposite trend comparing to the middle of the stack. For Li compounds there is no clear trend, which makes it challenging to obtain a reliable relation between different positions.

Generally, for both cycled cells, the differences due to sample location are in the same range as the differences between the two cells. These findings can be correlated to the visual observation of several variations in color and thickness along the electrode stack and across each electrode tape.

The XPS spectra and fitting results for the cathodes extracted from fresh and aged cells are compared in Figure 10. Table 7 reports the averaged atomic concentration for the different components. Due to the uncertainties caused by the overlap between the Fe 2p and Fe 3p core level spectra and the fluorine plasmon and, in the Li 1s core spectra, the overlap between Fe and Li contributions, the corresponding spectra were not used for cathode analysis. All samples investigated were taken at the center of electrode at the middle of the stack.

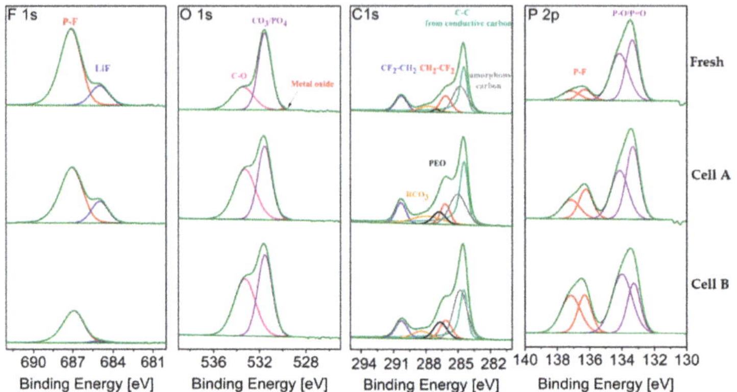

Figure 10. F 1s, O 1s, C 1s and P 2p core peaks XPS spectra of LiFePO4 electrode from fresh and aged cells (MC samples). Non-sputtered samples.

Table 7. Average atomic percentages (% at.) determined from three measurement spots on fresh and aged cathodes.

Region	Components	FreshMC	Cell A	Cell B
			MC	MC
F 1s	**P-F**	12.810	13.433	15.393
	LiF	2.751	3.704	0.866
O 1s	C-O	4.171	6.512	6.761
	CO_3/PO_4	9.101	6.508	7.043
	Metal Oxide	0.053	0.043	0.043
C 1s	Conductive carbon	15.365	18.827	16.033
	CH_2-CF_2	4.783	3.812	4.575
	CF_2-CH_2	4.463	3.809	4.268
	RCO_3	2.720	4.885	2.964
	PEO	0.451	1.823	4.401
	amorphous carbon	9.657	8.814	14.020
P 2p	P-F	0.643	0.787	1.131
	P-O/P = O	1.135	0.762	0.792

The F 1s spectra show traces of LiF at 685 eV on the fresh and aged LFP electrodes. As found for the anode, LiF represents a higher fraction of the cathode electrolyte interphase (CEI) layer in the cell cycled at 0.3C (Cell A) than at 2C (Cell B).

In the O 1s region, the amount of signal attributed to oxygen atoms in the phosphate group $(PO_4)^{3-}$ of LFP decreases upon cycling (Figure 10). This peak is well pronounced for the fresh cathode, indicating the very thin nature of the CEI layer. The peak at higher binding energy (C–O) is attributed to the oxygenated species at the electrode surface and increases in aged cathodes. Upon ageing, the difference between these two peaks decreases, as a result of the growth of the CEI layer. For Cell B the $(PO_4)^{3-}$ contribution remains dominant (see Table 7) while the fraction of oxygenated species in Cell A equals that calculated from the LFP feature.

In the C 1s spectrum, the peak at 284.5 eV corresponds to conductive carbon, which conceals the peak of amorphous carbon related to carbon coating [16]. The two other peaks toward higher energy at 286.3 eV and 290.4 eV are attributed to CF_2CH_2 and CH_2CF_2, corresponding to the PVdF binder [17]. Moreover, two small peaks are due to polyethylene glycol (PEO) and RCO_3 [7]. PEO and RCO_3 have been proved to be part of both the SEI and the CEI [18,19]. It can be seen that both contributions increase upon cycling in both cases.

In the P 2p region, the peak of PO_4 (at 133.3–133.5 eV, labeled P–O/P = O) declines with ageing, which indicates for the CEI growth. The higher amount of phosphate feature in the cell cycled at 2 C is in good agreement with results from the O 1s region, indicating thinner CEI layer. The higher amount of P–F compound ($LiPF_6$ decomposition products) is detected in the cell cycled at higher C-rate, the same result was obtained for graphite electrode.

Figure 11 compares the atomic percentages at the center and edge of the electrode tape from different positions of the stack (beginning and middle of the stack) of the aged cathodes. In both cells, a higher LiF amount is detected at the edge of each electrode tape, which means more $LiPF_6$ degradation. Overall, the amount of LiF is higher at any positions in the cell cycled at 0.3C. However, higher amounts of LiF are found in the middle of the stack, while the opposite is observed for the cell cycled at 2C. Nonetheless, the total amount of fluoride species are found to be higher in the middle of the stack for both cells.

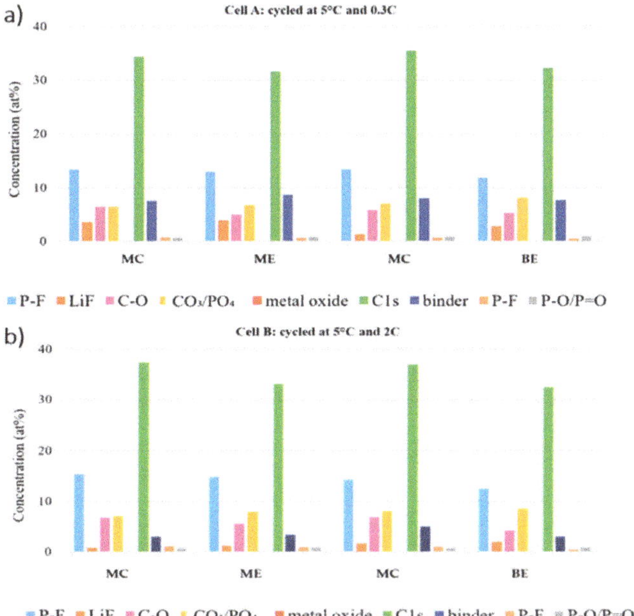

Figure 11. Atomic concentrations of cathode electrolyte interphase (CEI) components at the center and edge of the electrode tape in different position of the stack of (**a**) Cell A and (**b**) Cell B.

In O 1s and P 2p regions, the amount of oxygenated species (C–O and P–F) is lower at the edge of each stack, while the LFP feature stays more pronounced at all the edges. Thus, we can conclude that the formed CEI is thinner at the electrode edge in both cells. Moreover, the LFP feature (PO_4 and P–O/P = O) in O 1s and P 2p region is more pronounced in the beginning of the stack, which implies that the CEI at the middle of the stack is thicker.

Generally, for both cycled cells greater differences were found in electrodes of different stacks. However, the trend of the features in the F 1s, O 1s and P 2p regions suggest that the dissimilarities caused by varying the C rates are as marked as the differences between different locations in the cells.

3. Conclusions

The post-mortem analysis conducted on 16 Ah graphite//$LiFeO_4$ cells after formation and aged at 5 °C by cycling either at 0.3C (Cell A) or 2C (Cell B) revealed inhomogeneous ageing of the electrodes along the stack and from the center to the edge of the electrode tapes. Differences in the residual capacity values and $LiFePO_4/FePO_4$ phase distribution were detected. Although a clear trend could not be identified between C-rates, one of the possible reasons for the non-uniform ageing can be related to differences in the pressure experienced by the electrodes, which are higher in the middle of the stack and at the center of the electrode. Furthermore, the inhomogeneities are more marked for the cell cycled at low rate (Cell A) than for that cycled at 2C (Cell B). The cell cycled at 2C (Cell B) performed higher number of equivalent cycles compared to that cycled at 0.3C (Cell A). It is reasonable to assume that the higher current also results in more heat generated, therefore the temperature experienced by the cell could have been higher than the 5 °C of the environmental temperature, and has positively influenced the cell performance. From XPS analysis it appears that more electrolyte decomposition took place at the edge of the electrodes and at the outer part of the cell stack independently of the ageing conditions. Therefore, in these areas, more Li is consumed by SEI formation and side reactions, which contributed to the inhomogeneity in the cathode residual capacity values. In both aged cells the most evident signs of graphite electrode ageing are the increased thickness of the SEI and the increase of salt decomposition products (compared with fresh cells) and they are more pronounced in Cell A. Overall, the cathode is less affected by cycling and able to recover the initial capacity for both aged cells. On the other hand, the anode of the cell cycled at 2C (Cell B) displayed graphite exfoliation in the central parts of the electrode and the presence, in the same areas of Fe contamination.

4. Materials and Methods

The 16 Ah pouch cells were manufactured by CEA-LITEN. The stack contained 38 anodes 76 separators and 37 cathodes. The stacks with tabs as terminal were placed between two half shells based of aluminium and heat sealed before the electrolyte filling.

The ageing tests on the 16 Ah pouch cells were carried on in a custom-made climate chamber using a PEC battery tester with a maximum of 5 V and 50 A per channel.

Prior the opening, the cells have been discharged to Vmin (2.5 V) using a constant current (C/2) step followed by a constant voltage step (10 h or C/50). The cells were opened inside the glove box under Ar atmosphere (H_2O < 1 ppm, O_2 < 1 ppm) cutting the long side of the bag with ceramic scissor and scalpel avoiding to damage the stack or to induce a short circuit.

The adhesion strength of the coated electrodes on the current collector is measured via the 90° peel test carried out in ambient conditions at 20 mm/min crosshead speed on 20 × 90 mm (Width × Length) electrode strips cut from the ~6 × 9 cm samples to obtain a peeling strength value ($N \cdot m^{-1}$).

The morphological analysis of the cell components was conducted using a Scanning Electron Microscope (SEM) using an inert transfer chamber to protect the sample from the external atmosphere.

X-Ray Diffraction measurements were carried out directly on electrodes using a Bruker D8 Discover diffractometer equipped with a Cu Kα source (λ = 0.154 nm) where 2θ range was between 10 to 85°. Refinements of the cathode materials diffraction patterns were performed by the Rietveld Method using the FullProf program [20].

XPS measurements were conducted on an Axis UltraDLD (Kratos, U.K.) equipped with a monochromatic X-Ray source (Al Kα, filament current and voltage 15 mA and 15 kV respectively, with charge neutralizer to compensate for the charging of samples, and 20 eV pass energy for core spectra) The investigated sample area was ca. 700 μm × 300 μm. The XPS equipment is equipped with an antechamber, preventing atmospheric exposure when loading samples. Samples were not washed for XPS investigation. Sputter depth profiling (sputter times 60, 120 and 600 s) was done using an Argon ion gun (a coronene ion source with a filament voltage of 0.5 kV and an emission current of 8 mA, with a sputter crater diameter set to 1.1 mm and the incident angle between the sample surface and the ion gun beam at 45°. Measurements were done in the field of view 2 with a 110 μm aperture and a pass energy of 40 eV). The fitting of the spectra was done with the CasaXPS software (Version 2.3.16 PR 1.6, Casa Software Ltd., Teignmouth, UK). Core peaks were analyzed using a nonlinear Shirley-type background. The peak positions and areas were optimized by a weighted least-square fitting method using 70% Gaussian, 30% Lorentzian peak shapes. The intense C 1s peak at 284.5 eV was used as reference. For each XPS sample, at least 3 spots per sample were measured to test reproducibility.

The electrochemical tests were carried out using coin cells 2032 assembled in argon-filled glove box (H_2O < 0.1 ppm, O_2 < 0.1 ppm). Half-cells were assembled using the pristine electrodes (never assembled into a cell) and aged electrodes. The glass fibre separator (Whatman, GF/D) was soaked with the electrolyte 1 M $LiPF_6$ in EC:PC:DMC (1:1:3 vol.) + 2 wt. % VC. One side of the coated electrode was removed to eliminate artefacts at low current density [21]. In the argon-filled glove box, the cathode layer, made with PVDF binder, was removed using a paper tissue wet with *N*-methyl-2-pyrrolidone (NMP) solvent while the anode layer was removed using a sand paper. The removal of the coating from one side of the electrode resulted in heavy damage of the anode. Afterwards, small discs (area = 1.13 cm^2) were punched out of the electrode sheet and dried under vacuum in order to remove any residues of solvent and electrolyte. Galvanostatic cycling was carried out at a constant temperature of 20 ± 0.1 °C (Binder KB 400) using a battery tester (MACCOR 4300) following the test protocols described in Table S2 in Supplementary Materials. A C-rate of 1 C corresponds to 2.3 Ma cm^{-2}.

Supplementary Materials: The following are available online at http://www.mdpi.com/2313-0105/5/2/45/s1, Figure S1. Results of 16 Ah soft prismatic cells cycled at different temperatures: 5 °C (blue), 25 °C (green) and 45 °C (red) at 1 C discharge rate and 0.3 C (circle), and 2 C (square) in charge. Table S1. Summary of measurements performed in different stack's positions. Figure S2. SEM images of LFP electrode extracted from different parts of the stack of Cell A. Figure S3. SEM images of LFP samples. Figure S4. EDX spectra of aged cathodes. Figure S5. EDX analysis of the fresh graphite electrode. Figure S6. EDX spectra of negative electrode extracted from Cell B. Figure S7. SEM of separators from Fresh Cell, Cell A and Cell B. Figure S8. Voltage profile of the first cycle of a pristine cathode. Figure S9. Comparison of first de-lithiation of aged cathode in freshly re-assembled half-cells. Table S2. C-rate test protocol for half-cells with aged cathodes. Figure S10. Examples of Rietveld refinement of XRD patterns of aged cathodes. Figure S11. XRD patterns (focus on 001 peak at 26.8°) of pristine, fresh and aged anodes.

Author Contributions: Conceptualization, A.M., E.P., I.d.M., W.P., K.T. and N.E.; methodology, A.M., E.P., I.d.M., W.P. and K.T.; investigation, D.V.C., N.E., D.B.-B., J.-B.D., A.E.-B. and I.d.M.; writing—original draft preparation, A.M., N.E., E.P., I.d.M. and J.-B.D.; writing—review and editing, A.M., N.E., S.P., I.d.M., E.P., J.-B.D. and K.T.; supervision, A.M. and S.P.; funding acquisition, S.P., I.d.M. and E.P.

Funding: This research was funded by the European Union, under the grant agreement No. 653373 (SPICY—Silicon and polyanionic chemistries and architectures of Li-ion cell for high energy battery).

Conflicts of Interest: The authors declare no conflict of interest.

References

1. Mulder, G. White Paper on Test Methods for Improved Battery Cell Understanding. Available online: https://www.batterystandards.info/literature (accessed on 15 February 2019).
2. Kalhoff, J.; Eshetu, G.G.; Bresser, D.; Passerini, S. Safer Electrolytes for Lithium-Ion Batteries: State of the Art and Perspectives. *ChemSusChem* **2015**, *8*, 2154–2175. [CrossRef] [PubMed]
3. Xu, K. Nonaqueous Liquid Electrolytes for Lithium-Based Rechargeable Batteries. *Chem. Rev.* **2004**, *104*, 4303–4418. [CrossRef] [PubMed]
4. An, S.J.; Li, J.; Daniel, C.; Mohanty, D.; Nagpure, S.; Wood, D.L. The state of understanding of the lithium-ion-battery graphite solid electrolyte interphase (SEI) and its relationship to formation cycling. *Carbon* **2016**, *105*, 52–76. [CrossRef]
5. Waldmann, T.; Hogg, B.-I.; Wohlfahrt-Mehrens, M. Li plating as unwanted side reaction in commercial Li-ion cells—A review. *J. Power Sources* **2018**, *384*, 107–124. [CrossRef]
6. Petzl, M.; Kasper, M.; Danzer, M.A. Lithium plating in a commercial lithium-ion battery—A low-temperature ageing study. *J. Power Sources* **2015**, *275*, 799–807. [CrossRef]
7. Koltypin, M.; Aurbach, D.; Nazar, L.; Ellis, B. On the Stability of LiFePO4 Olivine Cathodes under Various Conditions (Electrolyte Solutions, Temperatures). *Electrochem. Solid-State Lett.* **2007**, *10*, A40–A44. [CrossRef]
8. Evertz, M.; Horsthemke, F.; Kasnatscheew, J.; Börner, M.; Winter, M.; Nowak, S. Unraveling transition metal dissolution of Li1.04Ni1/3Co1/3Mn1/3O2 (NCM 111) in lithium ion full cells by using the total reflection X-ray fluorescence technique. *J. Power Sources* **2016**, *329*, 364–371. [CrossRef]

9. Qi, Y.; Harris, S.J. In Situ Observation of Strains during Lithiation of a Graphite Electrode. *J. Electrochem. Soc.* **2010**, *157*, A741–A747. [CrossRef]

10. Williard, N.; Sood, B.; Osterman, M.; Pecht, M. Disassembly methodology for conducting failure analysis on lithium–ion batteries. *J. Mater. Sci. Mater. Electron.* **2011**, *22*, 1616–1630. [CrossRef]

11. Waldmann, T.; Iturrondobeitia, A.; Kasper, M.; Ghanbari, N.; Aguesse, F.; Bekaert, E.; Daniel, L.; Genies, S.; Gordon, I.J.; Löble, M.W.; et al. Review—Post-Mortem Analysis of Aged Lithium-Ion Batteries: Disassembly Methodology and Physico-Chemical Analysis Techniques. *J. Electrochem. Soc.* **2016**, *163*, A2149–A2164. [CrossRef]

12. Wang, S.; Wang, C.; Ji, X. Towards understanding the salt-intercalation exfoliation of graphite into graphene. *RSC Adv.* **2017**, *7*, 52252–52260. [CrossRef]

13. Ciosek Högström, K.; Hahlin, M.; Malmgren, S.; Gorgoi, M.; Rensmo, H.; Edström, K. Ageing of electrode/electrolyte interfaces in LiFePO$_4$/graphite cells cycled with and without PMS additive. *J. Phys. Chem. C* **2014**, *118*, 12649–12660. [CrossRef]

14. Ciosek Högström, K.; Malmgren, S.; Hahlin, M.; Rensmo, H.; Thébault, F.; Johansson, P.; Edström, K. The Influence of PMS-Additive on the Electrode/Electrolyte Interfaces in LiFePO$_4$/Graphite Li-Ion Batteries. *J. Phys. Chem. C* **2013**, *117*, 23476–23486. [CrossRef]

15. Andersson, A.M.; Henningson, A.; Siegbahn, H.; Jansson, U.; Edström, K. Electrochemically lithiated graphite characterised by photoelectron spectroscopy. *J. Power Sources* **2003**, *119–121*, 522–527. [CrossRef]

16. Dedryvère, R.; Maccario, M.; Croguennec, L.; Le Cras, F.; Delmas, C.; Gonbeau, D. X-Ray Photoelectron Spectroscopy Investigations of Carbon-Coated Li$_x$FePO$_4$ Materials. *Chem. Mater.* **2008**, *20*, 7164–7170. [CrossRef]

17. Castro, L.; Dedryvère, R.; El Khalifi, M.; Lippens, P.E.; Bréger, J.; Tessier, C.; Gonbeau, D. The spin-polarized electronic structure of LiFePO$_4$ and FePO$_4$ evidenced by in-lab XPS. *J. Phys. Chem. C* **2010**, *114*, 17995–18000. [CrossRef]

18. Qian, Y.; Niehoff, P.; Börner, M.; Grützke, M.; Mönnighoff, X.; Behrends, P.; Nowak, S.; Winter, M.; Schappacher, F.M. Influence of electrolyte additives on the cathode electrolyte interphase (CEI) formation on LiNi1/3Mn1/3Co1/3O2 in half cells with Li metal counter electrode. *J. Power Sources* **2016**, *329*, 31–40. [CrossRef]

19. Zhuang, G.; Chen, Y.; Ross, P.N. The Reaction of Lithium with Dimethyl Carbonate and Diethyl Carbonate in Ultrahigh Vacuum Studied by X-ray Photoemission Spectroscopy. *Langmuir* **1999**, *15*, 1470–1479. [CrossRef]

20. Rodriguez-Carvajal, J. Recent advances in magnetic structure determination by neutron powder diffraction. Fullprof Program. *Physica B* **1993**, *192*, 55–69. [CrossRef]

21. Zhou, G.; Wang, Q.; Wang, S.; Ling, S.; Zheng, J.; Yu, X.; Li, H. A facile electrode preparation method for accurate electrochemical measurements of double-side-coated electrode from commercial Li-ion batteries. *J. Power Sources* **2018**, *384*, 172–177. [CrossRef]

Article

EIS Study on the Electrode-Separator Interface Lamination

Martin Frankenberger *, Madhav Singh, Alexander Dinter and Karl-Heinz Pettinger

Technology Center for Energy, University of Applied Sciences Landshut, Wiesenweg 1, D-94099 Ruhstorf, Germany; madhav@iitdalumni.com (M.S.); Alexander.Dinter@haw-landshut.de (A.D.); Karl-Heinz.Pettinger@haw-landshut.de (K.-H.P.)
* Correspondence: martin-frankenberger@mytum.de; Tel.: +49-8531-914044-0

Received: 30 September 2019; Accepted: 11 November 2019; Published: 17 November 2019

Abstract: This paper presents a comprehensive study of the influences of lamination at both electrode-separator interfaces of lithium-ion batteries consisting of $LiNi_{1/3}Mn_{1/3}Co_{1/3}O_2$ cathodes and graphite anodes. Typically, electrode-separator lamination shows a reduced capacity fade at fast-charging cycles. To study this behavior in detail, the anode and cathode were laminated separately to the separator and compared to the fully laminated and non-laminated state in single-cell format. The impedance of the cells was measured at different states of charge and during the cycling test up to 1500 fast-charging cycles. Lamination on the cathode interface clearly shows an initial decrease in the surface resistance with no correlation to aging effects along cycling, while lamination on both electrode-separator interfaces reduces the growth of the surface resistance along cycling. Lamination only on the anode-separator interface shows up to be sufficient to maintain the enhanced fast-charging capability for 1500 cycles, what we prove to arise from a significant reduction in growth of the solid electrolyte interface.

Keywords: lithium-ion battery; lamination; electrochemical impedance spectroscopy; fast-charging capability; lifetime

---

1. Introduction

Lithium-ion battery (LIB) technology has grown to a market leader in the field of rechargeable batteries in the last decades. LIBs are used as the energy source for portable devices and electric vehicles (EVs), as well as for stationary energy storage systems to ensure grid stability upon fluctuations from renewable energy sources. To improve the fast-charging capability as well as the travelling distance of the EVs, ongoing research mainly addresses the basic cell components like active materials [1,2], electrolyte [3], separator [4–7], and manufacturing steps. Different manufacturing techniques, such as ultra-thick electrodes [8–11], calendering process [12], controlled stack pressure [13,14], laser structuring [15–20] and lamination [21], have been applied to increase the power density, energy density, lifetime and for cost reduction of LIBs. Typically, the calendering process improves the contact situation between the active material particles [14], which leads to an increase of the rate capability as well. On the other hand, extensive calendering can break the active material particles and block the lithium-ions at the electrode-electrolyte interface which makes the fast-charging capability problematic [12,22].

The electrode-separator lamination technique is known for the simplification of the stacking process upon reducing the probability of stack component slipping in the anode-separator-cathode compound [23], as well as accelerating the manufacturing speed. Besides, it can also improve the fast-charging capability and reduce the capacity fade at high C-rates [21]. This leads to the assumption that the detailed mechanisms that drive the fast-charging capability upon electrode-separator lamination are not completely understood so far, which inspired this study.

A recent study of lithium metal anodes showed that surface treatments for smooth lithium surfaces can significantly suppress lithium dendrite growth by modifying the surface topography and local surface chemistry, therefore lowering the solid electrolyte interface (SEI) growth losses during cycling [24]. Similarly, it was shown that electrochemical polishing on alkali metal anodes, which yield ultra-smooth surfaces, provide ultra-thin SEI layers which possibly suppress dendrite growth along cycling [25].

In our previous study, cross-section images gained by scanning-electron-microscopy (SEM) clearly showed pore size reduction at both electrode-separator interfaces upon lamination [21]. Hence, the lithium-ion diffusion paths can be expected to be shortened and homogenized at both electrode-separator interfaces. As a result, the ion current density distribution along the active area of the full cell gets equalized according to the ion path length homogenization on the electrode-separator interfaces. Equalizing the current density distribution reduces the possibility to locally reach exceptional high ohmic overpotentials, high enough to undergo the lithium deposition potential and cause local lithium dendrite formation, especially at high charging rates. Similar effects were found by Müller, S. et al., who recently reported a clear correlation for graphite anode inhomogeneities at different scale lengths to possibly cause local overpotentials high enough to undergo the lithium deposition potential [26].

To overcome some of the mentioned problems, the electrode-separator lamination technique provides proper contacts of separator and electrode at micro level. Suppression of ridges and wrinkles as well as reducing the probability to form cracks during cycling, analogously holds the potential to prevent dendrite formation and SEI growth. Validity of this assumption can be reasonably studied using electrochemical impedance spectroscopy (EIS).

To appropriately study the contribution of each electrode within a full cell using EIS, typically three electrode geometries are designed and used, which unfortunately requires special modifications to any kind of full cell design [27,28]. In case of two-electrode impedance spectroscopy, the distinction of each electrode contribution is challenging, but possible [27]. The EIS signal responses for $LiNi_{1/3}Mn_{1/3}Co_{1/3}O_2$ (NMC) cathodes [22] or related structures [29] and graphite anodes [30,31] have yet been studied in detail. Using this information, the EIS signal of a graphite/NMC full cell could be considered well understood so far. Nevertheless, in literature the interpretation of the surface resistance semicircle reflects several further aspects. While the signal is often assigned to mainly depend on the SEI [32,33], there are also studies that show separate influences arising from the electrode-current collector contact situation [34]. Therefore, reliable studies of SEI effects emerging from the surface resistance signal have to deal with a proper separation of the signal contributions.

To overcome this difficulty, EIS measurements are either driven at exceptional low temperatures [32], which is problematic when aiming for aging studies purely based on cycling effects, or by introducing reference electrodes into the cell geometry [27,28].

In this study, we will show the ability of the lamination technique to provide this signal separation even in a two-electrode geometry, and use the correlation to prove the enhanced fast-charging capability of laminated cells to arise from a reduction in SEI growth.

2. Results and Discussion

2.1. Cathode-Separator Lamination Effects

Single cells consisting of a NMC cathode, separator and graphite anode were studied via EIS in non-laminated/partially laminated/fully laminated state along varied state of charge (SOC) after three formation cycles (C/10 rate) and two initial cycles at 1C rate. Figure 1 shows the impedance spectra gathered upon different lamination modes.

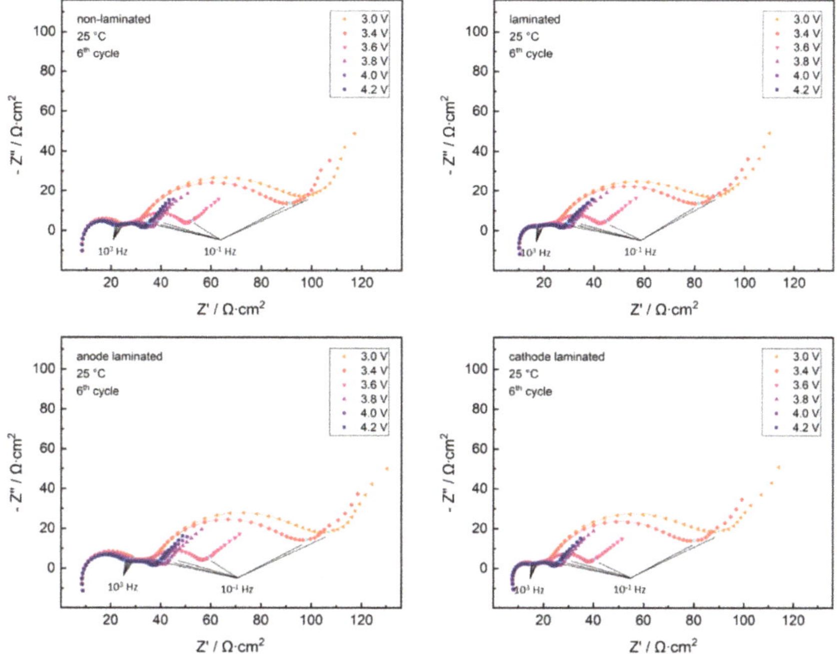

Figure 1. Impedance measurements of laminated/partially laminated/non-laminated cells: EIS (100 kHz–10 mHz) along charging step, 6th cycle; fitting curves indicated as solid lines; data points at 10^3 Hz and 10^{-1} Hz highlighted in pale blue.

The Nyquist plots uniquely show inductive behavior at the high-frequency regime, followed by three semicircles and the typical Warburg behavior at the lowest frequencies. For further studies, we characterize the three semicircles and the Warburg regime using the equivalent circuit model shown in Figure 2.

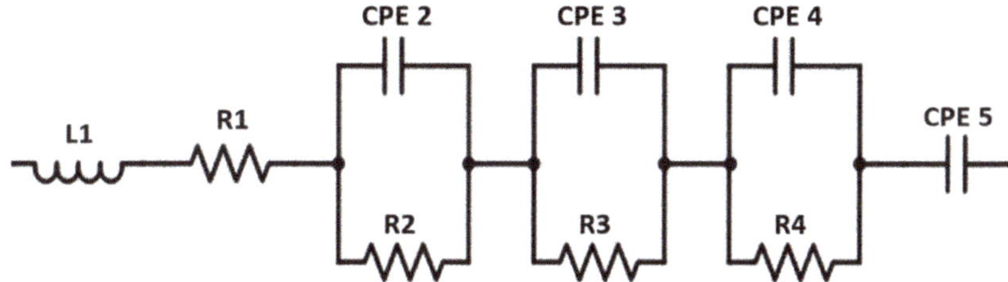

Figure 2. Equivalent circuit model used for EIS fit analysis.

This common equivalent circuit model is frequently used in literature [32] to describe NMC/graphite full cells. In the highest frequency regime, the signal is dominated by inductive phenomena arising from the impedance measurement environment. Both inductance element L1 and the electronic resistance R1, considered to mainly arise from the electrolyte resistance, contain side influences from the setup, such as cell tabs, welding points, impedance channel contacts and cables. The first semicircle represents the surface resistance phenomena. The second and third semicircle are driven by the charge-transfer reactions of graphite anode and NMC cathode, respectively. Low frequency phenomena arise from solid state diffusion characteristics, that can be split into closed and open Warburg regime [35], and therefore sometimes occurring as a consecutive series of varying slopes in the Nyquist presentation of the lowest frequency responses. As described above, signals from both charge-transfer reactions tend to overlap in this reference system as arising from the sum of the graphite anode and NMC cathode charge-transfer contributions [29–31,36]. For proper signal separation of these charge-transfer contributions, typically three-electrode cell designs are used involving reference electrodes. But in the case of NMC/graphite full cells, given knowledge on the separate EIS signal trends along SOC for the NMC cathode [37] or related composite materials [29], and graphite anode [30,31] can be used to identify the individual full cell signal contributions to the charge-transfer resistance. It is well accepted that in case of LIB full cells the sum of the signal contributions of the separate electrodes define the EIS response in a two-electrode geometry [36,38]. Jimenez Gordon et al. described the signal contribution of graphite anodes in LIBs [30,31]. While the surface resistance contribution, arising from porosity aspects and SEI characteristics, stays independent from the SOC of the cell [30], the charge-transfer signal of graphite decreases with increasing SOC, following a characteristic trend [31]. Liu et al. clarified the EIS trends of composite cathodes with respect to the amount and ratio of the conductive additive and PVDF binder, identifying separate mechanisms to drive the raise in impedance depending both on the ratio and on the total content of binder and conductive additive [29]. In the case of a ratio of 0.8:1 for acetylene black: PVDF, both for the total contents of 3.6% and 9%, they found a characteristic increase of the charge-transfer resistance with increasing depth of discharge (DOD). A drastic increase of the slope was found approaching 100% DOD [29].

The highlighted data point set at 10^3 Hz in Figure 1 lies close to the minimum between the first and second semicircle for all shown Nyquist datasets. As its relative position does not change with SOC for any cell, the surface resistance, that characterizes the first semicircle, can be considered independent from SOC. The second highlighted data point series at 10^{-1} Hz lies within the Warburg regime at high SOC for all cells, shifting closer to the minimum between the third semicircle and Warburg regime with decreasing SOC. This behavior indicates a shift to lower time constants of the cathode charge-transfer process with decreasing SOC, which is a known effect on the charge-transfer resistance of NMC based full cells [37].

Using the described correlations of the EIS signal response, the equivalent circuit analysis of the signals shown in Figure 1 is unambiguous. As for the structure of the chosen equivalent circuit, the open Warburg behavior was excluded from the data fit. Results are presented in Figures 3 and 4.

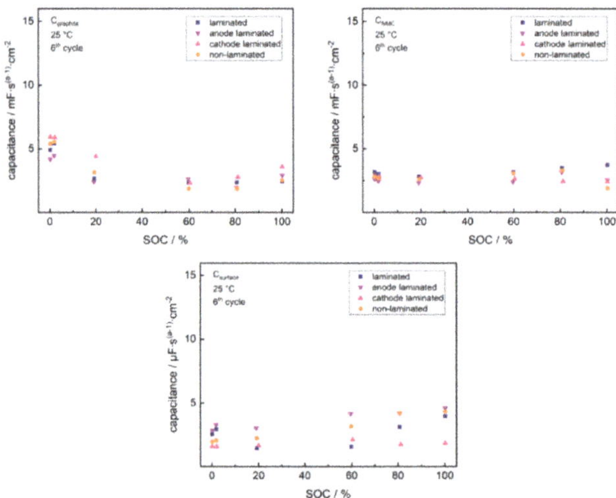

Figure 3. EIS capacitance fit parameters of laminated/partially laminated/non-laminated cells along charging step, sixth cycle; data normalized to geometric electrode area.

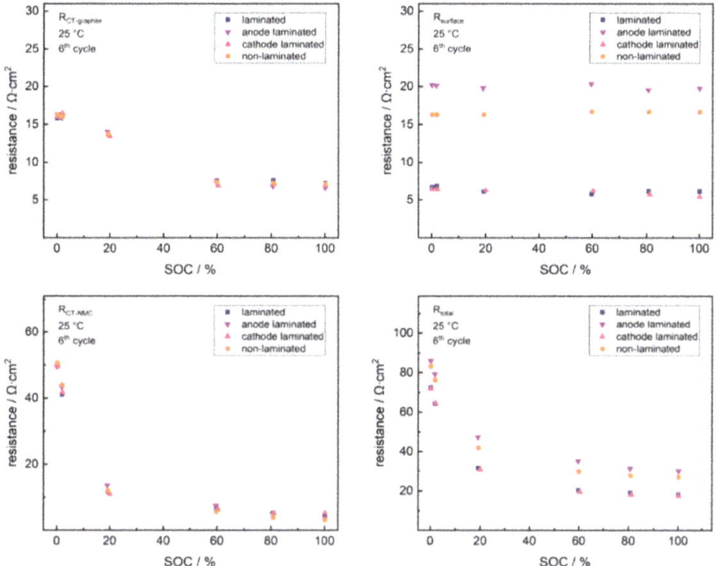

Figure 4. EIS resistance fit parameters of laminated/partially laminated/non-laminated cells along charging step, sixth cycle; data normalized to geometric electrode area.

As shown in Figure 3, the capacitance fit parameters for the graphite anode and NMC cathode lie around 3–4 mF·s $^{(a-1)}$·cm^{-2} with negligible dependence on SOC, which is in the range of typically reported values for EIS capacitance fit parameters, normalized to the geometric electrode area, of graphite anodes and composite cathodes in non-aqueous electrolytes [39–41]. The capacitance fit parameters of the surface resistance semicircle show no significant correlation to the SOC and lie around 3–5 µF·s $^{(a-1)}$·cm^{-2} for all cells, which is in the range of double layer capacitances of non-aqueous electrolytes [39]. Further information can be extracted from the resistance fit parameters, shown in Figure 4.

For all cells, the charge-transfer resistance signal for the graphite anode decreases initially down to ~14 Ω·cm^2 at 20% SOC. For SOC higher than 60%, the charge-transfer signals lie at a constant lower plateau at around 7 Ω·cm^2. This trend of the graphite anode charge-transfer resistance was well-described by Jiménez Gordon, I. et al. [30,31]. As it can be seen, there is no significant difference between different lamination modes.

Analogously, a logarithmic drop of the NMC charge-transfer resistance along increasing SOC is found for all cells, with no correlation to any lamination mode. This logarithmic trend of the NMC charge-transfer signal is well-known [29,42].

The first significant difference between the lamination modes can be found in the trends of the surface resistance. Both non-laminated cell (~16 Ω·cm^2) and anode-laminated cell (~20 Ω·cm^2) show relatively high surface resistance signals, whereas cathode-laminated and fully laminated cells both drop to a surface resistance signal around 6 Ω·cm^2, especially at higher SOC. Zheng, H. et al. found a similar drop of the surface resistance purely upon calendering NMC-cathodes to different porosities, where also lowest resistances at around 5 Ω·cm^2 are reached [22]. This recognizable drop arises from porosity changes and contact optimization at the cathode side. As both lamination and calendering technique are applied previous to final cell assembly, this surface resistance drop can act as a normalization for the starting conditions of the full cell surface resistance. After minimizing all well-known NMC cathode influences on the surface resistance as in the cathode-laminated and fully laminated state previous to cycling studies, any further changes on the surface resistance upon cycling uniquely arise from changes in the SEI.

2.2. Anode-Separator Lamination Effects

Figure 5 presents the results of non-laminated/partially laminated/fully laminated cells in the cycling test, when charging at 5C (CCCV mode) and discharging at 1C (CC mode). Recent studies on electrode-separator lamination showed a recognizable reduction in capacity fading upon fast-charging cycles at fully laminated state [21].

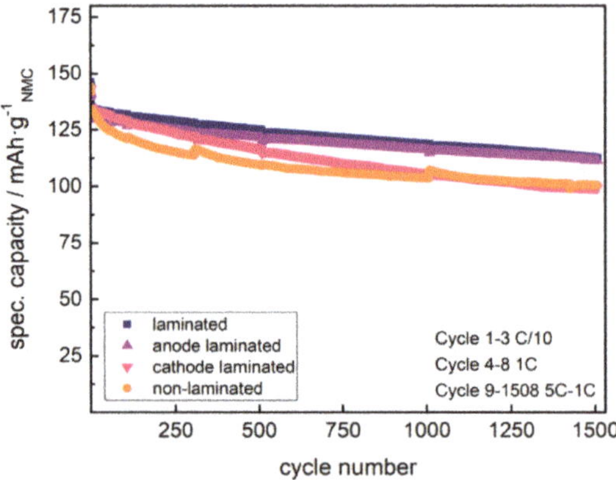

Figure 5. Discharge capacity data of laminated/partially laminated/non-laminated cells in fast-charging cycle test; charge at 5 C CCCV mode, discharge at 1 C CC mode.

Both the non-laminated cell and cathode-laminated cell reveal severe capacity fading, decreasing from 133.0 mAh·g^{-1} (100%) and 133.4 mAh·g^{-1} (100%) at the 9th cycle, down to 100.4 mAh·g^{-1} (75.5%) and 99.3 mAh·g^{-1} (74.6%) at the 1508th cycle, respectively. Additionally, both laminated and anode-laminated cells show a reduced capacity fading, decreasing from 133.2 mAh·g^{-1} (100%) and 129.6 mAh·g^{-1} (100%) at the 9th cycle, down to 112.5 mAh·g^{-1} (84.6%) and 111.4 mAh·g^{-1} (83.7%) at the 1508th cycle, respectively. Capacity fade trends are summarized in Table 1.

Table 1. Capacity fade of laminated/partially laminated/non-laminated cells in the fast-charging cycle test.

Cycle Number	Laminated	Anode-Laminated	Cathode-Laminated	Non-Laminated
9	133.2 mAh·g^{-1}	129.6 mAh·g^{-1}	133.4 mAh·g^{-1}	133.0 mAh·g^{-1}
1508	112.5 mAh·g^{-1}	111.4 mAh·g^{-1}	99.3 mAh·g^{-1}	100.4 mAh·g^{-1}

From the cycling test results, there arise two unique conclusions. First, minimization of the surface resistance via cathode lamination cannot ensure a permanent reduction in capacity fading upon fast-charging cycles, as is shown from the cathode-laminated cell. Second, yet the partial lamination on the anode interface is sufficient to generate the well-known reduction of capacity fade during fast-charging cycles, as shown from the discharge capacity trend of the anode-laminated cell in Figure 5. The data therefore show clearly that the fast-charging capability arises only from lamination at the anode interface, while the cathode interface does not affect the fast-charging aging mechanisms.

The cells studied at the fast-charging cycle test were also characterized with EIS along the cycling. Figure 6 shows the trends of the respective datasets in the Nyquist plots.

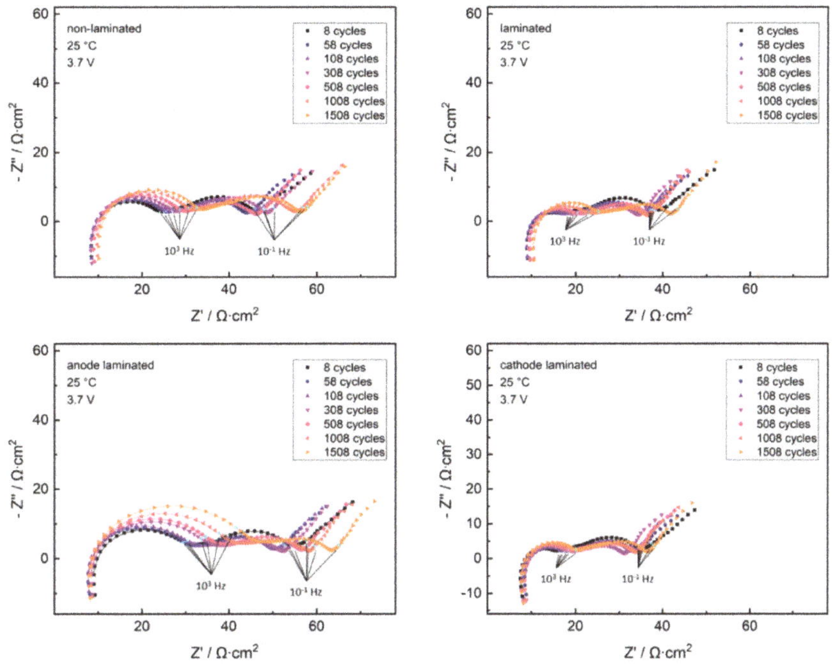

Figure 6. Impedance measurements of laminated/partially laminated/non-laminated cells: EIS (100 kHz–10 mHz) along fast-charging cycles; fitting curves indicated as solid lines; data points at 10^3 Hz and 10^{-1} Hz highlighted in pale blue.

Similar to Figure 1, the highlighted data point set at 10^3 Hz in Figure 6 lies close to the minimum between the first and second semicircle for all datasets, while the second highlighted dataset series at 10^{-1} Hz lies close to the minimum between the third semicircle and Warburg regime. Both benchmark frequency datasets do not change in relative position within the Nyquist datasets along cycling, indicating a negligible change of the time constant with cell aging at the chosen conditions. This correlates well to findings by Waag, W. et al. on NMC based full cells at moderate SOC [37].

To exclude side influences from cathode interface phenomena to the surface resistance signal, EIS aging studies focused on SEI effects require a minimization of the surface resistance starting condition. As discussed above, lamination at the cathode-separator interface minimizes the surface resistance previous to cycling influences. Initially, both the cathode-laminated and fully laminated cell have smaller surface resistance semicircles, while anode-laminated and non-laminated cell reveal enlarged surface resistance semicircles. All cells reveal, that the surface resistance semicircle increases with rising cycle number, while no clear trend for the charge-transfer semicircles arises along cycling for any cell. Further insights can be extracted by studying the semicircles in the equivalent circuit fit. Again, due to the structure of the chosen equivalent circuit, the open Warburg regime was excluded from the data fit. Figure 7 shows the trends of the resistances and capacitances calculated upon the fitting.

As both charge-transfer signals have drastic overlap, only the sum of the fitted charge-transfer resistances can be studied. No clear trend on the charge-transfer resistance signal along cycling is found for any lamination mode. Although all surface resistance signals differ in starting values, as indicated in the Nyquist plots, they all increase upon cycling. Both the capacitance fit parameters for the graphite anode and NMC cathode (3–4 mF·s $^{(a-1)}$·cm^{-2}) and the capacitance signals of the surface resistance semicircles (~3 μF·s $^{(a-1)}$·cm^{-2}) reveal no significant trend along cycling.

As discussed above, focusing on the SEI trends arising from the lamination modes requires minimization of the cathode influences on the surface resistance. Therefore, only the surface resistance signals of the cathode-laminated and fully laminated cell deliver unpersuaded information on the SEI changes.

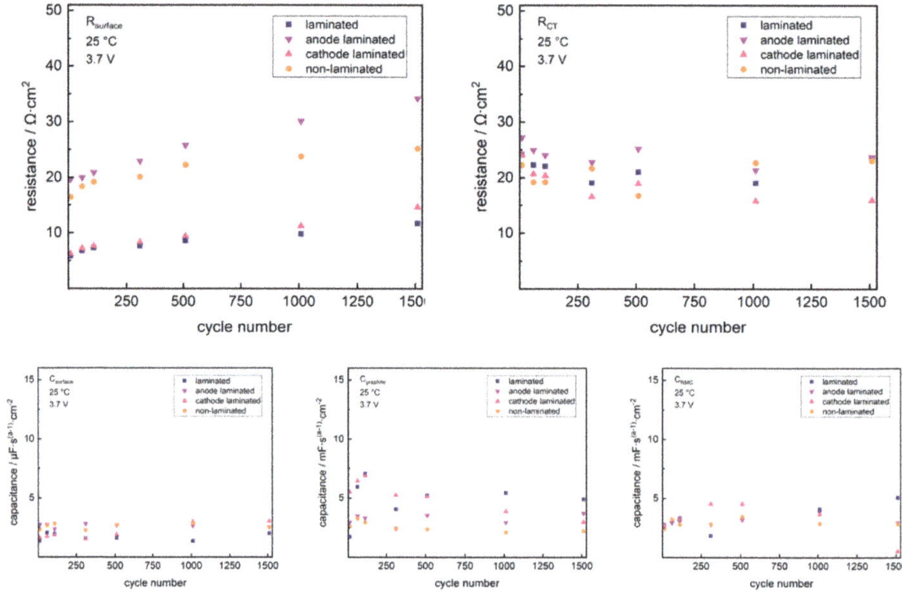

Figure 7. EIS resistance fit parameters of laminated and non-laminated single cells along fast-charging cycles; data normalized to geometric electrode area.

As shown in Figure 8, both cathode-laminated and fully laminated cells have a minimized surface resistance of 6.2 $\Omega \cdot cm^2$ and 5.8 $\Omega \cdot cm^2$ at the eigth cycle. The cathode-laminated cell increases in surface resistance up to 14.6 $\Omega \cdot cm^2$ after 1508 cycles, while the laminated cell dampens the surface resistance growth to 11.7 $\Omega \cdot cm^2$ after 1508 cycles. This trend in surface resistance clearly proves a reduction in SEI growth, which is specifically due to the lamination at the anode-separator interface.

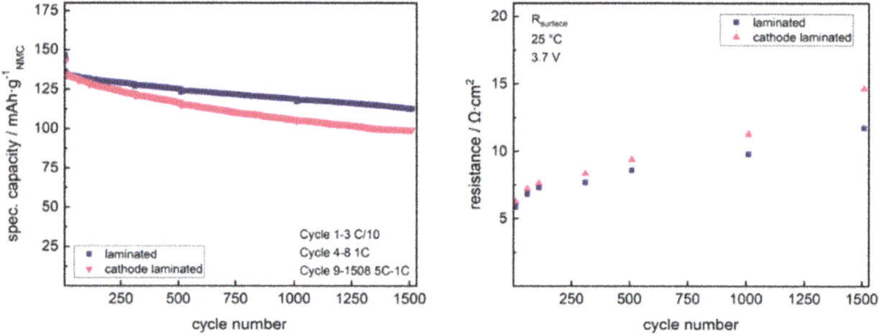

Figure 8. Decrease of discharge capacity vs. increase of surface resistance of laminated and cathode-laminated single cells along fast-charging cycles.

3. Materials and Methods

3.1. Electrode Preparation

For preparation of anode slurries, MAGE3 graphite (HITACHI CHEMICAL, Sakuragawa, Japan), Solef® 5130 polyvinylidene difluoride (PVDF, SOLVAY, Milan, Italy), Super C65 carbon (*IMERYS*, Bodio, Switzerland) and SFG6L graphite (IMERYS) were mixed in a ratio of 90/7/2/1 with N-methyl-pyrrolidone (NMP, Overlack, Mönchengladbach, Germany) in a planetary mixer (TX 2, INOUE, Isehara, Japan), while for cathodes, $LiNi_{1/3}Mn_{1/3}Co_{1/3}O_2$ (NMC, NM-3102 h, BASF TODA America, Battle Creek, USA), PVDF (SOLVAY), Super C65 carbon (IMERYS) and KS6L graphite (IMERYS) were mixed in a ratio of 93/3/3/1 with NMP. Anode and cathode slurries were coated on copper foil (15 µm, GELON LIB, Hong Kong, China) and aluminum foil (20 µm, GELON LIB), respectively, by single-side coating on a doctor-blade coater in a roll-to-roll process coating machine, including in-line drying in a two-step drying tunnel at the temperature range of 135–150 °C. The averaged active mass loadings of cathode and anode electrodes were ~8.4 mg·cm^{-2} (1.30 mAh·cm^{-2}) and ~4.2 mg·cm^{-2} (1.51 mAh·cm^{-2}), respectively. Cathodes and anodes were matched to have a capacity balancing factor of ~1:1.16 in all full cells.

3.2. Pouch Cell Preparation

Within the pouch cell, punched cathode, anode and separator (inorganic filled separator, 67% Al_2O_3 and 33% PVDF/HFP copolymer) sheets with the dimensions 5 × 8 cm^2, 5.4 × 8.4 cm^2 and 5.8 × 8.8 cm^2 were assembled. For the fully laminated state, stacks of cathode-separator-anode were laminated to form a single stack by using a lamination machine (BLE 282 D, MANZ Italy, former Arcotronics Italia, Bologna, Italy) at the roll speed of 1.4 m·min^{-1}, using a line force of 157 N·cm^{-1} in the temperature range of 100–120 °C. For separate electrode-separator lamination, stacks of cathode-separator-PE (polyethylene) carrier and anode-separator-PE carrier were laminated at identical parameters. Pre-assembled pouch cell stacks were dried under vacuum at 110 °C for 12 h. 1 M $LiPF_6$ in ethylene carbonate (EC): ethylmethylcarbonate (EMC) 3:7 *w/w* (Selectilyte LP57, BASF, Florham Park, USA) and vinylene carbonate (VC, Vinylene Carbonate E, BASF, Florham Park, USA), mixed in a ratio of 98/2, was used as electrolyte. The pre-assembled stacks were filled with 1000 µL electrolyte within an argon filled glovebox (MB20, H_2O and O_2 content <0.1 ppm, MBraun, Garching, Germany) and sealed under vacuum. Before starting the electrochemical characterization, wetting of all pores was ensured by keeping the cells at room temperature for 24 h previous to starting the formation cycles.

3.3. Electrochemical Characterization

Electrochemical characterization was done with a battery tester (CTS-LAB, BaSyTec, Asselfingen, Germany) and a potentiostat (PGSTAT204, METROHM, Filderstadt, Germany). Cells were cycled between 3.0 V and 4.2 V, using a CCCV protocol for charging (constant current protocol followed by constant voltage protocol) with a CV termination below 0.05C rate, and CC protocol for discharging. Formation was done by applying three cycles at 0.1C, using the nominal capacity of the NMC in each cell, calculated from the specific NMC capacity of 155 mAh·g^{-1}, given by the supplier. After formation, the discharge capacity of the third formation cycle was taken as the nominal capacity for C-rate calculation of all following steps.

For EIS analysis along SOC, cells were first discharged to 3.0 V at 0.2C rate after the fifth cycle, to then charge the cell up to each point of investigation at 0.2C rate. For EIS analysis along cycling, cells were charged to 3.7 V at 1C rate previous to EIS measurements at each specific cycle. Comparable temperature (25 °C) in EIS measurements was ensured by measuring in a cooled incubator (INCU-Line® IL 68 R, VWR, Ismaning, Germany). Cells rested for 2 h at OCV prior to each EIS measurement. EIS measurements were carried out in the frequency range of 100 kHz–10 mHz (potentiostatic mode) using an amplitude of 10 mV$_{rms}$. For EIS data fitting the Z-fit protocol, included in the BT-Lab software, was used (BT-Lab V1.55, BioLogic SAS via GAMEC, Illingen, Germany).

To ensure reproducibility of the study, for each cell assembly mode at least three cells were prepared and studied thoroughly. At each path, the performance of the respective cell with lowest initial impedance contributions and lowest capacity fade along 1500 fast-charging cycles is shown and discussed.

4. Conclusions

NMC/graphite full cells were studied in several lamination modes upon significant interface lamination effects revealed by EIS. Along variation of the SOC, both NMC cathode and graphite anode charge-transfer signals were found to stay independent from any lamination mode. The initial surface resistance gets minimized upon lamination at the cathode-separator interface previous to cycling influences.

Fast-charging cycling studies revealed a clear correlation of the reduction in capacity fade, to arise from lamination at the anode-separator interface. The surface resistance minimization via cathode-separator lamination was furthermore used to exclude cathode influences on the surface resistance signal evolution in cycling tests. Using this correlation, the cycling studies prove the fast-charging capability to arise from a reduction in SEI growth specifically arising from lamination at the anode-separator interface.

So, lamination at the cathode-separator interface is found to decrease the internal cell resistances, while lamination at the anode-separator interface reduces long term aging phenomena during fast-charging cycles.

Author Contributions: Conceptualization, M.F., M.S. and K.-H.P.; methodology, M.F. and M.S.; software, M.F.; validation, M.F. and A.D.; formal analysis, M.F. and A.D.; investigation, M.F.; data curation, M.F. and A.D.; Writing—Original draft preparation, M.F.; Writing—Review and editing, M.F., M.S., A.D. and K.-H.P.; visualization, M.F.; supervision, M.S. and K.-H.P.; project administration, M.F.; funding acquisition, K.-H.P.

Funding: This research was funded by BMWi (Federal Ministry for Economic Affairs and Energy, Germany), grant number 03ET6103C.

References

1. Geng, X.; Guo, H.; Wang, C.; Cheng, M.; Li, Y.; Zhang, H.; Huo, H. Surface modification of $Li_{1.20}Mn_{0.54}Ni_{0.13}Co_{0.13}O_2$ cathode materials with SmF3 and the improved electrochemical properties. *J. Mater. Sci. Mater. Electron.* **2018**, *29*, 19207–19218. [CrossRef]
2. Lee, M.-J.; Lho, E.; Oh, P.; Son, Y.; Cho, J. Simultaneous surface modification method for $0.4Li_2MnO_3$-$0.6LiNi_{1/3}Co_{1/3}Mn_{1/3}O_2$ cathode material for lithium ion batteries: Acid treatment and LiCoPO4 coating. *Nano Res.* **2017**, *10*, 4210–4220. [CrossRef]
3. Ming, J.; Li, M.; Kumar, P.; Lu, A.-Y.; Wahyudi, W.; Li, L.-J. Redox Species-Based Electrolytes for Advanced Rechargeable Lithium Ion Batteries. *ACS Energy Lett.* **2016**, *1*, 529–534. [CrossRef]
4. Fukutsuka, T.; Koyamada, K.; Maruyama, S.; Miyazaki, K.; Abe, T. Ion Transport in Organic Electrolyte Solution through the Pore Channels of Anodic Nanoporous Alumina Membranes. *Electrochim. Acta* **2016**, *199*, 380–387. [CrossRef]
5. Kannan, D.R.R.; Terala, P.K.; Moss, P.L.; Weatherspoon, M.H. Analysis of the Separator Thickness and Porosity on the Performance of Lithium-Ion Batteries. *Int. J. Electrochem.* **2018**, *2018*, 1–7. [CrossRef]
6. Xu, Q.; Kong, Q.; Liu, Z.; Zhang, J.; Wang, X.; Liu, R.; Yue, L.; Cui, G. Polydopamine-coated cellulose microfibrillated membrane as high performance lithium-ion battery separator. *RSC Adv.* **2014**, *4*, 7845–7850. [CrossRef]
7. Zhang, J.; Yue, L.; Kong, Q.; Liu, Z.; Zhou, X.; Zhang, C.; Xu, Q.; Zhang, B.; Ding, G.; Qin, B.; et al. Sustainable, heat-resistant and flame-retardant cellulose-based composite separator for high-performance lithium ion battery. *Sci. Rep.* **2014**, *4*, 3935. [CrossRef]

8. Danner, T.; Singh, M.; Hein, S.; Kaiser, J.; Hahn, H.; Latz, A. Thick electrodes for Li-ion batteries: A model based analysis. *J. Power Sources* **2016**, *334*, 191–201. [CrossRef]

9. Singh, M.; Kaiser, J.; Hahn, H. Thick Electrodes for High Energy Lithium Ion Batteries. *J. Electrochem. Soc.* **2015**, *162*, A1196–A1201. [CrossRef]

10. Singh, M.; Kaiser, J.; Hahn, H. A systematic study of thick electrodes for high energy lithium ion batteries. *J. Electroanal. Chem.* **2016**, *782*, 245–249. [CrossRef]

11. Singh, M.; Kaiser, J.; Hahn, H. Effect of Porosity on the Thick Electrodes for High Energy Density Lithium Ion Batteries for Stationary Applications. *Batteries* **2016**, *2*, 35. [CrossRef]

12. Kang, H.; Lim, C.; Li, T.; Fu, Y.; Yan, B.; Houston, N.; De Andrade, V.; De Carlo, F.; Zhu, L. Geometric and Electrochemical Characteristics of $LiNi_{1/3}Mn_{1/3}Co_{1/3}O_2$ Electrode with Different Calendering Conditions. *Electrochim. Acta* **2017**, *232*, 431–438. [CrossRef]

13. Barai, A.; Guo, Y.; McGordon, A.; Jennings, P. A study of the effects of external pressure on the electrical performance of a lithium-ion pouch cell. In Proceedings of the 2013 International Conference on Connected Vehicles and Expo (ICCVE), Las Vegas, NV, USA, 2–6 December 2013; IEEE: Piscataway, NJ, USA, 2013; pp. 295–299, ISBN 978-1-4799-2491-2.

14. Cannarella, J.; Arnold, C.B. Stress evolution and capacity fade in constrained lithium-ion pouch cells. *J. Power Sources* **2014**, *245*, 745–751. [CrossRef]

15. Habedank, J.B.; Kraft, L.; Rheinfeld, A.; Krezdorn, C.; Jossen, A.; Zaeh, M.F. Increasing the Discharge Rate Capability of Lithium-Ion Cells with Laser-Structured Graphite Anodes: Modeling and Simulation. *J. Electrochem. Soc.* **2018**, *165*, A1563–A1573. [CrossRef]

16. Kim, J.S.; Pfleging, W.; Kohler, R.; Seifert, H.J.; Kim, T.Y.; Byun, D.; Choi, W.; Lee, J.K.; Jung, H.-G.; Jung, H.-G.; et al. Three-dimensional silicon/carbon core–shell electrode as an anode material for lithium-ion batteries. *J. Power Sources* **2015**, *279*, 13–20. [CrossRef]

17. Lim, D.G.; Chung, D.-W.; Kohler, R.; Pröll, J.; Scherr, C.; Pfleging, W.; Garcia, R.E. Designing 3D Conical-Shaped Lithium-Ion Microelectrodes. *J. Electrochem. Soc.* **2014**, *161*, A302–A307. [CrossRef]

18. Mangang, M.; Seifert, H.; Pfleging, W. Influence of laser pulse duration on the electrochemical performance of laser structured LiFePO4 composite electrodes. *J. Power Sources* **2016**, *304*, 24–32. [CrossRef]

19. Proll, J.; Kim, H.; Piqué, A.; Seifert, H.; Pfleging, W. Laser-printing and femtosecond-laser structuring of $LiMn_2O_4$ composite cathodes for Li-ion microbatteries. *J. Power Sources* **2014**, *255*, 116–124. [CrossRef]

20. Smyrek, P.; Pröll, J.; Seifert, H.J.; Pfleging, W. Laser-Induced Breakdown Spectroscopy of Laser-Structured Li (NiMnCo) O_2 Electrodes for Lithium-Ion Batteries. *J. Electrochem. Soc.* **2016**, *163*, A19–A26. [CrossRef]

21. Frankenberger, M.; Singh, M.; Dinter, A.; Jankowksy, S.; Schmidt, A.; Pettinger, K.-H. Laminated Lithium Ion Batteries with improved fast charging capability. *J. Electroanal. Chem.* **2019**, *837*, 151–158. [CrossRef]

22. Zheng, H.; Tan, L.; Liu, G.; Song, X.; Battaglia, V.S. Calendering effects on the physical and electrochemical properties of $Li[Ni_{1/3}Mn_{1/3}Co_{1/3}]O_2$ cathode. *J. Power Sources* **2012**, *208*, 52–57. [CrossRef]

23. Pettinger, K.-H. Fertigungsprozesse von Lithium-Ionen-Zellen. In *Handbuch Lithium-Ionen-Batterien*; Korthauer, R., Ed.; Springer: Heidelberg/Berlin, Germany, 2013; ISBN 978-3-642-30652-5.

24. Meyerson, M.L.; Sheavly, J.K.; Dolocan, A.; Griffin, M.P.; Pandit, A.H.; Rodriguez, R.; Stephens, R.M.; Bout, D.A.V.; Heller, A.; Mullins, C.B.; et al. The effect of local lithium surface chemistry and topography on solid electrolyte interphase composition and dendrite nucleation. *J. Mater. Chem. A* **2019**, *7*, 14882–14894. [CrossRef]

25. Gu, Y.; Wang, W.-W.; Li, Y.-J.; Wu, Q.-H.; Tang, S.; Yan, J.-W.; Zheng, M.-S.; Wu, D.-Y.; Fan, C.-H.; Hu, W.-Q.; et al. Designable ultra-smooth ultra-thin solid-electrolyte interphases of three alkali metal anodes. *Nat. Commun.* **2018**, *9*, 1339. [CrossRef] [PubMed]

26. Muller, S.; Eller, J.; Ebner, M.; Burns, C.; Dahn, J.; Wood, V. Quantifying Inhomogeneity of Lithium Ion Battery Electrodes and Its Influence on Electrochemical Performance. *J. Electrochem. Soc.* **2018**, *165*, A339–A344. [CrossRef]

27. Song, J.; Lee, H.; Wang, Y.; Wan, C. Two- and three-electrode impedance spectroscopy of lithium-ion batteries. *J. Power Sources* **2002**, *111*, 255–267. [CrossRef]

28. Solchenbach, S.; Pritzl, D.; Kong, E.J.Y.; Landesfeind, J.; Gasteiger, H.A. A Gold Micro-Reference Electrode for Impedance and Potential Measurements in Lithium Ion Batteries. *J. Electrochem. Soc.* **2016**, *163*, A2265–A2272. [CrossRef]

29. Liu, G.; Zheng, H.; Kim, S.; Deng, Y.; Minor, A.M.; Song, X.; Battaglia, V.S. Effects of Various Conductive Additive and Polymeric Binder Contents on the Performance of a Lithium-Ion Composite Cathode. *J. Electrochem. Soc.* **2008**, *155*, A887–A892. [CrossRef]

30. Gordon, I.J.; Grugeon, S.; Takenouti, H.; Tribollet, B.; Armand, M.; Davoisne, C.; Débart, A.; Laruelle, S. Electrochemical Impedance Spectroscopy response study of a commercial graphite-based negative electrode for Li-ion batteries as function of the cell state of charge and ageing. *Electrochim. Acta* **2017**, *223*, 63–73. [CrossRef]

31. Gordon, I.J.; Grugeon, S.; Debart, A.; Pascaly, G.; Laruelle, S. Electrode contributions to the impedance of a high-energy density Li-ion cell designed for EV applications. *Solid State Ion.* **2013**, *237*, 50–55. [CrossRef]

32. Momma, T.; Matsunaga, M.; Mukoyama, D.; Osaka, T. Ac impedance analysis of lithium ion battery under temperature control. *J. Power Sources* **2012**, *216*, 304–307. [CrossRef]

33. Stroe, D.I.; Swierczynski, M.; Stan, A.I.; Knap, V.; Teodorescu, R.; Andreasen, S.J. Diagnosis of lithium-ion batteries state-of-health based on electrochemical impedance spectroscopy technique. In Proceedings of the 2014 IEEE Energy Conversion Congress and Exposition (ECCE), Pittsburgh, PA, USA, 14–18 September 2014; Institute of Electrical and Electronics Engineers, Ed.; IEEE: Piscataway, NJ, USA, 2014; pp. 4576–4582, ISBN 978-1-4799-5776-7.

34. Gaberšček, M.; Moskon, J.; Erjavec, B.; Dominko, R.; Jamnik, J. The Importance of Interphase Contacts in Li Ion Electrodes: The Meaning of the High-Frequency Impedance Arc. *Electrochem. Solid-State Lett.* **2008**, *11*, A170–A174. [CrossRef]

35. Jossen, A.; Weydanz, W. *Moderne Akkumulatoren*, 2nd ed.; CUVILLIER VERLAG: Göttingen, Germany, 2019; ISBN 978-3-7369-9945-9.

36. Dollé, M.; Orsini, F.; Gozdz, A.S.; Tarascon, J.-M. Development of Reliable Three-Electrode Impedance Measurements in Plastic Li-Ion Batteries. *J. Electrochem. Soc.* **2001**, *148*, A851–A857. [CrossRef]

37. Waag, W.; Käbitz, S.; Sauer, D.U. Experimental investigation of the lithium-ion battery impedance characteristic at various conditions and aging states and its influence on the application. *Appl. Energy* **2013**, *102*, 885–897. [CrossRef]

38. Levi, M.D.; Dargel, V.; Shilina, Y.; Aurbach, D.; Halalay, I.C. Impedance Spectra of Energy-Storage Electrodes Obtained with Commercial Three-Electrode Cells: Some Sources of Measurement Artefacts. *Electrochim. Acta* **2014**, *149*, 126–135. [CrossRef]

39. Pritzl, D.; Landesfeind, J.; Solchenbach, S.; Gasteiger, H.A. An Analysis Protocol for Three-Electrode Li-Ion Battery Impedance Spectra: Part II. Analysis of a Graphite Anode Cycled vs. LNMO. *J. Electrochem. Soc.* **2018**, *165*, A2145–A2153. [CrossRef]

40. Landesfeind, J.; Pritzl, D.; Gasteiger, H.A. An Analysis Protocol for Three-Electrode Li-Ion Battery Impedance Spectra: Part I. Analysis of a High-Voltage Positive Electrode. *J. Electrochem. Soc.* **2017**, *164*, A1773–A1783. [CrossRef]

41. Ovejas, V.J.; Cuadras, A. Impedance Characterization of an LCO-NMC/Graphite Cell: Ohmic Conduction, SEI Transport and Charge-Transfer Phenomenon. *Batteries* **2018**, *4*, 43. [CrossRef]

42. Huang, J.; Li, Z.; Zhang, J.; Song, S.; Lou, Z.; Wu, N. An Analytical Three-Scale Impedance Model for Porous Electrode with Agglomerates in Lithium-Ion Batteries. *J. Electrochem. Soc.* **2015**, *162*, A585–A595. [CrossRef]

batteries 

Article

Lithium-Ion Capacitor Safety Testing for Commercial Application

Omonayo Bolufawi [1,2], Annadanesh Shellikeri [1,2] and Jim P. Zheng [1,2,*]

[1] Department of Electrical and Computer Engineering, Florida A&M University-Florida State University, Tallahassee, FL 32310-6046, USA; omonayo1.bolufawi@famu.edu (O.B.); ahs07e@my.fsu.edu (A.S.)

[2] Aero-Propulsion, Mechatronics and Energy Centre, Florida State University, Tallahassee, FL 32310-6046, USA

* Correspondence: zheng@eng.fsu.edu; Tel.: +1-8504106464

Received: 3 November 2019; Accepted: 2 December 2019; Published: 7 December 2019

Abstract: The lithium-ion capacitor (LIC) is a recent innovation in the area of electrochemical energy storage that hybridizes lithium-ion battery anode material and an electrochemical double layer capacitor cathode material as its electrodes. The high power compared to batteries and higher energy compared to capacitors has made it a promising energy-storage device for powering hand-held and portable electronic systems/consumer electronics, hybrid electric vehicles, and electric vehicles. The swelling and gassing of the LIC when subjected to abuse conditions is still a critical issue concerning the safe application in power electronics and commercial devices. However, it is imperative to carry out a thorough investigation that characterizes the safe operation of LICs. We investigated and studied the safety of LIC for commercial applications, by conducting a comprehensive abuse tests on LIC 200 F pouch cells with voltage range from 3.8 V to 2.2 V manufactured by General Capacitors LLC. The abuse tests include overcharge, external short circuit, crush (flat metal plate and blunt indentation), nail penetration test, and external heat test.

Keywords: abuse test; lithium-ion capacitor; safety; temperature; thermal runaway

---

1. Introduction

The last decade has seen increasing use of lithium-ion capacitor (LIC) in various applications due to its high power and energy density. They are also gaining traction as a power source in electric vehicles (EVs), hybrid electric vehicles (HEVs) and plug-in hybrid electric vehicles (PHEVs) because of the power and energy density and the ability of the LIC to charge and discharge fast. High demand for EVs, HEVs and PHEVs have made it imperative to investigate their safety and performance under various abuse condition. Lithium-ion capacitors are hybrid electrochemical energy-storage systems which, combine chemical reactions: Faradaic at the anode where intercalation occurs and non-Faradaic at the cathode where only surface adsorption-desorption occurs. The structure is made of lithium-ion battery anode materials (hard carbon) and electrochemical double-layer capacitor cathode (activated carbon) materials [1–4]. The electrolyte is made up of lithium $LiFP_6$ ethylene carbonate (EC) and dimethyl carbonate (DMC) [5]. Unfortunately, one of the challenges faced by lithium-ion capacitors is the difficulty of increasing the energy and power density simultaneously with enhanced safety benefit. Recent work reported by [5] has shown that the lithium-ion capacitor can achieve both high energy and power density, and good longevity. An approach to further improve the performance of the lithium-ion capacitor has been demonstrated, [6] that uses the graphitic porous carbon (GPC) and high-purity vein graphite (PVG) prepared from Sri Lanka graphite ore by KOH activation, and high-temperature purification. An electrochemical performance with a maximum energy density of 86 Whkg^{-1} at 150 Wkg^{-1}, and 48 Whkg^{-1} at a high-power density of 7.4 kWkg^{-1} was achieved at a relatively low cost. Another study [7] demonstrated a graphene-based LIC with reduced graphene oxide-carbon nanotube (rGO-CNT) film as capacitor-type electrode and pre-lithiated rGO-CNT film as battery-type electrode based on electrostatic spray deposition. Their approach delivered 114.5 Whkg^{-1} energy densities and maximum power density of 2569 Wkg^{-1}. However, as the electrochemical performance of lithium-ion capacitors increases, the safety risk associated with their use increases as well [8] which generates critical concern of which parameter (performance or safety) should be compromised for the other. This concern can be address during design state for specific use case and the safety could be accounted for by using a more chemical and thermally stable materials. An embedded cell-monitoring system is a plausible way of improving the safety, but the electrochemical performance will be sacrificed due to additional weight. There is also a critical need to improve the operating temperature of the lithium-ion capacitor in order to withstand cases of abuse conditions. Temperature is one of the critical parameters that affect the stability of lithium ion capacitor [9–12] which influence the aging degradation and response to abuse conditions. However, most studies are limited to lithium-ion batteries and the safety study analysis are relatively less common for lithium ion capacitor [13–16]. Simulations and experimental studies were conducted for both external and internal short circuits of lithium ion battery, where results indicated that external short circuit is worse for smaller size batteries due to their higher internal resistance. In the internal short-circuit test, there is a higher chance of failure found due to larger battery capacity with a rise in temperature and voltage drop. The simulation model shows the capability of estimating the start time of thermal runaway. The internal temperature and structural degradation of the lithium-ion cell was investigated [17] under a nail penetration test with different penetration position for lithium ion cells, where they found out that internal temperature is higher than the surface temperature. The variation between the inner temperature and surface temperature of a supercapacitor was investigated [10] with three dimensional symmetric thermal model based on the heating rate measured during cycling [9] The modeling of increased temperature effect was studied in relation to the charge–discharge current of the supercapacitor [18] which shows that the temperature response is dependent on the current applied. The effect of thermal charging on the supercapacitor by inducing an external heat condition was reported in [19]. The mechanical integrity of lithium ion battery was investigated at cell level with a finite element model for cell compression between flat plates, where the compressive and punch indentation test are usually conducted on a stack layer because single layer with small thickness will produce less accurate result [20]. Figure 1 shows the LIC sample used for the abuse test, manufactured by General Capacitor LLC. The schematic of typical abuse conditions and responses is illustrated in Figure 2 These abuse conditions can lead to a critical failure of lithium-ion capacitors by initiating a temperature increase wherein a chain exothermic reaction leads to swelling, gassing, thermal runaway and fire [21].

Figure 1. Lithium-ion capacitor (LIC) 200F pouch cell.

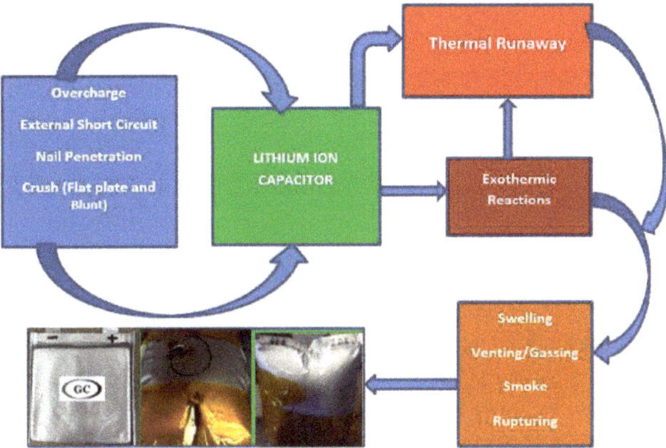

Figure 2. Overview of abuse conditions and response of a LIC.

Abuse tests are conducted to investigate the response of lithium-ion capacitors under conditions that exceed their normal operating mode and evaluate the thermal and chemical stability of the LIC under such conditions. The increase in temperature, as identified as major failure factor during abuse conditions is due to decomposition of the electrolyte through melting of the separator which in turn leads to exothermic reaction [22].

The previous study [23], investigated the performance of the same LIC manufactured by General Capacitor at wide temperature range electrolytes. We expanded further to study the safety performance of LIC pouch cells by investigating and monitoring their abuse response in terms of voltage, current, and temperature during the abuse testing.

2. Experimental Method

Lithium-ion capacitor pouch cells manufactured by General Capacitors were tested. The LIC cells were prepared using commercialized active materials for positive and negative electrodes. The positive and negative electrode consisted of activated carbon and graphite/hard carbon respectively. The cell specification is provided in Table 1.

Table 1. LIC specification.

Parameters	Specification
Dimension	Thickness = 4.5 mm, Heights = 58 mm, Width = 48 mm
Weight	16 g
Specific Power	6 kW/kg
Specific Energy	14 Wh/g
Voltage Range	2.2–3.8 V
Maximum Voltage	4.0 V
Capacitance	200 F

Abuse tests conducted include the following; external short circuit, overcharge, external heating, nail penetration, flat metal plate, and blunt indentation crush tests. The abuse tests were conducted in a transparent glass door safety box for personal protection and observation of physical real-time reaction. An Omega type K thermocouple (accuracy 1 °C) were attached to the cell's surface to data log the surface temperature. Likewise, the Arbin BT2000 test station was used to apply charge and discharge operation where necessary and to record the voltage and current during the abuse test. An external short circuit occurs as result of cell cathode and anode terminals being connected via a conducting path [24] which causes a high flow of current with rapid conversion of stored chemical energy to heat. The external short-circuit test was initiated using a low-resistance (<0.1 ohms) connecting wire. Overcharge is an electrical abuse condition that occurs when current is forced into the cell beyond its normal operating voltage limit [25] and the equation for energy balance during the overcharge condition is expressed in Equation (1):

$$Q_{overcharge} = i_{oc}^2 \frac{L_i}{\sigma_i} \tag{1}$$

where i_{oc}^2 is the overcharging current, L_i is the thickness of the cell materials, σ_i is the cell material conductivity.

Some of the reasons for overcharge includes power surge, faulty charger and battery management etc. [26–28]. Overcharge can be a severe abuse test since additional energy is added to the cell leading to chemical and thermal instability [29,30]. In the overcharge test, the LIC was charged at three times higher current beyond the voltage limit specified by the manufacturer. The nail penetration test is an important mechanical abuse test that illustrates the piercing of a LIC during a crash. A real-life occurrence was in the case of Tesla Model S battery pack that caught fire after a metallic object impact on the road [31]. Nail penetration induces an internal short circuit that can lead to thermal runaway after an exothermic reaction caused by heating [27,32]. Heat is generated by both the nail and the cell due to large current flowing through them such as ohmic heating [22,25] The LIC nail penetration test was conducted by a stainless-steel nail with full penetration at full state of charge. The position of the penetration was perpendicular to the electrode's surfaces of the LIC. The nail diameter was 0.3 cm, 7.8 cm in length. The nail joint was attached to a hydraulic press to drive the nail through the cell while ensuring the cell was correctly positioned in a holder to avoid movement during the penetration. The impact/crush test determines the ability of the LIC to withstand an impact or crush (flat metal plate and blunt indentation). The stress-strain compression relationship during a mechanical crush test is expressed in Equations (2) and (3) to describe the material behavior by the yield surface (Y) based on finite element modeling for detecting onset short circuit due to mechanical loads and deformation [33]:

$$Y = Y_c + \varepsilon_v , \, \delta_i > 0 \tag{2}$$

$$Y = Y_t, \, \delta_i < 0 \tag{3}$$

where Y_c is the compression stress cut of stress, ε_v is the volumetric strain, δ_i is the principal stress, Y_t is the tensile cut-off stress.

The LIC at full state of charge was crushed between two flat plates and indented by round metal of 2.47 cm diameter until the cell voltage drops to zero volts, or the cell is deformed. As the cell is indented, the indenter compresses the cathode, separator, and anode. External heat was applied on LIC to observe their thermal stability which could also be referred to as a thermal ramp test by placing in a thermostatically controlled oven (FO-19000 Series Forced Air Drying Oven). The oven microcontroller was used to regulate the internal temperature of the oven with an internal fan being used to circulate the oven air to ensure temperature uniformity around the cell. The LIC was placed inside the oven on an insulated surface to avoid conduction. The oven temperature was set to 300 °C until plausible thermal runaway occurrence.

3. Electrochemical–Thermal Reaction Mechanisms Governing Equation

The electrochemical–thermal reaction mechanism of lithium-ion cells are governed by the electrochemical and thermal equations expressed in Equations (3)–(6) due to exothermic reactions. Exothermic chemical reactions are related to thermal abuse mechanisms which occur inside a cell as the temperature rises. This may generate heat that accumulates inside the cell and accelerates the chemical reaction between the cell components, if the heat-generation rate exceeds the dissipation rate to the surroundings. External conditions for a temperature rise can be external heating, over-charging or over-discharging, nail penetration, and external short etc. In these cases, a thermal runaway can occur as a consequence with leakage, smoke, gas venting, flames etc., which leads to the destruction of the cells [34,35].

The current flow through the nail during the penetration can be determined from the ratio of the voltage drop and the nail resistance based on Ohm's law. The resistance of the nail can be estimated using Equation (4):

$$R = \frac{\sigma l}{A} \tag{4}$$

where σ is the electrical conductivity of the material, l; is the length and A; is the cross-sectional area of the nail.

The nail properties and the contact resistance affect the response of the cell and the internal shorting resistance created by the nail is expresses as:

$$R_s = R_{nai\ l} + R_{ct} \tag{5}$$

R_s is the internal shorting resistance, $R_{nai\ l}$ is the nail resistance and R_{ct} is the contact resistance

The electrical resistivity is the inverse of electrical conductivity. Therefore, according the Matthiessen's rule, [36] the temperature dependence of metal resistivity is expressed as shown in Equations (6) and (7):

$$\frac{1}{\sigma} = \frac{1}{\sigma_o}[1 + \beta(T - T_o)] \tag{6}$$

$$R = R_o[1 + \beta(T - T_o)] \tag{7}$$

where T is the metal temperature, β is the temperature coefficient of resistivity. σ_o and T_o are the electrical conductivity and the reference temperature.

The heat generation inside a lithium-ion cell is produced majorly from electrochemical operation, joule heating, entropy change and heat transferred to ambient conditions by convection [37–42]. The heat can be estimated based on the thermodynamic energy balance generally expressed in Equation (8):

$$m_{LIC}Cp\frac{dT_{LIC}}{dt} = I^2R + T_{LIC}\Delta S\frac{1}{nF} + Ah(T_{LIC} - T_{amb}) \tag{8}$$

where the first term is heat generation due to joule heating and the second term is the heat generation due to entropy changes and the third term is the heat transferred to ambient conditions by convection. The parameters in Equation (8) above are defined as: m_{LIC} is the *LIC* mass, Cp is the specific heat capacity, I is the current, R is the cell internal Resistance, T_{LIC} is the *LIC* temperature, ΔS is the entropy change, A is the surface area, h is the heat transfer coefficient, T_{amb} is the ambient temperature.

4. Result and Discussion

Figure 3 shows the temperature, voltage vs. time plot during the LIC external short-circuit test which indicates a sudden drop in the cell voltage because of excess current with a rise in temperature which leads to pressure on the ion flow and venting. The cell swells with an increase in temperature as well as an increase in the cell thickness due to internal rupturing and gassing. The possible solution for the cell gassing is to use electrolyte additives to enhance stability as studied [23]. A maximum surface temperature of 68 °C was reached due to the exothermic reaction caused by internal heat generation but does not result in thermal runaway.

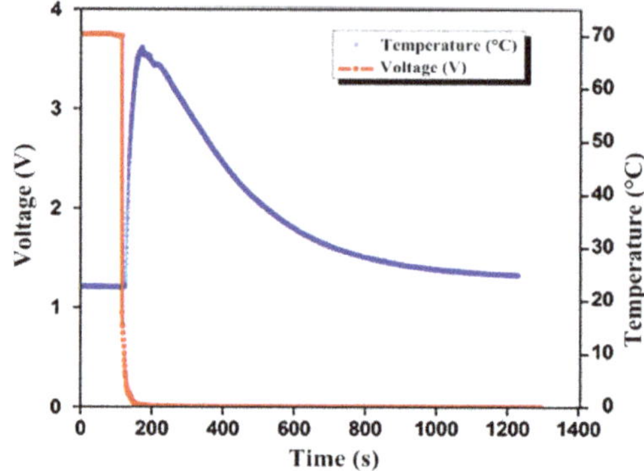

Figure 3. LIC external short-circuit test.

The overcharge abused response occurred at a condition beyond the normal charge state. The voltage, current and temperature were observed and plotted to describe their relationship during the overcharge condition. The LIC started overcharging when the rate of lithium insertion from cathode to anode increases due to increasing potential leading to lithium plating. Figure 4 shows the overcharged test result with a cell surface temperature plateau at 37 °C and a minimal weight loss observed because of minor venting of the electrolyte. The overcharge test led to swelling, gassing and temperature rise due to the exothermic reaction between the electrodes and electrolyte decomposition [43].

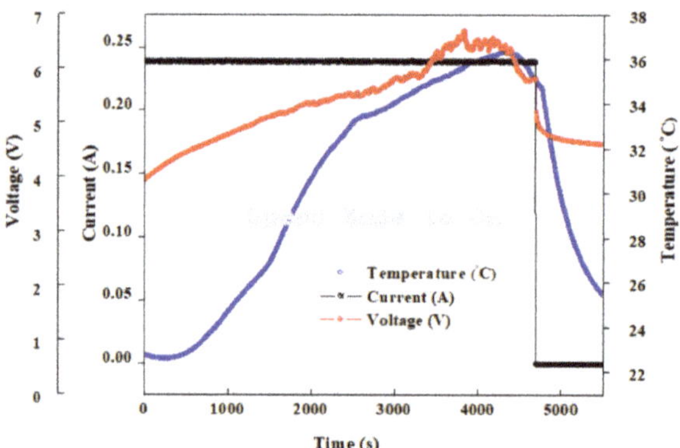

Figure 4. LIC overcharge test.

During cell overcharging, the lithium ions remaining in the cathode are removed and more lithium ions are intercalated in the anode. However, if the lithium insertion ability of the anode is small, lithium metal may be deposited on the anode [44]. The cell response, in this case, is not extremely catastrophic (resulting in only swelling of the cell due to internal gas formation) because of the lower charging current but a more catastrophic response (fire or explosion) could occur when overcharged at a higher current rate [18]. The heat generated by the applied current does not vary its behavior like the cell surface temperature but increases proportionally with current rates. Joule heating gives the relationship for the heat generated by the cell described in Equation (9).

$$Q = I^2 R \tag{9}$$

where Q is the Joule heating in J/s, I is the current, R is the cell internal resistance.

The overcharging current applied to the LIC is mainly responsible for the initial heat generation and not necessarily chemical reactions until the cell reaches a critical point beyond the cell range of operation. [43]. Figure 5 shows a heat generated over time estimated on experimental data and Equation (9) with constant current and varying overcharge voltage during the overcharge process with 1.6 J/s heat generated. The internal resistance was computed with Ohm's law by taking the ratio of the continuous overcharging voltage to the constant current thereby giving a varying resistance value. The profile indicated the amount of heat the cell generated due to rupturing and internal reaction; this reaction is believed to be a result of the decomposition of electrodes and electrolyte which affect the thermal stability of the lithiated electrode. Figure 6 shows the result of LIC nail penetration test with a sudden drop in the LIC voltage and a corresponding rapid increase in the cell surface temperature. The instantaneous voltage-drop during internal short circuit is due to loss of potential difference when the anode and cathode at different potentials are connected due to conductive stainless-steel nail material, resulting in large current flow. The start of nail penetration as illustrated in Figure 6, shows the trigger point for the instantaneous voltage drop (large current flow) followed by rise in temperature, which initiates the decomposition of the electrolyte [43], which are exothermic reactions. This generates more heat, which in turn triggers a thermal runaway, generating gaseous products.

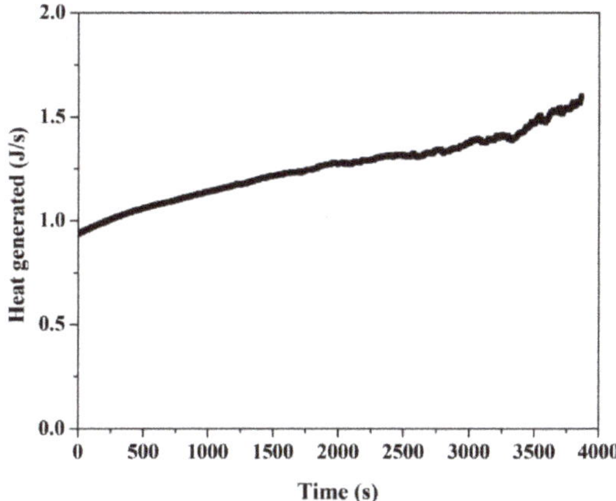

Figure 5. Heat output generation during overcharge test.

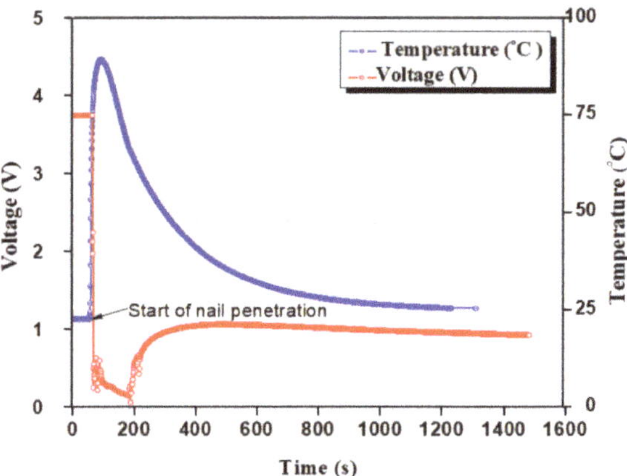

Figure 6. LIC nail-penetration test.

Upon nail removal, the cell temperature decreases, and the cell does not explode violently because the nail made a partial opening for the release of gaseous products. The cell temperature peaked at 90 °C with excessive gassing, smoke, and swelling. Figures 7 and 8 show the result for blunt indentation and flat metal plate crush test respectively. As the LIC was indented, due to mechanical impact, the indenter compresses the cathode, separator, and the anode thereby creating an internal short that led to a temperature rise to about 120 °C within a few seconds when crushed.

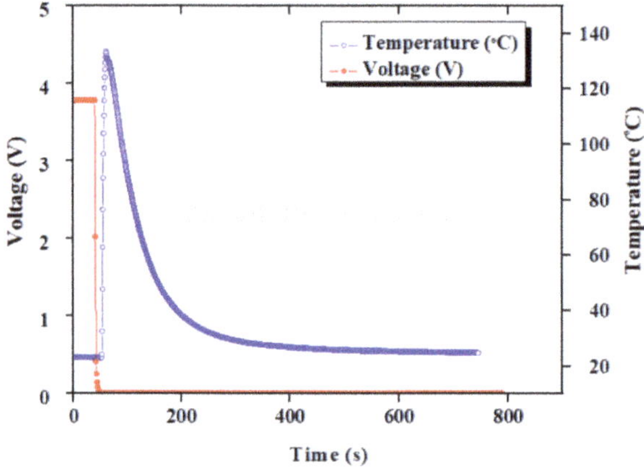

Figure 7. LIC blunt crush test.

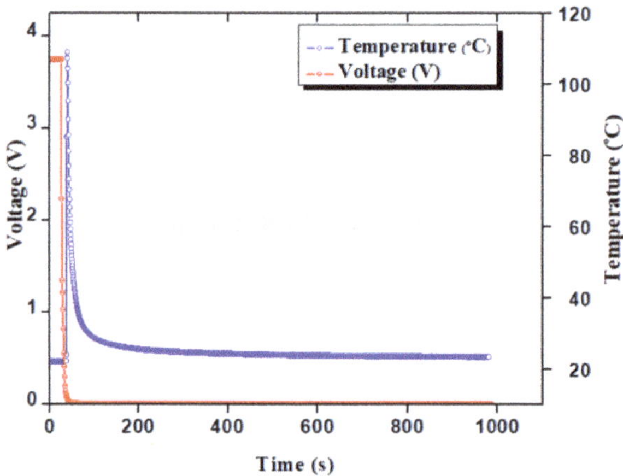

Figure 8. LIC flat-plate crush test.

Indenting the cell with a round metal piece created a more focused compression on the center of the cell making it like a nail penetration but with a wider contact surface area. The onset temperature for thermal runaway varied inversely with the degree of lithiation of the negative electrode [45] in which the thermal stability of the cell was reported to increase with increasing lithiation of the cathode.

The abuse tolerance of lithium-ion cells depends on the rate of generation and dissipation of heat from the cell. When a cell cannot transfer heat to its environment at a rate equal or higher than the rate of heat generation, the cell is subject to thermal runaway. Similarly, in the flat-plate crush test, a sudden increase in temperature was observed which was believed to be due to internal heat generation resulting from internal short-circuit of anode and cathode electrodes. There is the possibility of thermal diffusion between the flat metal plate and the cell in which the flat metal plate absorbs certain percent of the heat generated. However, the amount of heat absorbed by the flat metal plate was not accounted for in this study. Figure 9 shows the voltage and temperature with respect to time during the external heating test. A rapid surface temperature was observed with the cell gassing, venting and smoke but no fire. External heating of the LIC is at higher risk of fire since additional heat was applied to investigate its thermal stability tolerance. However, the cell did not result in a catastrophic fire but vented and smoked excessively. The maximum cell surface temperature reached 210 °C due to the exothermic reaction which is attributed to the decomposition of electrolyte and electrode thus resulting in thermal runaway. During the continuous heating, it took about 15 min for the cell surface temperature to reach 163 °C which was the point where the cell voltage terminals were disconnected due to physical damage as indicated at point A on Figure 9. The cell voltage drops to 0 V, as shown in Figure 9, resulted from the voltage terminal disconnection during the gas venting, which compromised the cell integrity. The gaseous products were a result of electrolyte decomposition, which was triggered by the rising cell temperature [46,47].

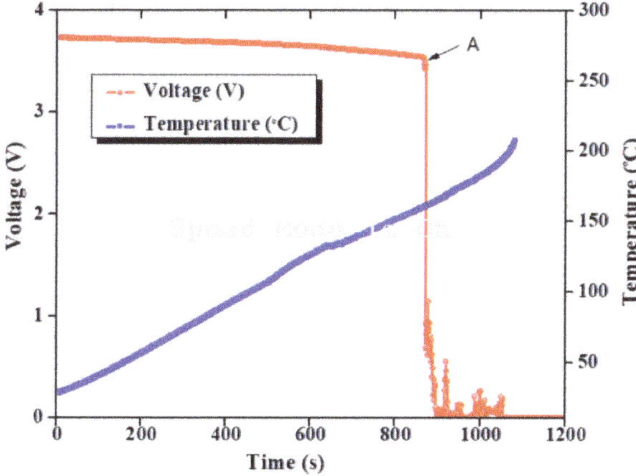

Figure 9. LIC external heat test.

The lithium-ion capacitor demonstrated a relatively safe, improved thermal and chemical stability because of their electrode materials composition which has minimal oxide content that could serve as catalyst for heat generation. The LIC is relatively safer than lithium-ion batteries when compared to a previous study [9] on lithium-ion batteries. From studies, the heat generation was reported to be due to reasons such as: the irreversible resistive heating, the reversible entropic heat, the heat change of chemical side reactions, and heat of mixing due to the generation and relaxation of concentration gradients [48]. Irreversible resistive heat loss occurred when current flows through internal resistance during charge and discharge which cause deviation of the cell potential from its equilibrium and the cell voltage difference is converted to heat. The heat by entropy change is relatively small compared to heat from resistive heating. The internal temperature of the LIC may be higher than the measured surface temperature because the heating begins inside the cell due to exothermic reaction before conducting outwards through the electrode layers towards the outer surface [17,49]. However, the present study accounts for the external surface temperature only. The nail penetration test demonstrated an instant internal shorting of the LIC cathode and anode electrodes, thus generating heat due to the exothermic reaction from decomposition of the electrode and electrolyte leading to a temperature increase. The temperature increased rapidly to about 90 °C in about 40 s before gradually decreasing because of the separator damage [47]. The temperature continued to decline until the cell was fully discharged with total energy dissipation. Due to the high specific surface area of the electrode, the heat generated was believed to have sufficiently distributed over the electrode surface which helps to reduce the temperature rise during abuse conditions. The LIC swelled, vented, gassed and smoked but there was no fire during overcharge as showed in Figure 10a.

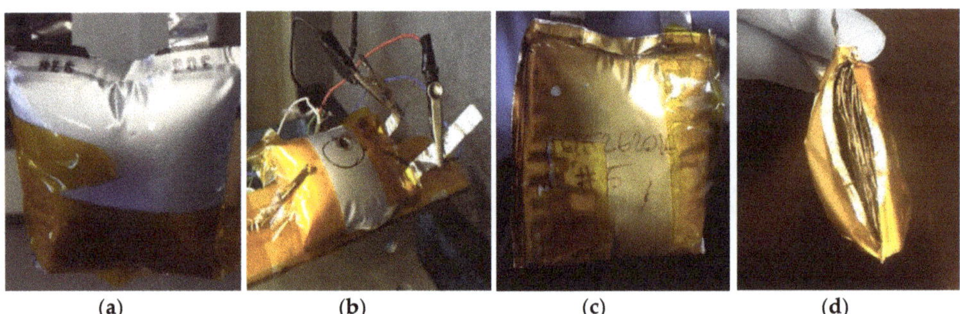

(a)   (b)   (c)   (d)

Figure 10. (**a**) A swollen overcharged cell, (**b**) penetrated cell, (**c**,**d**) externally heated cell.

These abuse responses portray LICs to have a promising safe operation in power electronics and commercial electronic device applications. As the cathode is completely de-lithiated during overcharge, the cell voltage overcharged above 6.5 V, the ionic conduction ceased, and the flow of current through the cell became ohmic resulting in joule heating. An overcharge abuse test can be very severe since additional energy is added to the cell. In the nail penetration result shown in Figure 6, a constant voltage and temperature were observed at the start of the experiment. However, the voltage dropped to zero and the surface temperature of the cells rose to about 100 °C, which is related to a high rate of current resulting in joule heating. The peak temperature shows that the intensity of the current during the nail penetration is believed to be high. The mechanism of nail penetration can be compared with the outgrown of dendrite in the cell that initiates internal shorting. The penetrated cell does not explode violently as the nail penetration made way for the release of gaseous products which reduce the internal pressure of the cell. In nail penetration, the nail makes direct contact between the anode and cathode as illustrated in Figure 10b. Contrary to the crush test, the flat metal plate was used to horizontally compress the cell which initiated internal short-circuiting of the electrodes, thereby giving room for the flow of current in the cell similar to the nail penetration with maximum surface temperature of 135 °C. During LIC external heating, the open-circuit voltage remained stable until about 150 °C measured on the cell surface as seen in Figure 9, then fell rapidly leading to excessive smoke and explosion as shown in Figure 10c,d. This effect was due to separator breakdown since the melting points of typical separator materials polyethylene and polypropylene are at about 130 °C [47]. The final voltage drops occurred at about 175 °C because of a delay in heat dissipation which occur when the cells generate heat more than it can dissipate. The state of charge is believed to influence cell behavior when abused and may vary according to the type of abuse conditions. The total energy release and emitted gas could not be accounted for during the explosion which also constitutes a fraction of the energy release.

5. Conclusions

The safety of lithium ion capacitor was investigated under different abuse tests with varying response observed for each test case. The high specific surface area of LIC electrode is believed to enhance the thermal and chemical stability with reduced thermal runaway and fire effect. Therefore, a lithium-ion capacitor provides safety benefits because it is less flammable making it a promising electrochemical energy storage device for safe applications in portable and consumer electronics.

Author Contributions: Experimental design, execution and original draft preparation, O.B.; Review, result and discussion, all authors.

Funding: This research was funded by NATIONAL SCIENCE FOUNDATION, grant number 0812121.

Conflicts of Interest: The authors declare no conflict of interest.

References

1. Cao, W.J.; Zheng, J.P. The effect of cathode and anode potentials on the cycling performance of Li-Ion capacitors. *J. Electrochem. Soc.* **2013**, *160*, A1572–A1576. [CrossRef]
2. Zheng, J.P. The limitations of energy density of battery/double-layer capacitor asymmetric cells. *J. Electrochem. Soc.* **2003**, *150*, A484–A492. [CrossRef]
3. Zheng, J.P. Theoretical energy density for electrochemical capacitors with intercalation electrodes. *J. Electrochem. Soc.* **2005**, *152*, A1864–A1869. [CrossRef]
4. Zheng, J.P. High energy density electrochemical capacitors without consumption of electrolyte. *J. Electrochem. Soc.* **2009**, *156*, A500–A505. [CrossRef]
5. Cao, W.J.; Zheng, J.P. Li-ion capacitors with carbon cathode and hard carbon/stabilized lithium metal powder anode electrodes. *J. Power Sour.* **2012**, *213*, 180–185. [CrossRef]
6. Gao, X.; Zhan, C.; Yu, X.; Liang, Q.; Lv, R.; Gai, G.; Shen, W.; Kang, F.; Huang, Z.H. A high performance lithium-ion capacitor with both electrodes prepared from Sri Lanka graphite ore. *Materials* **2017**, *10*, 414. [CrossRef]
7. Adelowo, E.; Baboukani, A.; Chen, C.; Wang, C. Electrostatically sprayed reduced graphene oxide-carbon nanotubes electrodes for lithium-ion capacitors. *J. Carbon Res.* **2018**, *4*, 31. [CrossRef]
8. Belov, D.; Yang, M.H. Failure mechanism of Li-ion battery at overcharge conditions. *J. Sol. State Electrochem.* **2008**, *12*, 885–894. [CrossRef]
9. Lee, D.H.; Kim, U.S.; Sim, C.B. Modelling of the thermal behaviour of an ultracapacior for a 42-V automotive electrical system. *J. Power Sour.* **2008**, *175*, 664–668. [CrossRef]
10. Gualous, H.; Louahlia-Gualous, H.; Gallay, R.; Miraoui, A. Supercapacitor thermal modelling and characterization in transient state for industrial applications. *IEEE Trans. Ind. Appl.* **2009**, *45*, 1035–1044. [CrossRef]
11. Monzer Al, S.; Gualous, H.; Van Mierlo, J.; Culcu, H. Thermal modeling and heat management of supercapacitor modules for vehicle applications. *J. Power Sour.* **2009**, *194*, 581–587.
12. Zubieta, L.; Bonert, R. Characterization of double-layer capacitors for power electronics applications. *IEEE Trans. Ind. Appl.* **2000**, *36*, 199–205. [CrossRef]
13. Hu, H.; Zhao, Z.B.; Zhang, R.; Bin, Y.Z.; Qiu, J.S. Polymer casting of ultralight graphene aerogels for the production of conductive nanocomposites with low filling Content. *J. Mater. Chem. A* **2014**, *2*, 3756–3760. [CrossRef]
14. Song, Z.; Sun, K. Adaptive backstepping sliding mode control with fuzzy monitoring strategy for a kind of mechanical system. *ISA Trans.* **2014**, *53*, 125–133. [CrossRef]
15. Guo, G.; Bo, L. Three-dimensional thermal finite element modeling of lithium-ion battery in thermal abuse application. *J. Power Sour.* **2010**, *195*, 2393–2398. [CrossRef]
16. Wang, K.; Li, L.; Zhang, T.; Liu, Z. Nitrogen-doped graphene for supercapacitor with long-term electrochemical stability. *Energy* **2014**, *70*, 612–617. [CrossRef]
17. Finegan, D.P.; Tjaden, B.; Heenan, T.M.; Jervis, R.; Di Michiel, M.; Rack, A.; Hinds, G.; Brett, D.J.; Shearing, P.R. Tracking internal temperature and structural dynamics during nail penetration of lithium-ion cells. *J. Electrochem. Soc.* **2017**, *164*, A3285–A3291. [CrossRef]
18. Wang, K.; Li, L.; Yin, H.; Zhang, T.; Wan, W. Thermal modelling analysis of spiral wound supercapacitor under constant-current cycling. *PLoS ONE* **2015**, *10*, e0138672. [CrossRef]
19. Al-zubaidi, A.; Ji, X.; Yu, J. Thermal charging of supercapacitors: A perspective. *Sustain. Energy Fuels* **2017**, *1*, 1457–1474. [CrossRef]
20. Kermani, G.; Sahraei, E. Characterization and modeling of the mechanical properties of lithium-ion batteries. *Energies* **2017**, *10*, 1730. [CrossRef]
21. Larsson, F.; Mellander, B. Abuse by external heating, overcharge and short circuiting of commercial lithium-ion battery cells. *J. Electrochem. Soc.* **2014**, *161*, A1611–A1617. [CrossRef]

22. Lee, C.W.; Venkatachalapathy, R.; Prakash, J. A Novel Flame-Retardant Additive for Lithium Batteries. *Electrochem. Sol. State Lett.* **2000**, *3*, 63–65. [CrossRef]
23. Cappetto, A.; Cao, W.J.; Luo, J.F.; Hagen, M.; Adams, D.; Shelikeri, A.; Xu, K.; Zheng, J.P. Performance of wide temperature range electrolytes for Li-Ion capacitor pouch cells. *J. Power Sour.* **2017**, *359*, 205–214. [CrossRef]
24. Arora, A.; Medora, N.K.; Livernois, T.; Swart, J. Safety of Lithium-Ion Batteries for Hybrid Electric Vehicles. In *Electric and Hybrid Vehicle*; Elsevier: Amsterdam, The Netherlands, 2010.
25. Spotnitz, R.; Franklin, J. Abuse Behavior of High-Power, Lithium-Ion Cells. *J. Power Sour.* **2003**, *113*, 81–100. [CrossRef]
26. Tobishima, S.; Yamaki, J. A consideration of Lithium Cell Safety. *J. Power Sour.* **1999**, *81–82*, 882–886. [CrossRef]
27. Biensan, P.; Simon, B.; Peres, J.P. On Safety of Lithium-Ion Cells. *J. Power Sour.* **1999**, *81–82*, 906–912. [CrossRef]
28. Saito, Y.; Takano, K.; Negishi, A. Thermal behaviors of lithium-ion cells during overcharge. *J. Power Sour.* **2001**, *97–98*, 693–696. [CrossRef]
29. Leising, R.A.; Palazzo, M.J.; Takeuchi, E.S.; Takeuchi, K.J. Abuse Testing of Lithium-Ion Batteries: Characterization of the Overcharge Reaction of LiCoO$_2$/Graphite Cells. *J. Electrochem. Soc.* **2001**, *148*, A838. [CrossRef]
30. Leising, R.A.; Palazzo, M.J.; Takeuchi, E.S.; Takeuchi, K.J. A study of the overcharge reaction of lithium-ion batteries. *J. Power Sour.* **2001**, *97–98*, 681–683. [CrossRef]
31. Abada, S.; Marlair, G.; Lecocq, A.; Petit, M.; Sauvant-Moynot, V.; Huet, F. Safety Focused Modeling of Lithium-Ion Batteries: A Review. *J. Power Sour.* **2016**, *306*, 178–192. [CrossRef]
32. Sahraei, E.; Meier, J.; Wierzbicki, T. Characterizing and modeling mechanical properties and onset of short circuit for three types of lithium-ion pouch cells. *J. Power Sour.* **2014**, *247*, 503–516. [CrossRef]
33. Zavalis, T.G.; Behm, M.; Lindbergh, G. Investigation of Short-Circuit Scenarios in a Lithium-Ion Battery Cell. *J. Electrochem. Soc.* **2012**, *159*, A848. [CrossRef]
34. Ouyang, D.; Chen, M.; Huang, Q.; Weng, J.; Wang, Z.; Wang, J. A review on the thermal hazards of the lithium-ion battery and the corresponding countermeasures. *Appl. Sci.* **2019**, *9*, 2483. [CrossRef]
35. Yamauchi, T.; Mizushima, K.; Satoh, Y.; Yamada, S. Development of a simulator for both property and safety of a lithium secondary battery. *J. Power Sour.* **2004**, *136*, 99–107. [CrossRef]
36. Melcher, A.; Ziebert, C.; Lei, B.; Zhao, W.; Luo, J.; Rohde, M.; Seifert, H.J. Modeling and Simulation of Thermal Runaway in Cylindrical 18650 Lithium-Ion Batteries. In *Meeting Abstracts the Electrochemical Society*; The Electrochemical Society: Pennington, NJ, USA, 2016; p. 425.
37. Matula, R.A. Electrical resistivity of copper, gold, palladium and silver. *J. Phys. Chem.* **1979**, *8*, 1147. [CrossRef]
38. Benger, R.; Wenzl, H.; Beck, H.; Jiang, M.; Ohms, D.; Schaedlich, G. 2009 Electrochemical thermal modeling of lithium-ion cells for use in HEV or EV application. *World Electr. Veh. J.* **2009**, *3*, 342. [CrossRef]
39. Jeon, D.H.; Baek, S.M. Thermal modeling of cylindrical lithium ion battery during discharge cycle. *Energy Convers. Manag.* **2011**, *52*, 2973–2981. [CrossRef]
40. Wang, F.; Li, M. Thermal performance analysis of the Lithium-ion Batteries. In Proceedings of the 2010 International Conference on Parallel and Distributed Computing, Applications and Technologies (PDCAT), Wuhan, China, 8–11 December 2010; pp. 483–486.
41. Soltani, M.; Ronsmans, J.; Kakihara, S.; Jaguemont, J.; Van den Bossche, P.; van Mierlo, J.; Omar, N. Hybrid battery/lithium-ion capacitor energy storage system for a pure electric bus for an urban transportation application. *Appl. Sci.* **2018**, *8*, 1176. [CrossRef]
42. Onda, K.; Ohshima, T.; Nakayama, M.; Fukuda, K.; Araki, T. Thermal behavior of small lithium-ion battery during rapid charge and discharge cycles. *J. Power Sour.* **2006**, *158*, 535–542. [CrossRef]
43. Ismail, N.H.F.; Toha, S.F.; Azubir, N.A.M.; Ishak, N.H.M.; Hassan, M.K.; Ibrahim, B.S.K. Simplified heat generation model for lithium ion battery used in electric vehicle. *IOP Conf. Ser. Mater. Sci. Eng.* **2013**, *53*, 012014. [CrossRef]
44. Aurbach, D.; Zinigrad, E.; Cohen, Y.; Teller, H. A short review of failure mechanism of lithium metal and lithiated graphite anodes in liquid electrolyte solutions. *Sol. State Ion.* **2002**, *148*, 405–416. [CrossRef]
45. Dahn, J.R.; Fuller, E.W.; Obrovac, M.; Von Sacken, U. Thermal stability of LixCoO$_2$, LixNiO$_2$ and λ-MnO$_2$ and consequences for the safety of Li-ion cells. *Sol. State Ion.* **1994**, *69*, 265–270. [CrossRef]

46. Chen, Y.; Evans, J.W. Thermal analysis of lithium-ion batteries. *J. Electrochem. Soc.* **1996**, *143*, 2708–2712. [CrossRef]
47. Kriston, A.; Pfrang, A.; Döring, H.; Fritsch, B.; Ruiz, V.; Adanouj, I.; Kosmidou, T.; Ungeheuer, J.; Boon-Brett, L. External short circuit performance of Graphite-LiNi$_{1/3}$Co$_{1/3}$Mn$_{1/3}$O$_2$ and Graphite-LiNi$_{0.8}$Co$_{0.15}$Al$_{0.05}$O$_2$ cells at different external. *J. Power Sour.* **2017**, *361*, 170–181. [CrossRef]
48. Orendorff, C.J. The Role of Separators in Lithium-Ion Cell Safety. *Electrochem. Soc. Interface* **2012**, *21*, 61–65. [CrossRef]
49. Jiang, J.R.D.J.; Dahn, J.R. ARC studies of the thermal stability of three different cathode materials: LiCoO$_2$; Li[Ni$_{0.1}$Co$_{0.8}$Mn$_{0.1}$]O$_2$; and LiFePO$_4$, in LiPF$_6$ and LiBoB EC/DEC electrolytes. *Electrochem. Commun.* **2004**, *6*, 39–43. [CrossRef]

Article

State-of-Charge Monitoring and Battery Diagnosis of NiCd Cells Using Impedance Spectroscopy

Peter Kurzweil [1,*] and Wolfgang Scheuerpflug [2]

[1] Electrochemistry Laboratory, University of Applied Sciences (OTH), Kaiser-Wilhelm-Ring 23, 92224 Amberg, Germany
[2] Diehl Aerospace GmbH, Donaustraße 120, 90451 Nürnberg, Germany; wolfgang.scheuerpflug@diehl.com
[*] Correspondence: p.kurzweil@oth-aw.de; Tel.: +49-9621-482-3317

Received: 14 December 2019; Accepted: 6 January 2020; Published: 9 January 2020

Abstract: With respect to aeronautical applications, the state-of-charge (SOC) and state-of-health (SOH) of rechargeable nickel–cadmium batteries was investigated with the help of the frequency-dependent reactance Im $\underline{Z}(\omega)$ and the pseudo-capacitance $C(\omega)$ in the frequency range between 1 kHz and 0.1 Hz. The method of SOC monitoring using impedance spectroscopy is evaluated with the example of 1.5-year long-term measurements of commercial devices. A linear correlation between voltage and capacitance is observed as long as overcharge and deep discharge are avoided. Pseudo-charge $Q(\omega)$ = $C(\omega)\cdot U$ at 1 Hz with respect to the rated capacity is proposed as a reliable SOH indicator for rapid measurements. The benefit of different evaluation methods and diagram types for impedance data is outlined.

Keywords: battery life testing; capacitance; state-of-charge determination; state-of-health; aging; impedance spectroscopy; pseudo-charge

---

1. Introduction

Emergency power supplies in aircrafts require high reliability. After parking for a longer period of time without electric supply, the state-of-charge (SOC) of airplane batteries drops by self-discharge. Scheduled take-offs may be delayed by this. According to the state-of-the-art, the complete procedure of capacity determination lasts several hours. As a costly precaution, freshly charged batteries must be held in stock. With respect to more extended maintenance intervals, a reliable method for fast battery diagnosis is required, which reflects at least the upper SOC range.

Based on preliminary work on SOC determination by impedance spectroscopy [1–7], we studied new and aged batteries. Since the frequency response depends on the cell chemistry, we focused on nickel–cadmium batteries [8–10] in this work. In the following, the significance of the imaginary part of impedance is evaluated with respect to aging time and state-of-charge.

1.1. State-of-Charge Indicators

Rated capacity [11] denotes the electric charge Q_N, which is stored by a new battery conditions. The actually available capacity $Q(t)$, however, is lower by the already consumed charge Q_1, and the capacity loss Q_L in the course of aging during the time t. Moreover, the capacity Q_0 of the fully charged battery depends on temperature and previous charge–discharge cycles at given C-rates.

$$Q(t) = Q_N - Q_1 - Q_L = \alpha \cdot Q_0 = \alpha \cdot \beta \cdot Q_N$$
$$SOC = Q/Q_0 = \alpha \tag{1}$$

α, the state-of-charge (SOC) [12], describes the ratio between the actually momentary available capacity $Q(t)$ and maximum total available capacity Q_0 at the previous full charge. $\alpha = 1$ (100% SOC) represents the full charge, and $\alpha = 0$ (0% SOC) is an empty battery. For SOC determination [13,14], voltage measurements were general practice since the 1930s. Since the mid-1970s, impedance spectroscopy [15], coulomb counting [16], bookkeeping methods [17,18], and look-up tables [19] came along, which were complemented by fuzzy logic, Kalman filters, learning algorithms and predictive methods [14], and the analysis of relaxation times [20] over the past decades. The C-rate is defined by the current–capacity ratio. The indication 1C says that the battery was fully charged or discharged within one hour. 5C requires 0.2 h (12 min).

The state-of-health (SOH) [12] considers the actually available residual capacity in aged batteries Q_0 with regard to the nominal rated capacity Q_N of the new battery.

$$\text{SOH} = Q_0/Q_N = \beta \tag{2}$$

1.2. Aging Phenomena

Aging starts immediately after the battery has left the fabrication facilities. The degradation rate depends on cycling stress, temperature, charging method, overcharge, and deep-discharge. Some aging phenomena are reversible, but most are irreversible. Calendar aging during storage with time happens in the battery at rest at any temperature and state-of-charge, independently of the power load. Cyclic aging depends on current (C-rate), temperature, depth-of-discharge, power demand, and the load profile. Repeated charge–discharge events at 0%–20% SOC and 80%–100% SOC are more harmful to the battery than continuous operation at medium SOC levels.

The memory effect [21] or lazy battery effect is a special type of aging phenomenon with rechargeable nickel–cadmium (NiCd) chemistry that causes the battery to hold less charge (Figure 1). The battery gradually loses its maximum energy capacity when it is repeatedly recharged after being only partially discharged [22]. The battery seems to remember the previous state of charge, and causes an undesired early voltage drop when charged again. Due to crystal formation at the anode, the stored energy becomes available only at a lower voltage than before. Unfortunately, the memory effect is exceedingly difficult to reproduce in model experiments. In modern NiCd batteries, the resulting loss of capacity is partly compensated by a discharge reserve. The memory effect can be repaired by a complete charge–discharge cycle, which recovers the original capacity (except for calendar and cycle aging).

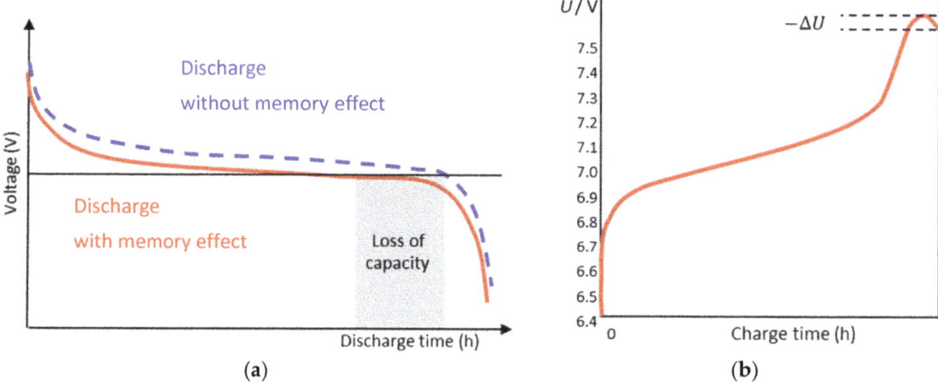

Figure 1. (**a**) Qualitative discharge characteristics of a nickel–cadmium battery suffering the memory effect. (**b**) End-of-charge determination by minus-delta-U cutoff. U = voltage in volts.

The voltage cutout is a widely used to prevent overcharge. The NiCd battery is continuously charged until the voltage drops by 0.01–0.02 V per cell although energy is supplied (Figure 1b).

1.3. Impedance Spectroscopy

The observed *ac* resistance of every electrochemical cell is caused by the electrolyte resistance and the kinetic inhibitions of the electrode processes, which act like non-ideal resistors and capacitances. With electrochemical impedance spectroscopy (EIS) [23], a periodically changing voltage (at constant cell voltage) or current excitation signal (at constant discharge current), respectively, is applied to the battery. The resulting phase shift ϕ between input signal and cell response is recorded frequency by frequency. The amplitude of the usually sinusoidal input signal must be small, so that the steady-state condition of the cell is not disturbed.

Commercial frequency response analyzers (FRA) deliver the frequency-dependent complex impedance $\underline{Z}(j\omega)$ or admittance $\underline{Y}(j\omega)$ in various mathematical formats, with respect to angular frequency $\omega = 2\pi f$, resistance $R = \mathrm{Re}\,\underline{Z}$ (real part of impedance), reactance $X = \mathrm{Im}\,\underline{Z}$ (imaginary part of impedance), modulus $Z = |\underline{Z}|$, and phase shift $\phi = \phi_U - \phi_I$ between *ac* voltage and *ac* current.

$$\underline{Z}(j\omega) = \frac{\underline{U}(j\omega)}{\underline{I}(j\omega)} = \mathrm{Re}\,\underline{Z}(j\omega) + j\,\mathrm{Im}\,\underline{Z}(j\omega) = |\underline{Z}(j\omega)| \cdot e^{j\varphi} = |\underline{Z}(j\omega)| \cdot [\cos\varphi + j\sin\varphi] \tag{3}$$

$$|\underline{Z}(j\omega)| = \frac{U_{eff}(j\omega)}{I_{eff}(j\omega)} = \sqrt{(\mathrm{Re}\,\underline{Z}(j\omega))^2 + (\mathrm{Im}\,\underline{Z}(j\omega))^2}$$

$$\varphi(j\omega) = \arctan\frac{\mathrm{Im}\,\underline{Z}(j\omega)}{\mathrm{Re}\,\underline{Z}(j\omega)} = \mathrm{sgn}[\mathrm{Im}\,\underline{Z}(j\omega)] \cdot \arccos\frac{\mathrm{Re}\,\underline{Z}(j\omega)}{|\underline{Z}(j\omega)|}$$

$$\underline{Y}(j\omega) = \mathrm{Re}\,\underline{Y}(j\omega) + j\,\mathrm{Im}\,\underline{Y}(j\omega) = \frac{1}{\underline{Z}(j\omega)} = \frac{\mathrm{Re}\,\underline{Z}(j\omega)}{|\underline{Z}(j\omega)|^2} + j\,\frac{-\mathrm{Im}\,\underline{Z}(j\omega)}{|\underline{Z}(j\omega)|^2}$$

Figure 2 illustrates these quantities in the complex plane and frequency domain.

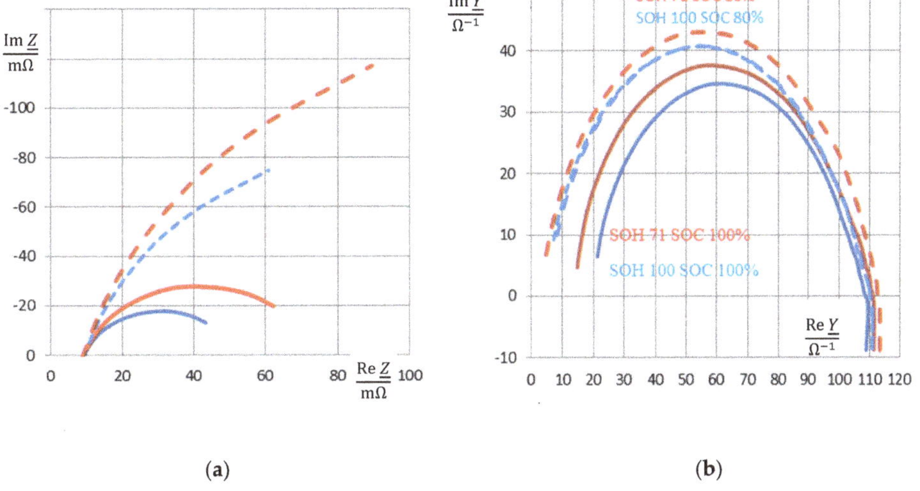

(**a**)   (**b**)

Figure 2. *Cont.*

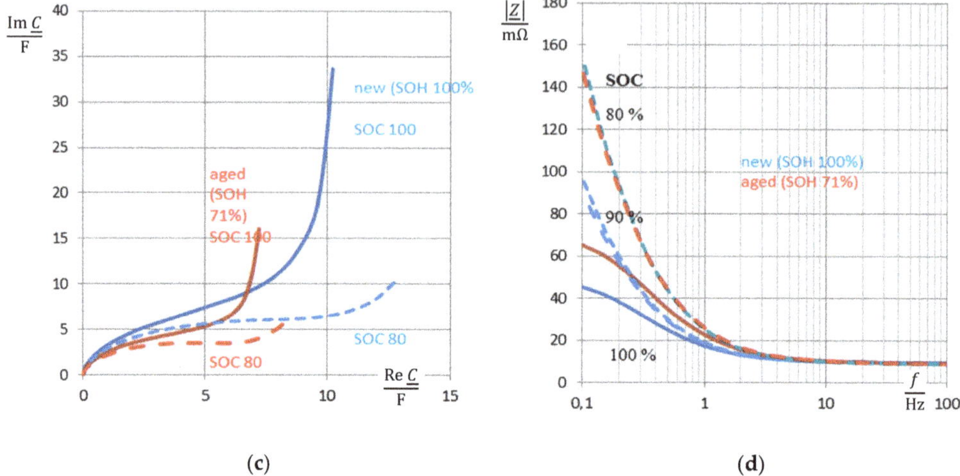

(c)   (d)

Figure 2. Impedance spectra of a nickel-cadmium (NiCd) battery (cell #5, new: State-of-health (SOH) 100% = 1.7 Ah, aged: SOH 71% = 1.21 Ah) at full charge (state-of-charge (SOC) = 100%, solid line) and 80% state-of-charge (dashed): (**a**) complex plane plot of impedance $\underline{Z}$, so-called Nyquist plot, (**b**) admittance $\underline{Y}$ = 1/$\underline{Z}$ in the complex plane, (**c**) complex capacitance $\underline{C}$ = $\underline{Y}$/(jω), (**d**) frequency response of modulus $|\underline{Z}|(\omega)$, part of Bode diagram. Reactance Im $\underline{Z}$, susceptance Im $\underline{Y}$, and pseudocapacitance Re $\underline{C}$ reflect the state-of-charge more clearly than the ohmic resistance Re Z, conductance Im Y, and the phase angle ϕ (not shown here).

1.4. Capacitance and Time Constant

We added the pseudo-capacitance $C(\omega)$ [1,24,25] to the above list, as a unique measure for the activity of the electrode/electrolyte interface, and as a qualitative indicator for the state-of-charge of the battery. The frequency response of the real part of complex capacitance $\underline{C}$ = $\underline{Y}$/(jω) is given by Equation (4).

$$C(\omega) = \frac{dQ}{dU} = \mathrm{Re}\,\underline{C}(\mathrm{j}\omega) = \frac{\mathrm{Im}\,\underline{Y}(\mathrm{j}\omega)}{\mathrm{j}\omega} = \frac{-\mathrm{Im}\,\underline{Z}(\mathrm{j}\omega)}{\omega \cdot |\underline{Z}(\mathrm{j}\omega)|^2} \tag{4}$$

At high frequencies ($\omega \to \infty$), pseudo-capacitance tends to the geometric double-layer capacitance C_D of the electrode/electrolyte interface. The ohmic resistance of the electrolyte solution, R_e = Re $\underline{Z}(\omega \to \infty)$, is found as the intersection of the complex plane plot with the real axis.

$$C_D = \lim_{\omega \to \infty} C(\omega) = \lim_{\omega \to \infty} \frac{-\mathrm{Im}\,\underline{Z}(\omega)}{\omega \cdot \left[[\mathrm{Re}\,\underline{Z}(\omega) - R_e]^2 + [\mathrm{Im}\,\underline{Z}(\omega)]^2 \right]} \approx \frac{-1}{\omega \cdot \mathrm{Im}\,\underline{Z}(\omega)} \tag{5}$$

The diagram of frequency-dependent capacitance $C(\omega)$ versus resistance R is useful for the direct comparison of battery capacities (see Section 3.5). The approximation in Equation (5) holds only for high frequencies, when the polarization resistance of the battery is negligible.

We think that the classification of aging phenomena is simplified by the apparent time constant of the charge-transfer reaction [26]. As a useful quantity for the comparison of new (at time t = 0) and aged batteries at any time t, we will consider below the relative growth of electrolyte resistance (at a given high frequency, e.g., 1 kHz) and the loss of capacitance (at a low frequency, e.g., 0.1 Hz).

$$\tau(t) = \frac{R(t) \cdot C(t)}{R_0 C_0} \tag{6}$$

2. Experimental Setup

In the course of a long time, with tests under real conditions as in an airplane, six NiCd battery packs were kept at a state-of-charge of 100% by trickle charge in order to balance self-discharge. With cycling, the packs were charged and discharged at a rate of C/2 between 100% SOC (7.5 V) to 80% SOC (6.0 V), separated by a rest period of 15 min, as shown in Figure 3 and Table 1. The ambient temperature was 50 °C during aging. Impedance spectra were measured at 25 °C after every 400 cycles, because (i) this temperature change is close to the practical application in the airplane, and (ii) helps to standardize the temperature-dependence of impedance, which affects the electrolyte resistance and the polarization resistance and the electrochemical reactions in a different way.

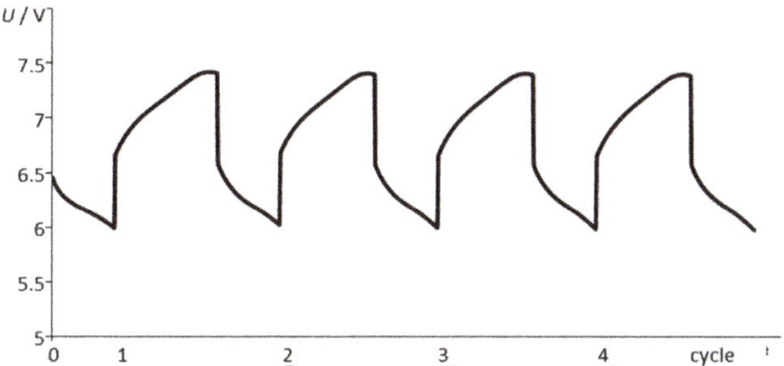

Figure 3. Voltage during continuous cycle testing at 50 °C (NiCd, 1.7 Ah, 0.5C) for 1200 cycles.

Table 1. Overview of experiments.

Test Method	Battery Pack: 7.5 V, 1.7 Ah, 5 Single Cells	A. Cycling (SOC) at 50 °C	B. Impedance Measurements During Discharge (SOC 1 → 0.7) After 400, 800, 1200 Cycles	C. Capacity After Full 0.5C Charge (Ah Counting)
1 Full discharge	(a) old (#1) (b) new (#4)	1C (1→ 0) 0.5C (1→ 0)	by 2% voltage steps	at cycle 400, 800, and 1200
2 Partial discharge	(a) old (#2) (b) new (#5)	1C (1→ 0.8) 0.5C (1→ 0.8)	by 0.19 Ah steps	at cycle 400, 800, and 1200
3 Partial discharge	(a) old (#3) (b) new (#6)	0.1C (1→ 0.8) 0.5C (1→ 0.8)	by 2% voltage steps	at cycle 400, 800 and 1200

With respect to the investigation of the memory effect, three types of experiments were defined:

1. Successive discharge. Pack #1 (fabricated in 2017) was cycled between SOC = 100% and SOC = 0% with 1C rate. Pack #4 (2018) was cycled between SOC = 100% and SOC = 80% at 0.5C rate. After every 400 cycles, the batteries were fully charged and then successively discharged by 2% steps, each characterized by impedance spectroscopy, until SOC = 70% was reached. Capacity was measured at the end of the endurance test, so that no memory effect was eliminated.

2. Rapid test. Pack #2 (2017) was cycled between SOC = 100% and SOC = 80% at 1C rate, and pack #5 of 2018 suffered at 0.5C. Every 400 cycles the batteries received (i) full charge, (ii) impedance measurement, (iii) discharge by 0.19 Ah, (iv) again impedance measurement, and (v) capacity measurement by ampere-hour counting. The memory effect was studied between successive tests.

3. Extensive test. Pack #3 (2017) was cycled between SOC = 100% and SOC = 80% at 0.1C rate and pack #6 (of 2018) at 0.5C. The analysis combined the above methods: (i) full charge, (ii) successive discharge by 2% and impedance measurement, until SOC = 70% was reached, (iii) full charge, and (iv) capacity determination by ampere-hour counting. The loss of capacity was studied for a large number of charge–discharge cycles.

3. Results and Discussion

3.1. State-of-Charge Monitoring

The stored residual electric charge Q (capacity) of a battery is expected to correlate with the pseudo-capacitance according to the definition $C = dQ/dU$. The pseudo-capacitance, according to Equation (4), is coined by the reactance, therefore the imaginary part of impedance is expected to show the state-of-charge as well. Unfortunately, we found no linear relationship between state-of-charge and capacity (Figure 4). The reactance at 0.1 Hz drops linearly between 80% and 20% SOC, whereas the edges at high and low SOC do not obey the trend.

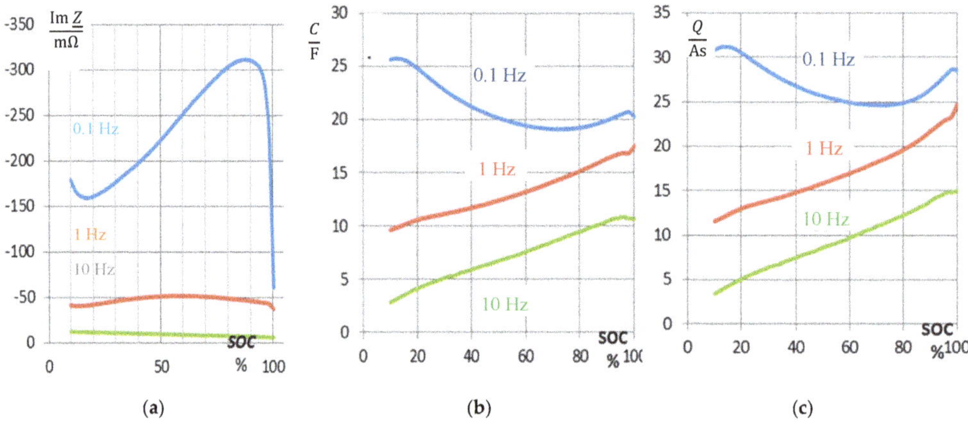

Figure 4. SOC monitoring by impedance spectroscopy of a used NiCd battery (cell 5 of pack #3 of 2017). (**a**) Reactance $X = \text{Im } \underline{Z}$ at different frequencies versus state-of-charge. (**b**) Pseudo-capacitance C and (**c**) calculated residual electric pseudo-charge $Q(t) = C\,U(t)$ at the momentary voltage U (SOC).

The absolute value of reactance $\text{Im } \underline{Z}$(1 Hz) shows a maximum around 65% SOC. In practice, $\text{Im } \underline{Z}$(10 Hz) shows a linear trend in the full SOC window. Figure 4 proves that capacitance $C(\omega)$ is a proper measure of the actual battery capacity. Low frequencies are useful for capacity determination.

As well, the pseudo-charge $Q = C{\cdot}U$ resolves the SOC characteristics more clearly than the imaginary part of impedance. As a compromise, between high signal level (at low frequencies) and linearity (at high frequencies), the imaginary part at 1 Hz might be appropriate for use in the airplane application, because low charge states are usually not reached in normal use.

3.2. Reactance during Aging and the Memory Effect

In long-term tests, all NiCd packs were aged by 1200 cycles. Impedance was recorded every 400 cycles. After 800 and 1200 cycles, respectively, the cells were refreshed in order to eliminate the memory effect.

Figure 5a compiles the reactance at 1 Hz versus the real state-of-charge, which was defined by ampere-hour counting. During aging, reactance drops due to the loss of available capacity (100% to 80% SOH). At the 800th cycle, 76% SOH is reached. The memory effect is eliminated by a single deep-discharge and the successive full charge. Indeed, capacity improves again to 91% SOH. After 1200 cycles, 97% SOH is restored and 77% SOH without removal of the memory effect.

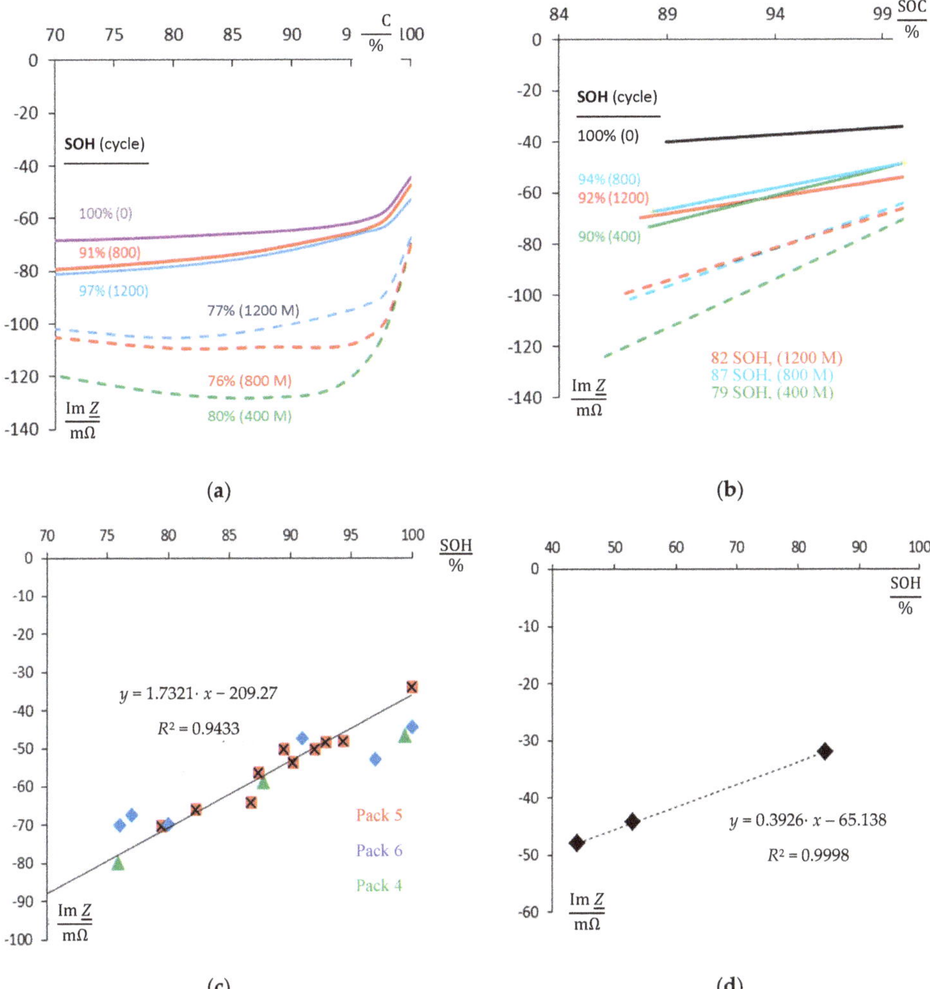

Figure 5. State-of-health monitoring (SOH) by impedance spectroscopy of NiCd batteries. (**a**) Reactance Im $\underline{Z}$(1 Hz) during 1200 charge–discharge cycles (cell of new pack #6, C/2, 50 °C). (**b**) Rapid method with 0.19 mAh discharge and capacity measurement by ampere-hour counting (cell of new pack #5). (**c**) Im $\underline{Z}$(1 Hz) of fully charged battery packs versus SOH. (**d**) Cell 5 of pack #1 after 400 pre-cycles. The state-of-health SOH = Q_0/Q_N (ratio of actual and rated available capacity Q) correlates quite well with the reactance Im $\underline{Z}$.

The rapid method (Section 2, Figure 5b) considers the difference between the full battery, and the state-of-charge after discharge of exactly 190 mAh (independent of SOH). The battery was fully discharged after 400, 800, and 1200 cycles. Obviously, the imaginary part of impedance reflects the state-of-health (capacity) during aging. Again, the memory effect can be eliminated by a single deep-discharge and a full charge and the reactance Im $\underline{Z}$(1 Hz) displays the real state-of-charge. The results of battery packs #1 and #4 are qualitatively identical with packs #5 and #6. The 'old' packs #2 and #3 were handled in the same way as the 'new' packs #5 and #6 except for the changed C-rate, respectively. However, due to previous history with extended dwell times under overcharge, the impedance and aging status do not clearly correlate.

In Figure 5c, the reactance, Im $\underline{Z}$(1 Hz), of three fully charged NiCd packs is compared at different aging states, including and excluding the memory effect. According to the residual capacity between 100% and 76% SOH, reactance lies between −40 mΩ and −80 mΩ. Low capacity correlates with highly negative reactance. This relation holds even if the memory effect is taken into account. Different battery packs show the same trend in the same order of magnitude.

In Figure 5d, the old battery pack #1, which suffered from 400 charge–deep discharge cycles before the impedance tests, shows the most serious loss in capacity. After 1200 cycles of the test procedure, residual capacity drops to 44% SOH (including memory effect), and 53% SOH (after refresh of the memory effect), respectively. Again, the reactance shows the state-of-health.

3.3. Capacitance as an Aging Indicator

Pseudo-capacitance $C(\omega)$ at different frequencies was investigated at different cells between beginning of life (BoL) and end of life (EoL, SOH = 71%) at state-of-charge values between 70% and 100%. In Figure 6, capacitance displays the state-of-health for the charged batteries between the rated voltage U_0 and 0.9 U_0. Pseudo-capacitance at 0.1 Hz shows the difference in the SOH of new and aged batteries more clearly than the measurement at 0.1 Hz. C calculated by Equation (4) contains the internal resistance of the battery, whereas the approximation C_D requires a higher frequency to give a reliable correlation with SOH.

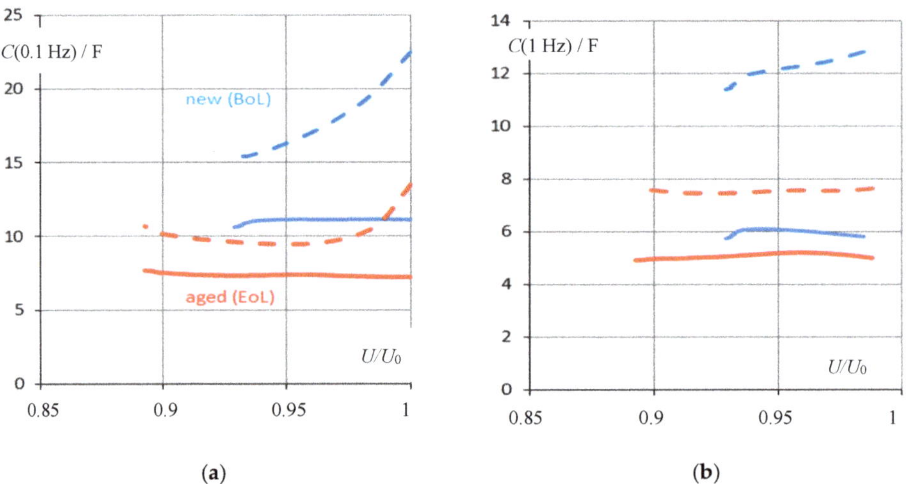

Figure 6. Capacitance-based state-of-health monitoring with respect to terminal voltage U/U_0 at (**a**) 0.1 Hz and (**b**) 1 Hz. BoL = beginning of life (1.7 Ah), EoL = end of life (1.2 Ah) of cell #5 of pack #6. Solid: $C = \text{Im } \underline{Y}/(j\omega)$, according to Equation (4). Dashed: Approximation $C_D = C(\omega \to \infty)$.

In any case, capacitance allows distinguishing between full charge and partial charge new and aged batteries, whereas the message at low state-of-charge is less reliable.

3.4. Pseudo-Charge as a SOH Indicator

Pseudo-charge, $Q(\omega) = C(\omega)/U$, which was calculated by the measured capacitance C and the momentary cell voltage U, correlates more obviously with the state-of-charge along the battery life. For the old battery pack #6 with respect to the new battery, Figure 7 compiles the relative pseudo-charge at different state-of-charge values.

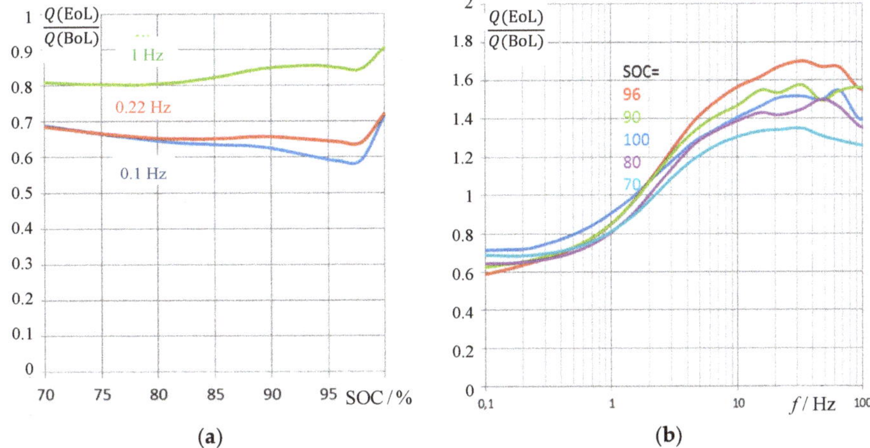

Figure 7. Aging study. Relative pseudo-charge $Q(\omega) = C(\omega)\cdot U(\text{SOC})$ of an aged 1.7 Ah battery (pack #6, EoL = end of life) with regard to the new battery (BoL = beginning of life). (**a**) Impedance measurements at selected frequencies versus the actual state-of-charge SOC = Q/Q_0 received by genuine Ah counting. (**b**) Frequency response of relative pseudo-charge, which is determined by the internal resistance of the battery below 1 Hz, and the surface capacitance above 1 Hz.

It becomes obvious that the impedance-based pseudo-charge correlates both with the actual Ah-based state-of-charge SOC between 70% and 96%. The fully charged battery (SOC = 100%) creates an outlier, because of the overcharge phenomenon that distorts the impedance-based capacitance measurement by parasitic electrochemical reactions. As an alternative to Figure 7, the relative loss of charge between new and old batteries can be drawn (i.e., ((Q(EoL) − Q(BoL))/Q(BoL)), which shows the same trend in the reverse direction. Useful frequencies for practical SOH determination lie below 1 Hz or below 100 Hz whether or not the internal resistance of the battery is considered as a proper aging criterion with calculating the pseudo-capacitance, $C(\omega) = \text{Re } \underline{C}$, according to Equation (5). Above 500 Hz, the spiral roll of the cell generates inductive impedance values, which are not useful for state-of-charge determination.

3.5. Separation of SOC and SOH

For the comparison of old and new batteries after a given operation time t, the pseudo-charge Q is normalized into a range between 0 and 1, which unlinks the SOC from the SOH.

$$Q_n(\omega,t) = \frac{Q(\omega,t)}{Q_N} = \frac{C(\omega,t)/U(t)}{C(\omega,0)/U_0} \tag{7}$$

Q_N is the rated capacity of the new battery (SOH = 1, SOC = 1). Q_0 is the residual capacity of the aged battery (SOC = 1, SOH <1). U_0 is the rated voltage and $U(t)$ is the momentary cell voltage.

The normalized pseudo-charge correlates excellently with the real state-of-charge, which was determined by ampere-hour counting (Figure 8). Advantageously, Q(1 Hz) is determined very quickly.

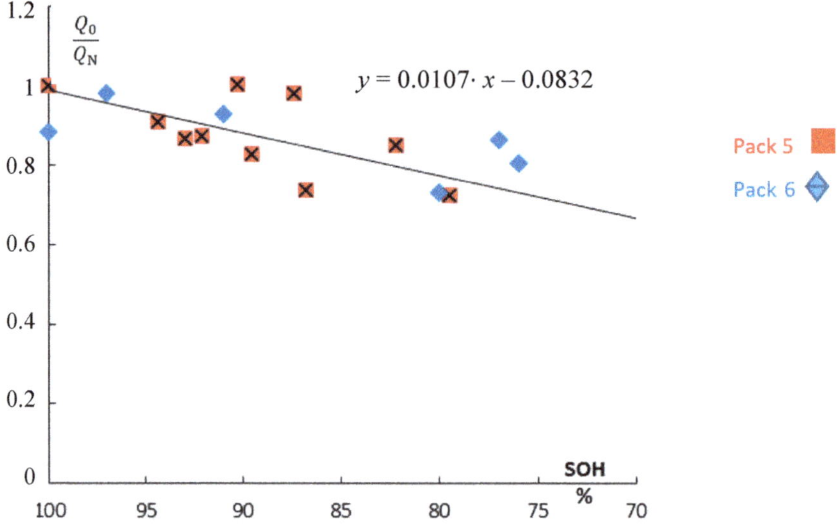

Figure 8. The impact of aging. Ratio of the available electric pseudo-charge $Q_0(t)$ of used NiCd batteries with respect to the rated value of the new battery Q_N (pack #5 and pack #6). Data are taken from Figure 5. Impedance-based pseudo-charge $Q_0 = C(1\text{ Hz})\,U$ (divided by the rated capacity Q_N) correlates well with the true SOH values from Ah measurements.

3.6. Impedance-Based Aging Indicators

Diagrams of capacitance versus resistance $C(R)$ have been used for the characterization of supercapacitors, whereby the rated capacitance is reached at low frequencies [1]. Aging generates a growing resistance and a loss of pseudo-capacitance. In Figure 9, however, a constant value of *dc* pseudo-capacitance is not quickly reached with NiCd batteries at frequencies down to 0.1 Hz. Especially at low SOC, aged batteries show a less steep slope $\Delta C/\Delta R$ compared with new cells. In any case, the $C(R)$ diagram estimates the extent of aging quickly. We propose this diagram for SOH monitoring.

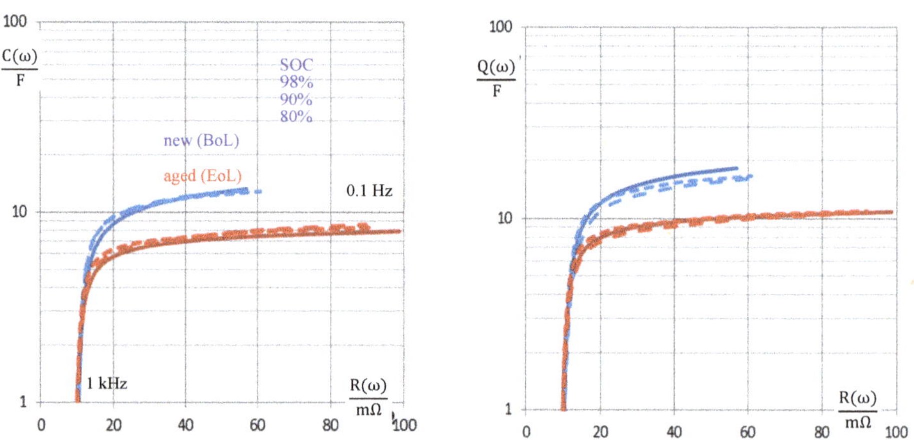

Figure 9. Aging characteristics of NiCd cell #5 of pack #6 (new: 1.7 Ah, aged: 1.3 Ah, SOH = 76%) in the plot of pseudo-capacitance $C(\omega)$ and pseudo-charge $Q(\omega) = C(\omega)\,U$ versus the internal resistance (real part of impedance). $U(\text{SOC})$ = actual cell voltage at the time of measurement.

The reactance (imaginary part of impedance) versus SOC shows a local maximum (Figure 10). This quantity is therefore less useful for SOH determinations. In contrast to that, relative pseudo-capacitance C/C_0 increases with SOC the same way as cell voltage U/U_0 does (Figure 10). This proves why we suggest using capacitance for impedance-based SOC determination.

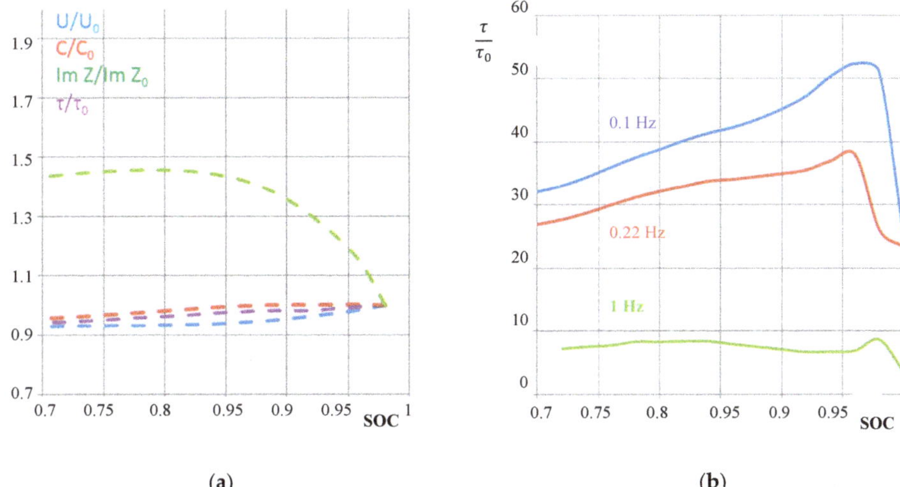

(a)   (b)

Figure 10. Aging characteristics of a NiCd battery (cell #5 of pack #6). (**a**) Different normalized state-of-charge quantities with respect to voltage U/U_0, pseudo-capacitance C/C_0 at 0.22 Hz, imaginary part of impedance at 0.22 Hz, and relative time constant τ/τ_0 at 0.22 Hz against the actual state-of-charge received from Ah counting. (**b**) Relative time constant between the old battery $\tau = R(1\ \text{kHz}){\cdot}C(0.1\ \text{Hz})$ and the new battery τ_0 at different frequencies according to Equation (6).

The relative time constant τ according to Equation (5) was determined by the pseudo-capacitance at 0.1 Hz. With this, new and aged batteries can qualitatively be distinguished by a single quantity. A large value of τ/τ_0 means that the aged battery suffers from high electrolyte resistance and low interfacial capacitance.

4. Conclusions

For NiCd batteries used in airplanes at SOC values above 70%, this work proves the impact of cycle aging on the available electric charge (capacity) at different state-of-charge values (SOC) by the help of impedance spectroscopy. During aging, the cell resistance grows and the pseudo-capacitance drops (Figure 9), and the imaginary part of impedance rises (Figure 5). For rapid impedance measurements, we propose a frequency around 0.2 Hz. For practical use, the pseudo-capacitance according to Equation (4) best reflects the real state-of-charge (Figure 10). The imaginary part of impedance is less useful for SOC and SOH monitoring. Thanks to modern electronics, impedance measurements can be performed in airplanes throughout the pre-departure process with little expense and effort.

In addition, we found that pseudo-charge $Q(\omega) = C(\omega){\cdot}U$ at 1 Hz is a proper measure for the real state-of-charge of a battery, where U is the momentary voltage of battery at the time of impedance measurement. With this approach, there is no need for any model descriptions or equivalent circuits, which are often unclear and complicate the system analysis during operation. For SOH monitoring, the pseudo-charge of the aged battery with regard to the new battery is considered (Figure 8).

Author Contributions: Writing—original draft preparation, review and editing, P.K. and W.S. All authors have read and agreed to the published version of the manuscript.

Funding: This work was by supported by Diehl Aerospace GmbH.

Conflicts of Interest: The authors declare no conflict of interest.

References

1. Kurzweil, P.; Shamonin, M. State-of-Charge Monitoring by Impedance Spectroscopy during Long-Term Self-Discharge of Supercapacitors and Lithium-Ion Batteries. *Batteries* **2018**, *4*, 35. [CrossRef]
2. Dubarry, M.; Truchot, C.; Liaw, B.Y. Synthesize battery degradation modes via a diagnostic and prognostic model. *J. Power Sources* **2012**, *219*, 204–216. [CrossRef]
3. Hammouche, A.; Karden, E.; De Doncker, R.W. Monitoring state-of-charge of Ni–MH and Ni–Cd batteries using impedance spectroscopy. *J. Power Sources* **2004**, *127*, 105–111. [CrossRef]
4. Xiong, R.; Cao, J.; Yu, Q.; He, H.; Sun, F. Critical Review on the Battery State of Charge Estimation Methods for Electric Vehicles. *IEEE Access* **2017**, *6*, 1832–1843. [CrossRef]
5. Sarmah, S.B.; Kalita, P.; Garg, A.; Niu, X.; Zhang, X.-W.; Peng, X.; Bhattacharjee, D. A Review of State of Health Estimation of Energy Storage Systems: Challenges and Possible Solutions for Futuristic Applications of Li-Ion Battery Packs in Electric Vehicles. *J. Electrochem. Energy Convers. Storage* **2019**, *16*, 040801. [CrossRef]
6. Zou, Y.; Hu, X.; Ma, H.; Li, S. Combined State of Charge and State of Health estimation over lithium-ion battery cell cycle lifespan for electric vehicles. *J. Power Sources* **2015**, *273*, 793–803. [CrossRef]
7. Rodrigues, S.; Munichandraiah, N.; Shukla, A.K. A review of state-of-charge indication of batteries by means of a.c. impedance measurements. *J. Power Sources* **2000**, *87*, 12–20. [CrossRef]
8. Bounaga, A. Device and Method for Measuring the Charge State of a Nickel-Cadmium Accumulator. U.S. Patent 5,650,937A, 22 July 1997.
9. Latner, N. Method for Determining the State of Charge of Nickel Cadmium Batteries by Measuring the Farad Capacitance Thereof. U.S. Patent 3,562,634A, 9 February 1971. Available online: https://worldwide.espacenet.com/searchResults?ST=singleline&locale=de_EP&submitted=true&DB=&query=US3562634 (accessed on 8 January 2020).
10. Sathyanarayana, S.; Venugopalan, S.; Gopikanth, M.L. Impedance parameters and the state-of charge. I. Nickel-cadmium battery. *J. Appl. Electrochem.* **1979**, *9*, 125–139. [CrossRef]
11. Wenzl, H. Capacity. In *Encyclopedia of Electrochemical Power Sources*; Garche, J., Dyer, C.K., Moseley, P., Ogumi, Z., Rand, D., Scrosati, B., Eds.; Elsevier: Amsterdam, The Netherlands, 2009; Volume 1, pp. 395–400.
12. Waag, W.; Sauer, D.U. State-of-Charge/Health. In *Encyclopedia of Electrochemical Power Sources*; Garche, J., Dyer, C.K., Moseley, P., Ogumi, Z., Rand, D., Scrosati, B., Eds.; Elsevier: Amsterdam, The Netherlands, 2009; Volume 4, pp. 793–804.
13. Piller, S.; Perrin, M.; Jossen, A. Methods for state-of-charge determination and their applications. *J. Power Sources* **2001**, *96*, 113–120. [CrossRef]
14. Bergveld, H.J.; Danilov, D.; Notten, P.H.L.; Pop, V.; Regtien, P.P.L. Adaptive State-of-charge determination. In *Encyclopedia of Electrochemical Power Sources*; Garche, J., Dyer, C.K., Moseley, P., Ogumi, Z., Rand, D., Scrosati, B., Eds.; Elsevier: Amsterdam, The Netherlands, 2009; Volume 1, pp. 450–477.
15. Dowgiallo, E.J. Method for Determining Battery State of Charge by Measuring A.C. Electrical Phase Angle Change. U.S. Patent 3,984,762A, 5 October 1975.
16. Finger, E.P.; Sands, E.A. Method and Apparatus for Measuring the State of Charge of a Battery by Monitoring Reductions in Voltage. U.S. Patent 4,193,026A, 11 March 1978.
17. Kikuoka, T.; Yamamoto, H.; Sasaki, N.; Wakui, K.; Murakami, K.; Ohnishi, K.; Kawamura, G.; Noguchi, H.; Ukigaya, F. System for Measuring State of Charge of Storage Battery. U.S. Patent 4,377,787A, 22 March 1983.
18. Seyfang, G.R. Battery State of Charge Indicator. U.S. Patent 4,949,046, 14 August 1990.
19. Peled, E.; Yamin, H.; Reshef, I.; Kelrich, D.; Rozen, S. Method and Apparatus for Determining the State-of-Charge of Batteries Particularly Lithium Batteries. U.S. Patent 4,725,784A, 16 February 1988.
20. Hahn, M.; Schindler, S.; Triebs, L.C.; Danzer, M.A. Optimized Process Parameters for a Reproducible Distribution of Relaxation Times Analysis of Electrochemical Systems. *Batteries* **2019**, *5*, 43. [CrossRef]
21. Sato, Y.; Takeuchi, S.; Kobayakawa, K. Cause of the memory effect observed in alkaline secondary batteries using nickel electrode. *J. Power Sources* **2001**, *93*, 20–24. [CrossRef]
22. Linden, D.; Reddy, T.B. *Handbook of Batteries*; McGraw-Hill: New York, NY, USA, 2002.
23. Barsoukov, E.; Macdonald, J.R. *Impedance Spectroscopy: Theory, Experiment, and Applications*; John Wiley Sons: Hoboken, NJ, USA, 2018.

24. Kurzweil, P.; Ober, J.; Wabner, D.W. Method for Correction and Analysis of Impedance Spectra. *Electrochim. Acta* **1989**, *34*, 1179–1185. [CrossRef]
25. Kurzweil, P.; Fischle, H.J. A new monitoring method for electrochemical aggregates by impedance spectroscopy. *J. Power Sources* **2004**, *127*, 331–340. [CrossRef]
26. Kurzweil, P.; Hildebrand, A.; Weiss, M. Accelerated Life Testing of Double-Layer Capacitors: Reliability and safety under Excess Voltage and Temperature. *ChemElectroChem* **2015**, *2*, 150–159. [CrossRef]

Review

Degradation and Aging Routes of Ni-Rich Cathode Based Li-Ion Batteries

Philipp Teichert [1], Gebrekidan Gebresilassie Eshetu [2,3], Hannes Jahnke [1,*] and Egbert Figgemeier [2,4,*]

[1] Volkswagen AG, Group Innovation, Letterbox 011/17774, 38436 Wolfsburg, Germany; philipp.teichert1@volkswagen.de

[2] Rheinisch-Westfälische Technische Hochschule Aachen, Institut für Stromrichtertechnik und Elektrische Antriebe, Jägerstraße 17-19, 52066 Aachen, Germany; Gebrekidan.Eshetu@isea.rwth-aachen.de

[3] Department of Material Science and Engineering, Mekelle Institute of Technology—Mekelle University, Mekelle 1632, Tigrai, Ethiopia

[4] Helmholtz Institute Münster (HI MS), IEK-12, Forschungszentrum Jülich, Corrensstrasse 46, 48149 Münster, Germany

* Correspondence: hannes.jahnke@volkswagen.de (H.J.); e.figgemeier@fz-juelich.de (E.F.)

Received: 20 November 2019; Accepted: 14 January 2020; Published: 22 January 2020

Abstract: Driven by the increasing plea for greener transportation and efficient integration of renewable energy sources, Ni-rich metal layered oxides, namely NMC, Li $[Ni_{1-x-y}Co_yMn_z]$ O_2 ($x + y \leq 0.4$), and NCA, Li $[Ni_{1-x-y}Co_xAl_y]$ O_2, cathode materials have garnered huge attention for the development of Next-Generation lithium-ion batteries (LIBs). The impetus behind such huge celebrity includes their higher capacity and cost effectiveness when compared to the-state-of-the-art $LiCoO_2$ (LCO) and other low Ni content NMC versions. However, despite all the beneficial attributes, the large-scale deployment of Ni-rich NMC based LIBs poses a technical challenge due to less stability of the cathode/electrolyte interphase (CEI) and diverse degradation processes that are associated with electrolyte decomposition, transition metal cation dissolution, cation–mixing, oxygen release reaction etc. Here, the potential degradation routes, recent efforts and enabling strategies for mitigating the core challenges of Ni-rich NMC cathode materials are presented and assessed. In the end, the review shed light on the perspectives for the future research directions of Ni-rich cathode materials.

Keywords: Li-Ion battery; Ni-rich cathode; degradation; cathode-electrolyte interphase; electro mobility

---

1. Introduction

There is a growing necessity for reducing CO_2 emissions to prevent further upsurge of the global temperature and, thereby, safeguard the fate of our planet and its occupants [1]. This calls for an urgent transition from the limited, as well as polluting, fossil fuels and internal combustion vehicles (ICEs) to renewable (e.g., solar, wind, etc.) energy sources and electro mobility (electric vehicles, xEVs) respectively [2,3]. 80% of the transportation in USA and Germany is linked to on-road vehicles and, thus, moving towards electric vehicles (xEVs), offers the potential for emission saving. However, green energy sources are intermittent in nature and their proper utilization demands the use of highly efficient and durable electrochemical energy storage devices [4]. Similarly, xEVs necessities the use of high energy density batteries that are capable of negating the existing "driving range anxiety" [5]. Amid existing electrochemical energy storage devices, lithium-ion batteries (LIBs) have attracted huge attention as one of the most versatile and enabler devices for use in a wide range of applications. However, further improvements in the energy density of LIBs is urgently needed to meet the ever-growing stringent requirements for emerging large-scale systems. For instance, the electrical propulsion demands battery devices with high energy density allowing for the ~500 km driving range ([energy density of > 400 Wh kg^{-1}, > 500 Wh L^{-1} and cost < 125 US\$ kWh^{-1}]) [6] requirements. The energy density (E_{cell}) of a given LIB cell is collectively determined by the discharge cell voltage (V_{cell}) and Li-storage capacity (C_{cell}), as shown in Equation (1) [7,8], inferring an increase in the energy density of LIBs requires increasing both the cell capacity and maximizing the potential difference between the anode and cathode electrodes.

$$E_{Cell} = \int V_{Cell} dC_{Cell} \tag{1}$$

Owing to the fact that the anode materials offer a higher Li-ion storage capacity than the cathode, the latter presents to be the most important and limiting factor for the energy density of LIBs [9–12]. Following Sony's introduction of lithium cobalt dioxide (LiCoO$_2$)-graphite cell in 1991, it has become the most dominant energy storage device in the market. LiCoO$_2$ (LCO) has been the most investigated electrode material from the time of its earliest discovery by Goodenough et al. [13]. However, despite its successful commercialization, LCO suffers from several drawbacks, e.g., structural degradation and oxygen release at highly de-lithiated states (Li$_{1-x}$CoO$_2$ where $x > 0.5$) [10,14] and low specific capacity of 140 mAh g^{-1} (i.e., ~half of the theoretical capacity, 274 mAh g^{-1}) [15]. Besides, cobalt (Co) is very expensive and toxic, which leads to an increase in the carbon footprint of the cathode active materials [16]. These practical challenges have been the stimulus behind the search for other alternative crystallographic systems. Lithium iron phosphate (LiFePO$_4$, LFP) is another interesting cathode material of high structural and thermal stabilities (i.e., no oxygen release) due to the strong Fe-O covalent bonds and cycling stability; however, it operates at a lower voltage, thus reducing its power density when compared to LCO [17].

On the contrary, despite the high energy and power densities and relatively high operating voltage (~4.1 V vs. Li/Li$^+$) of high voltage spinel-like Li$_2$MnO$_4$ (LMO) cathode materials, they suffer from accelerated capacity fading during prolonged cycling, storage, and, at elevated temperatures, attributed to Mn dissolution [14].

Ni-rich layered cathodes have been materialized as next generation materials because of their relatively high discharge capacity (i.e., 180–230 mAh g^{-1}), lower overall cost, and so forth in the search for high-performance devices. Recently, the research interest in Ni-rich cathodes has been exponentially augmented (Figure 1a,b), once again proving the huge attention that is given by the academia, governments and industries. However, Ni-rich cathode materials are beset with detrimental challenges, hindering their large-scale integration into the evolving xEVs market.

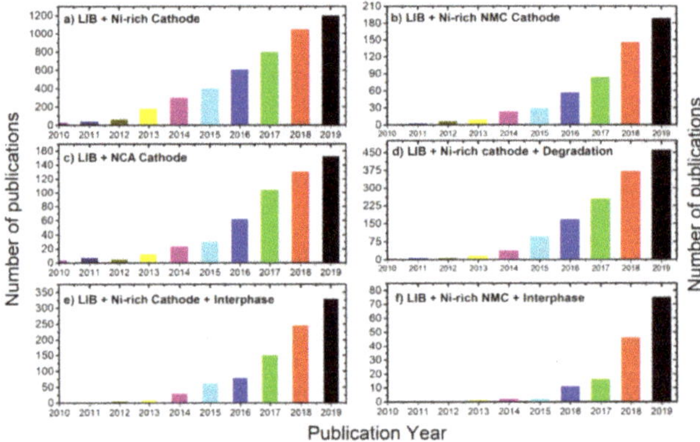

Figure 1. Overall research activity on Ni-rich cathode based lithium-ion batteries (LIBs) (Last updated 20th October 2019). The key search words used in Scopus are, "Lithium-ion battery", Lithium-ion battery + Nickel rich cathode" and "Lithium-ion battery + Nickel rich cathode + degradation or interphase".

Among others, issues that are related to the Ni-rich cathode/electrolyte interphase (CEI) and complex degradation origins as well as routes have received significant attention (Figure 1c–f). Herein, we review the critical impediments for the large scale commercialization of Ni-rich cathode based LIBs and recent progress in the search of enabling mitigating strategies. For the sake of readers, the chemistry, major challenges enlisting the unstable CEI and degradation phenomena as well as proposed solutions and research directions are addressed in Sections 2–4, respectively. Finally, a conclusive remark shedding light on the future research directions is precisely presented.

2. Chemistry of Ni-Rich Cathode Materials

The fundamental features of Ni-rich cathodes, (Li $[Ni_xM_{1-x}]$ O_2, ($x \geq 0.6$, M = transition metal), are predominantly dictated by the chemistry, i.e., elemental composition, namely nickel (Ni), cobalt (Co), manganese (Mn), aluminum (Al), and any other incorporated dopants (if exist) [18]. Transition metal cations and their chemistries play a major role in the aging and degradation processes in layered cathode based LIBs. The composition and oxidation state of each transition metal cation contribute to the electrochemical performance, thermal stability, overall material cost, toxicity, etc. of the Ni-rich cathodes. In general, nickel (Ni) serves as a source of valence states of +2 (Ni^{2+}) and +3 (Ni^{3+}) in Ni-rich electrodes. However, at certain conditions, such as at a very low lithiation state, the high oxidative +4 (Ni^{4+}), whose compounds are soluble in the electrolyte, is observed [19]. Unlike LCO, $LiNiO_2$ (LNO) is more challenging due to the instability resulting in the spontaneous reduction of Ni^{3+} to Ni^{2+}, and thereby its migration within the Li layers. The substitution of stable Co^{3+} for the unstable Ni^{3+} minimizes the formation of Ni^{2+} species, favoring the development of Li (Ni, Co) O_2 solid solution, i.e., $LiNi_{1-y}Co_yO_2$ to prevent such undesired transformation. Such Co-doping improves the structural integrity and promotes better stability in the charged state. Interestingly, the substitution of Ni with electrochemically inactive Al appeared to be interesting, which further increases the electrode thermal stability due to the stronger Al-O linkage when compared to Ni-O bonding. However, the improvement in the cathode electrode stability at high voltage and cycling life decreases, owing to its electrochemical inertness. Later, dual substitution of Co and Al, aiming at harvesting the synergistic benefits of both elements, pursued the development of $Li[Ni_{1-x-y}Co_xAl_y]O_2$, coined as NCA [20]. Amid several NCA varieties, $Li[Ni_{0.80}Co_{0.15}Al_{0.05}]O_2$ and $Li[Ni_{0.81}Co_{0.10}Al_{0.09}]O_2$ have been ascribed as the most promising versions, which satisfy safety criteria without compromising the energy density, power or cost benefits of Ni-based cathodes. Another possibility to substitute Ni is with Mn, hoping to enhance the operating voltage while keeping the structural integrity and lowering the cost and toxicity of the resulting material. Doping Ni with both Co and Mn leads to the development of very exciting crystallographic systems, called NMC materials, $Li[Ni_xCo_yMn_z]O_2$ ($x + y + z = 1$), with specific variants, such as $x{:}y{:}z$ = 1:1:1 (NMC111), 5:3:2 (NMC532), 6:2:2 (NMC622), 8:1:1 (NMC811), 85:7.5:7.5 (NMC857575), and so forth, where the numbers denote the ratio of Ni, Co, and Mn on a mole fraction basis.[6] The role of each transition metal in NMC is detailed in Figure 2a,b, clearly depicting their splendid effect on the discharge capacity, capacity retention, and thermal stability. Increasing Ni content clearly depicts an increase in the discharge capacity, but at the expense of thermal stability and capacity retention.

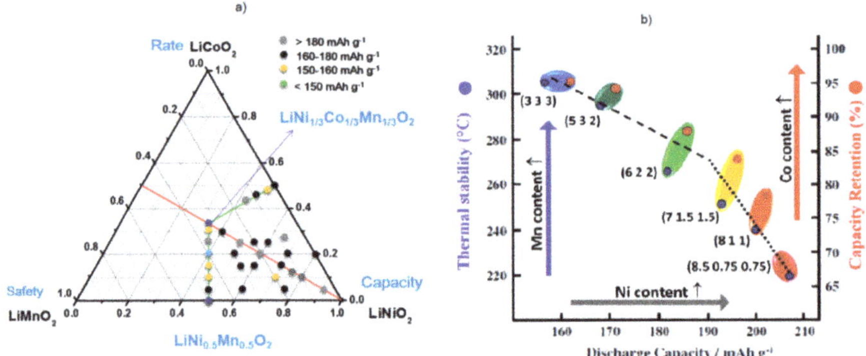

Figure 2. (**a**) Compositional phase diagrams of lithium stoichiometric-layered transition-metal oxide: $LiCoO_2$-$LiNiO_2$-$LiMnO_2$. The positions indicated by dots represent the described $LiNi_{1-x-y}Co_xMn_yO_2$, reproduced from Ref. [21] with permission, Copyright 2015, John Wiley and Sons; (**b**) Sketch on the effect of transition metals on discharge capacity (black), thermal stability (blue) and capacity retention (red) of $Li[Ni_xCo_yMn_z]O_2$ compounds with number in brackets corresponding to the composition (Ni, Mn, Co). The positive effect of Ni on capacity, Mn on thermal stability and Co on capacity retention and rate performances is underlined. Reproduced with permission from Ref. [22], Copyright 2015, Electrochemical Society.

3. Challenges and Origins Associated with Ni-Rich Cathode Materials

Despite the fact that Ni-rich cathode materials have reaped enormous courtesy as next generation cathode materials, their commercialization is still limited to lower Ni content, Ni ≤ 60% for NMC and Ni = 80% (by molar fraction) for NCA. Low Ni content-NMC/LMO blended cathode in Chevrolet Volt (PHEV), BMWi3 (BEV), as well as Nissan Leaf (BEV), pure NMC in Daimler Smart EV (BEV) and NCA in Daimler S class hybrid (HEV), Tesla Model S (BEV), and the BMW active Hybrid7 (HEV) have been commercialized for applications in the electro mobility [23,24]. However, these success stories cannot meet the harsh requirements of future generation batteries in terms of energy density (> 400 Wh kg^{-1}, > 500 Wh L^{-1}) and cost (< 125 US\$ kWh^{-1}) [25,26]. The cathode of choice of the automotive industry for the "near-future" xEVs versions is directed to Ni-rich NMC and NCA, combined with access to high voltage operations [27]. However, the practical application with Ni content over 60% is severely hindered due to the presence of various inter-linked challenges. Ni-rich layered oxides generally suffer from two major problems, i.e., performance degradation and safety hazards, especially during storage or operation near the fully delithiated state and/or at high temperature. Generally, the origins for the performance degradation are linked to residual lithium compounds (RLC), Ni/Li cationic mixing (disordering), oxygen evolution reaction (OER), irreversible phase transition, transition metal dissolution, micro cracking of secondary particle structure, and so forth. Thus, an in-depth evaluation of the existing obstacles and accompanying origins of NMC and NCA cathode materials in general and those Ni-rich one in particular are presented in the following sub-sections.

3.1. Lithium Residual Compounds from Synthesis

There exist various synthesis methods that lead to the production of Ni-rich cathode materials, such as atmospheric plasma spray pyrolysis [28], pulse combustion [29], sputter deposition [30], and co-precipitation [31], to name a few. The latter is the most popular synthesis route, which firstly involves the formation of precursor in a continuous stirring tank reactor (CSTR) by co-precipitation. The divalent transition metal ions are then introduced into the CSTR under inert atmosphere, e.g., as sulfate salts. Furthermore, sodium hydroxide and ammonium hydroxide are added to control the pH-value and as chelating agent, respectively. The stoichiometry of the transition metal hydroxide precursor determines the stoichiometry of the final product in terms of the Ni: Mn: Co ratio. Afterwards, the precursor is dried and mixed with a stoichiometric equivalent amount of Li_2CO_3 or LiOH by ball milling and, finally, sintered in air. During the synthesis route, a large number of parameters can affect the final product, such as pH-value, amount of ammonia, stirring speed, calcination temperature, and so forth [32].

For example, NMC111 sub-micron particles were obtained at an elevated pH-value, low ammonia content, and relatively low calcination temperature [33]. While the nucleation and its growth rate are pH dependent, the agglomeration of primary to secondary particles depends on the amount of ammonia that is added during the synthesis process. In short, the synthesis history plays a huge role in the degradation phenomenon of Ni-rich cathode materials. For instance, the synthesis conditions (e.g., nature and concentration of precursors and other reagents, sintering and co-precipitation temperatures, pH value, stirring speed, etc.) and synthesis protocols profoundly affect the particle size, crystallinity, morphology, structural stability, and cation ordering. This eventually plays a decisive role in the degradation process (e.g., by creating active surface driven side reactions and others) and, thus, in the electrochemical performance of the final products. Tian et al. [34] evidenced liquid electrolyte penetration limits, which are consequences of the morphological appearance of the NMC particles, leading to heterogeneous state-of-charge (SoC) distribution within the cathode active material particles. This, in turn, leads to an inhomogeneous aging within the particles.

Unreacted lithium ingredients can often remain on the surfaces of the active materials, ostensibly in the oxide form of Li_2O, because an excessive amount of lithium is compulsory to produce decidedly crystalline Ni-rich layered materials. The presence of these residual lithium compounds (RLCs), which are inescapably available in Ni-rich cathodes with their amount increasing with the Ni content (i.e., $\sim 6 \times 10^3$ to 25×10^3 ppm for Ni $\geq$ 60%), are highly detrimental. RLCs can be originated via two routes: (i) excess amount of LiOH introduced during the synthesis for recompensing the loss of Li_2O, by sublimation, and subduing the Ni/Li mixing [35], and (ii) via reactions with H_2O ($LiNiO_2 + xH_2O \rightarrow Li_{1-x}H_xNiO_2 + xLiOH$) and/or CO_2 ($2LiNiO_2 + xH_2O + xCO_2 \rightarrow 2Li_{1-x}H_xNiO_2 + xLi_2CO_3$) in air during storage [35,36]. Moreover, RLCs compounds can cause the gelation of the slurry during electrode preparation that is associated with dehydrofluorination of polyvinylidene fluoride (PVDF) binder by LiOH [$(CH_2\text{-}CF_2)_n + LiOH \rightarrow (CH=CF)_n + LiF + H_2O$] [37]. Moreover, the strong alkalinity nature of RLCs can catalyze the transesterification and, thus, cascading degradation of carbonate based solvents, resulting in battery swelling and/or gassing [38]. The electrochemical oxidation of Li_2CO_3 at a potential > 4.3 V vs. Li/Li$^+$ also results in the formation of oxygen and CO_2 ($2Li_2CO_3 \rightarrow 4Li^+ + 4e^- + O_2 + 2CO_2$), eventually causing battery swelling [39–41]. The potential reaction of Li_2CO_3 with protic species (e.g., $Li_2CO_3 + 2H^+ \rightarrow CO_2 + H_2O + 2Li^+$), which can inevitably be generated during cell cycling, can also lead to severe gassing. In general, RLCs in the form of mixtures of Li_2O, LiOH, and Li_2CO_3 play a key role in the degradation of Ni-rich cathode based LIBs and their ratio varies with storage conditions, such as humidity level in the air, time, etc. Thus, the amount of RLCs should kept lower than 3000 ppm for an acceptable level, owing to their detrimental effect [20]. Besides to employing tailored synthesis protocols, surface modification, for instance, with H_3PO_4 to produce a Li_3PO_4 coating layer upon heating and thus trapping three lithium atoms and thereby decreasing the impurities, has been proved to be an effective strategy [42]. The resulted Li_3PO_4 later can trap not only HF, but also H_2O as $Li_3PO_4 + H_2O \rightarrow Li_xH_yPO_4$ (or PO_xH_y) $+ Li_2O$, thus improving the electrochemical performances. Figure 3 depicts the surface change of Ni-rich cathode materials and accompanying changes in the cathode-electrolyte interphases [43].

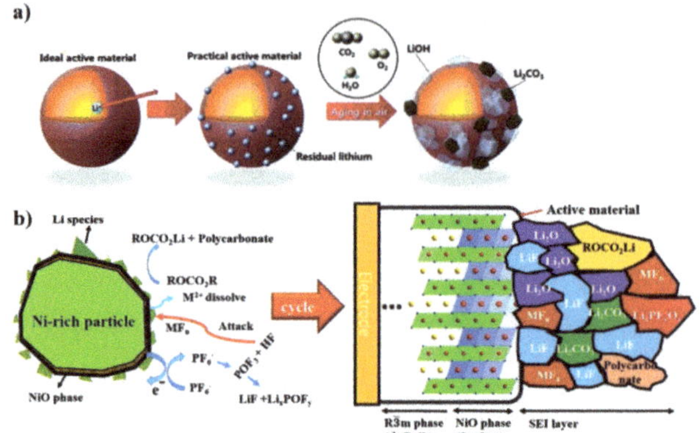

Figure 3. Illustration of the surface changes of Ni-rich cathode materials. (**a**) Surface change of Ni-rich materials after exposure in air, taken from Ref 37. (**b**) The microstructure and composition of the cathode-electrolyte interface at the surface of Ni-rich cathode materials, Reproduced from Ref. [20] with permission, Copyright 2015, John Wiley and Sons.

3.2. M/Li Cationic Mixing

Among the principal cations in Ni-rich cathode materials, i.e., Ni^{2+}, Co^{2+}, and Mn^{2+}, Ni^{2+} ions have the strongest propensity to mix with Li^+ ions (Figure 4), which results in the reduction of both capacity and Li mobility (thus conductivity) and at the same time transforming the layered over spinel crystal to NiO-like rock salt phase, both during preparation and application [43–46]. The mixing tendency is assumed to be favored from the similarity in the ionic radius of Ni^{2+} (0.69 Å) and Li^+ (0.76 Å) [47,48]. It is noteworthy to mention that the Ni^{2+}/Li^+ mixing occurs not only during the synthesis process, but also over the whole life of the battery. Its degree increases with increasing the Ni content, ratio of Ni^{2+} to Ni^{3+}, state-of-charge (SoC), and temperature. Instead, Doeff and co-workers proposed a super-exchange mechanism for Ni-rich NMC materials, where Ni^{3+} has priority to exchange with Li^+ [49]. A change in the valance state of Ni ion ($Ni^{3+} \rightarrow Ni^{2+}$) occurs in combination with oxidation of the nearest Co^{3+} ion ($Co^{3+} \rightarrow Co^{4+}$) for charge compensation, where oxygen ion has a charge carrier role.

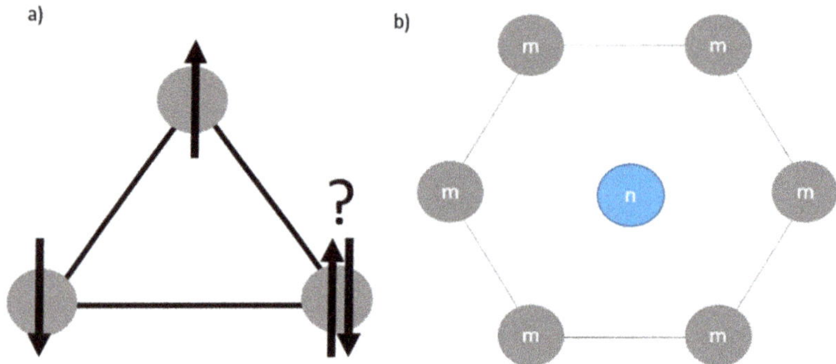

Figure 4. (**a**) Illustration of geometrical frustration where three ions with magnetic moments are arranged in trigonal planar, the sum of all moments cannot equal zero. (**b**) Rearrangement of ions within NMC to release geometric frustration, ions: m (grey) with magnetic moment, i.e., manganese and nickel form a hexagon with a non-magnetic, ion n (blue), i.e., cobalt or lithium, in its center.

Koyama et al. [50] recently reported that cation mixing is a result of frustration (Figure 4a). In the TM layer of NMC, Ni, Co, and Mn are arranged in trigonal planar. Hence, there is a potential for magnetic frustration, since the minimization of interaction energy of all nearest-neighbor spins is impossible. The frustration is released by a rearrangement of the TM elements from triangular coordination towards a hexagonal one, where Co is at the center of a hexagon of Ni and Mn, in the case of NMC111 with a high amount of nonmagnetic Co^{3+} (Figure 4b). This implies that lowering the Co content (i.e., increasing Ni) favors cationic mixing, which infers that cation-mixing leads to a thermodynamically preferred state of energy.

One of the possibilities for identifying and quantifying cation mixing is the utilization of X-ray diffraction (XRD), i.e., from the dependency of the ratio of the intensities of the (003) and (104) peaks on the amount of antisite Ni and Li. Patterns that were calculated using the WinPlotr tool of the FullProf Suite [50] show that, in low cation mixing (Figure 5a), the ratio of the peaks is 1.2, however, as the amount of cation mixing increases, the ratio I_{003}/I_{104} is decreased (Figure 5b).

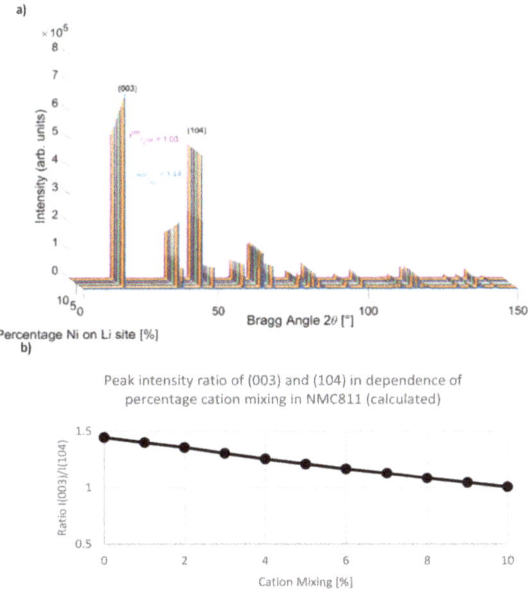

Figure 5. (**a**) Calculated X-ray diffraction (XRD) plot of NMC811 between $2\theta = 0 \dots 150°$ with indexed peaks (003) and (104) in case of cation mixing $0\% \dots 10\%$, with an intensity ratio of (003) and (104) peak of 1.44 in case of no cation mixing and a ratio of 1 in case of 10% antisite Nickel. (**b**) Calculated on dependency between peak intensity ratio of (003) to (104) and percentage cation mixing in the case of NMC811. Pattern calculation was carried out with WinPlotr tool of FullProf Suite.

Strategies for reducing cation mixing includes: (i) over-lithiation: where the ratio between the lithium species and the transition metal hydroxide precursor is higher than one. The positive influence of this strategy is explainable with the additional insertion of lithium into the TM layer and thereby reducing the frustration. However, the introduction of excess lithium species also lead to a greater amount of Li residues after calcination, which results with issues during electrode fabrication, storage, and cycling, (ii) increasing the partial pressure of oxygen deriving the reaction, $[Li_xNiO_2$ (layered) $\rightarrow$ $3NiO$ (rock salt) $+ 3x/2\ Li_2O + 6-3x/4O_2]$, towards the layered oxide, (iii) increasing the thermodynamic stability of the layered oxides, (iv) lowering Ni content and limiting SoC, by decreasing the lower and upper cut off voltages, but this is at the expense of the cell capacity and materials cost, (v) cation and/or anion doping, and (vi) surface coating and so forth.

3.3. Safety of Ni-Rich Cathodes

The commercialization of Ni-rich cathode materials austerely needs the full-scale appraisal of their safety-induced risks. It has been proven that the thermal stability of NMC decreases with increasing its Ni content, i.e., against increasing capacity and decreasing material cost [51]. From Figure 6a, it is clear that increasing the Ni content results in decreasing the onset decomposition temperature and increasing the strength of the exothermic peak (i.e., degree of heat release as can be evaluated by the enthalpy, ΔH) as well as pressure. This is attributed to the lower binding energy of Ni^{3+}-O bond and, thus, oxygen get released during calcination from the layered lithium transition metal oxide structure, which leads towards shifting in the ratio between Ni^{2+} and Ni^{3+} and introduction of oxygen defects into the lattice favoring cation mixing. Figure 6b,c give the decomposition of the predominant electrolyte solvent (ethylene carbonate, EC) and liability towards gas evolution of alkyl carbonates in the presence of different electrode materials. It is attributed to the oxidation of the alkyl carbonates by the reactive oxygen that is released from the NMC lattice. The gassing kinetics of alkyl carbonate-based electrolytes on various cathode electrodes and conductive carbon (C65) shows that the onset potential for the evolution of gases (e.g., CO_2 and CO) heavily depends on the nature of the electrode materials and increases in the order of NMC811 < NMC111 ~ NMC622 < C65 ~ LNMO (Figure 6b,c) [52]. The difference in the gas evolution on NMC811 vs. the other NMC versions (i.e., NMC111 and NMC622) can be explained due to an increase in the portion of the highly reactive Ni^{3+}/Ni^{4+} in the total amount of Ni with increasing the nickel content ($\geq$ 80%). This again reaffirms that a due care and in-depth investigation with respect to the thermal stability is needed for the commercialization of Ni rich (with Ni $\geq$ 80%) cathode materials.

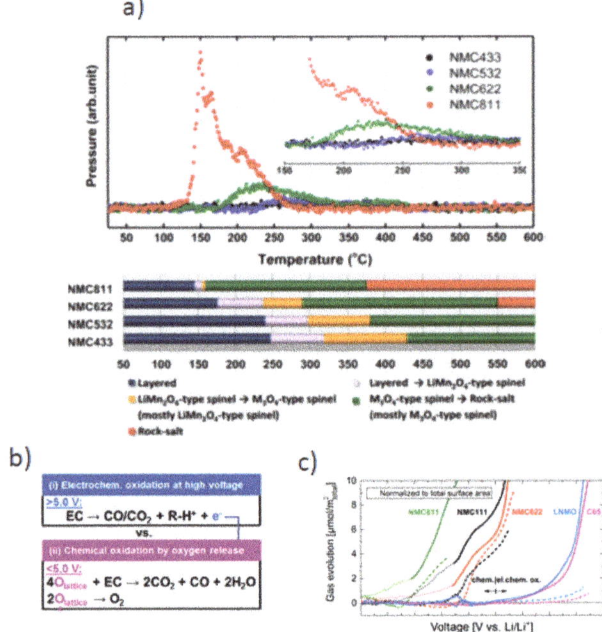

Figure 6. (**a**) Mass spectroscopy profiles for oxygen (O_2, m/z = 32) collected simultaneously during measurement of Time Resolved-X-Ray Difraction and the corresponding temperature region of the phase transitions for NMC samples (lower panel), reproduced from Ref [51] with permission, Copyright 2014, ACS, (**b**) gassing mechanisms in high-voltage LIBs, and (**c**) oxidative stability of carbonate electrolyte (1 M LiPF$_6$ in EC:EMC 3:7 wt/wt) on various electrode materials. (**b**) and (**c**) are reproduced from Ref. [52] with permission, Copyright 2017, ACS.

3.4. Effect of Electrode Manufacturing on Ni-Rich Cathode Ageing

For practical applications, a slurry made of Ni-rich cathode powder material; carbon black and PVDF are mixed and subsequently dissolved in N-methyl-2-pyrrolidone (NMP) or N-ethyl-2 pyrrolidone (NEP) solvent. Afterwards, the resulting slurry is coated onto aluminum (Al) foil current collector, followed by a drying step via evaporating the dissolving solvent.

This process of slurry making suffers from several drawbacks, such as cost and toxicity, because NMP is hazardous, teratogenic and irritating, and PVDF is mutagenic and teratogenic [53]. Hence, alternative electrode processing approaches are of high importance to enhance the ecological and cost reduction benefits of LIBs. In this regard, water-based green binders, e.g., sodium carboxymethyl cellulose (Na-CMC), are of paramount importance. In terms of cost, NMP is more expensive (1–3 \$ Kg^{-1}) when compared to ~ 0.015 \$ L^{-1} for water [53]. From polymeric binders perspective, the most commonly used aqueous-based binders, namely CMC (–5 \$ Kg^{-1}) and alginate (Alg, 8 \$ Kg^{-1}), are cheaper than PVDF (8–10 \$ Kg^{-1}). Moreover, more energy is required for evaporation, even though the latent heat of evaporation of water is four times higher than the one for NMP, since NMP has a higher boiling point than water (202 °C vs. 100 °C). Additionally, NMP solvent recovery is needed to avoid its dispersion into the atmosphere during drying (due to its higher toxicity) and a recovery system is also required to recycle the expensive solvent. However, despite the advantages of water as a solvent, water-based processing offers some challenges, such as the time-dependent dissolution of Li, materials compatibility (e.g., side reaction of Li with water), dispersion, and formulation challenges, which call for careful consideration. For instance, the reaction of NMC with water not only leads to dissolution of Li, but also increases the pH-value, which results in a more corrosive slurry against the Al current collector (i.e., leading to pitting corrosion). However, the formation of pitting can be lessened *via* controlling the pH-value with phosphoric acid to keep it below and by coating the current collector with an electronic conductive layer, e.g., carbon [54].

To overcome both drawbacks, i.e., metal dissolution and pitting, surface coating (e.g., making use of metal oxides, fluorides, phosphates, etc., as detailed in the later sections) of the NMC particles seems suitable, as it disables the direct interaction between the electrolytes constitutes (i.e., solvent and salt anions) and NMC electrode particles. Other challenges that come along with water-based binders can be solved easily, e.g., by adding isopropyl alcohol to avoid cracking (Figure 7) in thick electrodes [55] and an adjusted drying route to avoid residual water in the electrode.

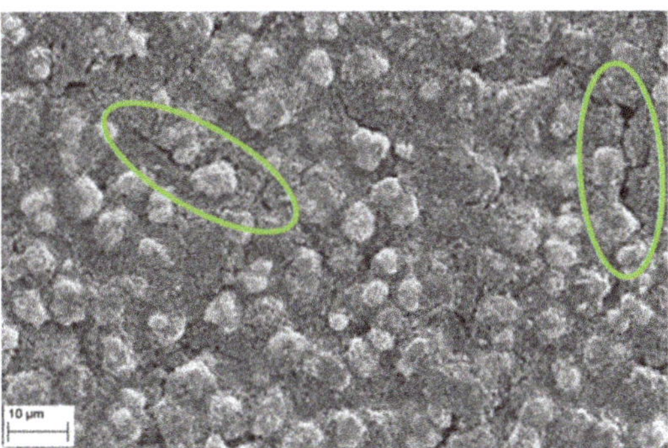

Figure 7. SEM image of cathode with cracks within binder structure but unimpaired cathode particles.

Another possibility for reducing the ecological footprint of LIB is the reusing and recycling of cell components, i.e.,finding second life application for cells. Since the end-of-life criteria for automotive cells is set to the state-of-health (SoH) at 80%, i.e., 80% of the initial capacity is remaining, they are still useable [56]. If a second life application is incapable, the recuperating of active material is possible in a destructive manner. After dissembling the LIB, one possibility would be the detachment of active material from the current collector by dissolving the binder [57]. The obtained particles may be unsuitable to be reused, as their active material is aged and binder residues diminish the electrochemical performance. Far more common is to leach out the transition metals and lithium. Hereby, in the case of the cathode, the Al current collector is dissolved in sodium hydroxide and NMC cathode in an acid. The obtained solution of transition metal ions and lithium ions can be treated and used for the synthesis of NMC [58–60]. Interestingly, Yang et al. [58] showed that the performance of NMC111 from recycled cathodes and pristine ones is comparable.

The importance of quality assurance to distinguish the sources of failure and ageing is highly recognized in the scientific community. Since binder degradation as a result of poor processing might lead to the detachment of active material and, thereby, capacity fade and increase of internal resistance, it can be confused with degradation phenomena of cathode active material, i.e., particle cracking and phase transformation. Hence, quality management might help to discriminate between the origins of both failures [60–62].

Defects can affect the performance of LIBs independent of the electrode-manufacturing route, and, in an extreme case, lead to failure of the whole cell [63]. Avoiding discard of electrodes and battery cells will lower fabrication costs, which makes LIB economical more attractive [64]. On the electrode's scale, Mohatny et al. [65,66] employed two methods for operando quality management. The first method monitors the uniformity of the coating with the help of a laser caliper. The second one, IR thermography, is able to detect manufacturing defects, such as agglomerates, pinholes, metal particle contaminates, and so forth. Furthermore, they investigated the influence of these defects on the LIB performance, revealing that defect prevention is essential while considering lifetime and efficiency.

3.5. Ni-Rich Cathode Ageing

Ageing leads to capacity and power fading in LIBs and it results from degradation phenomena. On a material level, it is mainly attributed to the loss of lithium inventory and that of active materials either on the anode or cathode compartments [67,68]. Such changes can occur in calendrical, i.e., time dependent, or cyclic, i.e., dependent on the amount of transferred charges. Various studies have evidenced that particle cracking and phase transformation are the degradation mechanisms for many of the NMC compositions, including NMC111 [69], NMC442 [70], NMC532 [71], and NMC811 [72]. In general, Ni-rich NMC face serious degradation at higher upper cut-off voltage and elevated temperatures [73,74]. Figure 8 summarizes the main performance deterioration routes, enlisting electrolyte decomposition, O_2 evolution, Ni^{2+} migration to the anode and precipitation, CEI dissolution, SEI contamination, and so forth in Ni-rich cathode materials [75].

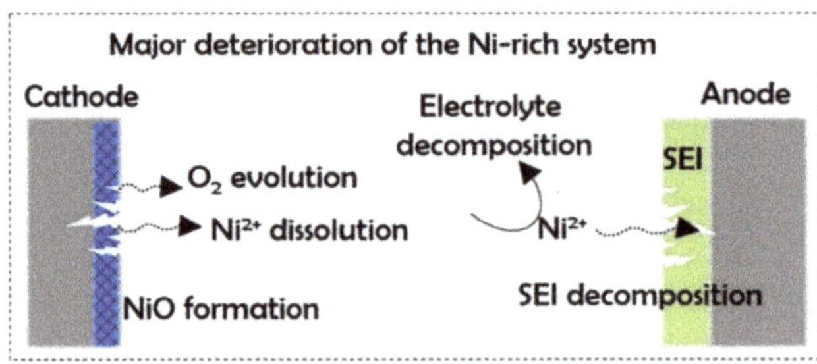

Figure 8. Schematic of the major deterioration of Ni-rich layered oxide system, adopted from Ref. [75] with permission, Copyright 2019, Elsevier.

3.5.1. Micro Cracking of Secondary Particle Structure

During charging, the voltage of the cathode electrodes is increased until a defined upper cut-off value leading to delithiated stage, i.e., extraction of Li ions from the layered structure [73,76,77].

The charge compensation is attributed to the transition metal, in particular, nickel ions and for higher voltage plateaus, cobalt [78]. XRD measurements revealed a volume change of the unit cell, which is attributed to changes of the lattice parameters a, and c. The shrinkage of parameter a is correlated with the oxidation of the transition metals and linked to a decrease in the bond length between the transition metal ions. It is also reported that a stronger volume change of the unit cell occurs as the nickel content increases, which might be due to the higher amount of Jahn—Teller distortions induced by Ni^{3+} ions [79].

In the case of the lattice parameter c, representing the distance of the transition metal layers to each other, the situation is more complex. At first, the unit cell expands in c-direction as lithium is extracted due to an increased electrostatic repulsion between the oxygen atoms. As charging continues, a shrinkage of c occurs. In NMC, there is a formation of a hybrid orbital consisting of transition metals with their characteristic d orbital and the oxygen p orbital. At higher degrees of delithiation, charge compensation is not solely provided by transition metal ions, but also by the hybrid orbitals, which leads to a charge transfer from the oxygen 2p orbital towards the partially filled nickel e_g orbital. Consequently, the repulsion between oxygen atoms decreases and the parameter c shrinks due to the increase in the covalency between the metal and oxygen, and, in this case, the NMC becomes more reactive, being associated to the release of highly reactive oxygen (e.g., singlet or atomic oxygen) from the NMC lattice [80]. This ultimately leads to further accelerated degradation phenomena. At first, the volume change induces stress in the secondary particles, leading to cracks between the primary particles, i.e., intergranular cracking (Figure 9) [81,82]. As a consequence, capacity fading occurs, which is linked to the detachment of primary particles from the secondary particle matrix [83]. Here, the resistivity of detached particles increases drastically and, in an extreme case, it is present as inert phase [84]. A recent investigation of NCA particles by Besli et al. [85] proved such correlation and the authors showed that a local increase of impedance and the rearrangement of the diffusion paths is a result of intergranular cracking within the NCA particle [85]. Furthermore, there is a correlation between the particle size distribution of NMC particles and state of health, which indicates that degradation is a result of particle cracking [57,86].

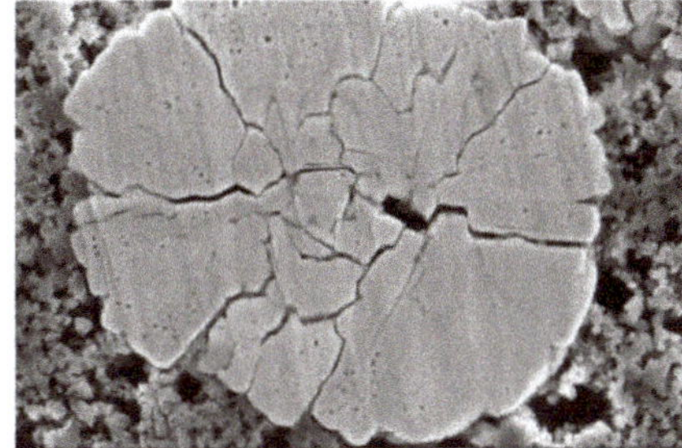

Figure 9. FIB-SEM secondary electron images of NMC111 after cycling under abusive conditions revealing crack evolution in secondary particles. (**Left**): magnification 30k, (**Right**): magnification 100k.

3.5.2. Electrolyte Degradation and Interphasial Reactions

The side reactions between the highly reactive delithiated cathode material and the electrolyte solution (i.e., electrolyte degradation and gas generation) is one of the challenges that need to be overcome for the massive deployment of Ni-rich cathode based LIBs [87]. The evolution of a highly detrimental surface reaction layer (ca. cathode-electrolyte interphase, CEI), which is comparable to the solid-electrolyte interface on the anode side but less known and investigated, occurs on the cathode surface. The electrolyte degradation also leads to transition metal reduction, in particular Ni^{4+} for Ni-rich layered oxides, which are attributed to its low-lying lower upper molecular orbital (LUMO) [88–90].

In the course of Ni reduction, the energy barrier for phase transformation is lowered, which makes Ni-rich cathodes more prone to structural reconstruction that is associated with a phase transformation from R-3m over spinel structure Fd-3m to the Fm-3m rock-salt structure [77,91]. The formation of rock-salt structure is more probable at high upper cut-off voltages and high Ni contents [92].

Most importantly, each phase transformation is accompanied by a volumetric change in the lattice structure, which facilitates crack formation. This process increases the electroactive surface, favoring further electrolyte invasion [77,91]. The phase transformation is more likely to occur at the particle's surface, whereas the outer part of the particle is enriched with oxygen vacancies. Due to this, the expansion in the outer part is larger compared to the core one. The resulting mechanical stress initiates and propagates radial cracks, which ultimately lead to intragranular voids and the breakdown of the particles [83]. Moreover, the disproportionation of Mn ions ($2Mn^{3+} \rightarrow Mn^{4+} + Mn^{2+}$) can possibly occur, which promotes further dissolution and formation of complexes between Mn^{2+} ions and carboxylate groups. Carboxylate groups result from electrolytic breakdown of the alkyl carbonate solvents, e.g., dimethyl carbonate or ethylene carbonate [93]. At this point, Mn^{2+} ions are capable of shuttling towards the anode and thereby contaminate it. Therefore, Mn^{2+} ions influence the SEI on the anode as they become deposited onto it. As a result, the loss of lithium inventory is accelerated, since a large amount of Li is trapped in the SEI in the presence of Mn-derived precipitated compounds, such as MnF_2 [94,95]. Thus, the electrocatalytic effect of manganese ions promotes electrolyte reduction, anode SEI growth, and, thereby, loss of lithium inventory.

3.5.3. Layered-Spinel to Rock Salt Phase Transition

Ni-rich cathode structural instability at the charged state is expressively connected to the oxygen gas evolution of the cell and Ni^{2+}/Li^+ mixing. The highly unstable cathode at a charged state can easily degrade by exothermic and endothermic phase transitions and this transformation occurs via multiple steps and is highly anisotropic by nature. The unstable Ni^{4+} can be reduced to the stable Ni^{2+} via forming oxygen-containing species, such as O_2^-, O^-, O_2^{2-}, and O_2. A larger transformation layer in the lithium diffusion direction is visible than perpendicular to it (Figure 10a). During the structural distortion, firstly cation mixing is induced during cycling. Hereby, primarily nickel is capable of migrating towards unoccupied Li sites in the delithiated NMC, [96] resulting in the clogging of the Li^+ diffusion paths, leading to a decrease in the ionic conductivity. The result is a lithium containing transition metal spinel (Figure 10b), where lithium can be found on the tetrahedral site. As the degradation proceeds, the residual lithium is mainly extracted and replaced by cobalt, which is capable of migrating in a highly oxidized state, likewise nickel, on the tetrahedral site [96]. In the end, the rock-salt structure (Figure 10c), which has poor Li kinetics, thus leading to capacity fading and rise in the internal resistance, is formed [35,97,98]. The latter is well detectable with the help of electrochemical impedance spectroscopy (EIS), as described elsewhere [99].

Mu et al. [70,100] proposed that the phase transformation is locally induced by abnormal changes in the valence states of the transition metal ions through modeling and experimental evidences, i.e., reduction of transition metal ions without the intercalation of Li. This could be linked to the interaction between electrolyte and delithiated NMC. As a result, the formation of nuclei of the new phase occurs. Each phase transformation is associated with the evolution of oxygen, owing to the hybridization of transition metal d and oxygen p orbitals [101]. Wandt et al. indicated that the release of singlet oxygen $^1O_2$, i.e., reactive oxygen species and this oxygen release is dependent of the phase transformation and vice-versa [98]. Figure 11 depicts this phenomenon and the incremental capacity analysis (ICA) of pseudo-OCV curves of NMC111, NMC622, and NMC811, showing a shift of the high potential redox peak from approx. 4.65 V vs. Li for NMC111 to approx. 4.2 V vs. Li for NMC811. The peak is linked to an oxygen redox process during the charge and discharge process [35].

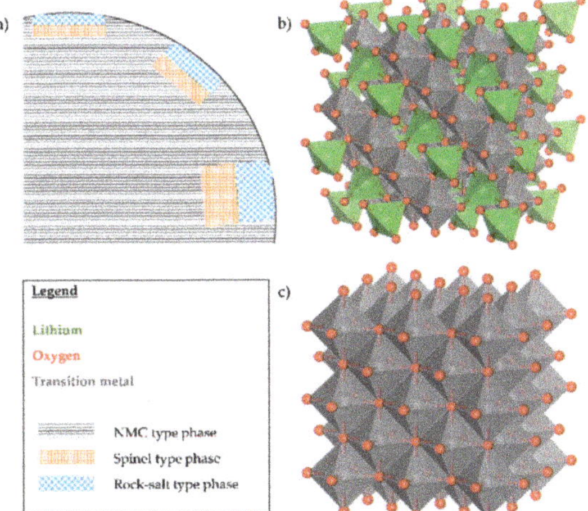

Figure 10. (**a**) Scheme of cross-sectional view in NMC particle revealing R-3m structure (layered) and exemplary regions of structural distortion, size of these sections is anisotropic, larger in direction of Li diffusion paths, orange section representing spinel type phase, blue section representing rock-salt type phase. (**b**) Fd-3m spinel type structure with Li (green) in tetrahedral coordination with oxygen (red) and transition metal (grey), mostly nickel in octahedral coordination representing first full phase transformation from layered structure towards distorted structures. (**c**) Fm-3m rock-salt structure with octahedral coordination of transition metal (grey), mostly nickel, with oxygen (red), structure is fully delithiated without Li ion conductivity. Structural models are plotted with VESTA.

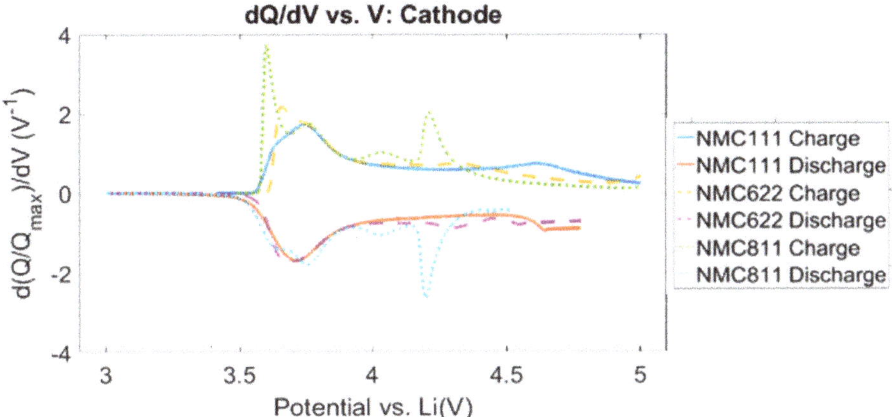

Figure 11. Incremental capacity analysis curve of NMC111 (solid line), NMC622 (dashed line) and NMC811 (dotted line) calculated from pseudo-OCV curves of the materials from C/25-charge and discharge. Highlighted peaks indicate redox peak with oxygen release characteristics.

The rate of phase transition is higher than oxygen transport and, thus, an accumulation of interstitial excess oxygen can occur to a certain degree of agglomeration due to phase transformation. During formation of the oxygen agglomerates, stress is evolved in the lattice and, finally, a crack is induced within the primary particle. Once the crack opened, oxygen is released, which might also result in additional electrolyte oxidation.

3.5.4. Transition Metal-Ion Dissolution and Anode-Cathode Dialogue

Similar to phase transition, transition metal ion dissolution is the result of the combined effect of the M/Li mixing and OER. They both lead to the formation of low valence transition metal oxides assisted by the loss of oxygen, leading to cations whose salts are more soluble in liquid electrolytes than those composed of high valence transition metal ions are. The oxygen assisted solvent decomposition [(e.g., $C_3H_4O_3$ (EC) + [O] $\rightarrow CO_2$ + CO + $2H_2O$)] results in the formation of H_2O, which, in turn, hydrolyzes the salt ($LiPF_6 \leftrightarrow LiF + PF_5$) in the electrolyte solution with the formation of HF ($LiPF_6$ + $H_2O \rightarrow POF_3 + 2HF + LiF$; $POF_3 + H_2O \rightarrow HF + HPO_2F_2$) [102]. This, in turn, attacks the metal oxide, which results in the formation of slightly more soluble MF_2 salts and H_2O.

4. Mitigating Strategies

Many promising strategies have been proposed to date in response to the multifaceted challenges associated with Ni-rich cathode materials. Some of the major and representative approaches are briefly discussed henceforward.

4.1. Tailoring Morphology of Ni-Rich Cathode Particles

The morphology of the Ni-rich active materials greatly influences the electrochemical performance of Ni-rich cathode based LIB cells [103,104]. Thus, one of the approaches is then to tailor the electroactive surface, i.e., the plane across which Li is inserted or extracted. A large amount of surface area will increase the power density and seemingly deliver higher specific capacity during constant current discharge, which is distorted by the decrease of internal resistance. Nevertheless, a large surface area is also prone to accelerated electrolyte degeneration. Wang et al. [105] reported an improvement on the cycle performance by decreasing the surface area via adjusting the precursor concentration. Peralta et al. [33] synthesized non-agglomerated sub-micron flake-shape particles with a small ratio of electroactive surface at the edge planes, i.e., a small area for lithium insertion, following the approach to diminish the amount of active surface area. Thus, the surface reconstruction, which primarily occurs at the insertion planes, is impeded. A similar approach was carried out by Dahn group with the synthesis of single crystal NMC particles: NMC532 [106], and NMC622 [107], which were found to outperform their polycrystalline counterparts [108]. Single crystal morphologies are also found to be advantageous in avoiding intergranular cracking and, thus, improving the intercalation of Li, which will otherwise be hindered by grain boundaries [109].

4.2. Foreign-Ion Doping

The incorporation of foreign-ions, which involve both cation and anion, is one of the simplest and most common strategy to augment the structural and thermal stabilities of Ni-rich layered oxides. In the case of cation doping, Li and transition metal sites are both considered. On the Li site, other alkali metals can be also introduced, i.e., besides to Li. The mechanism by which doping benefits the stabilization of Ni-rich cathodes is linked to: (i) integration of electrochemically inactive elements into the host structure, (ii) inhibition of undesired phase alterations, from layered over spinel to rock salt structure, and (iii) preferment of Li^+ mobility due to increased Li slab distance by the dopants. Electrochemically inactive ions, such as Na^+ [110], Mg^{2+} [111], Al^{3+} [112], and Ti^{4+} [113], etc., are among cation dopants that are recommended to boost the structural stability, reduce oxygen release and avoid cation mixing. The improvement is largely linked to the enlarged interlayer distance of O-Li-O, impeding cation mixing and thereby enhancing structural stability.

Overlithiation, which was presented earlier as an approach for reducing initial cation mixing during synthesis, can be considered as cation doping of the transition metal layer [114]. In this case, the resulting structure is comparable to Li-rich NMC, where Li is inserted into the transition metal layer of the layered R-3m structure. The beneficial features of this approach is the increase in capacity, albeit poor cycle life and fast capacity fading [115]. However, Wang et al. showed that there is an optimum limit for overlithiation of Ni-rich layered oxide cathodes, because, beyond this point, the performance of the material is diminished [116]. Another route of improving the material characteristic is the partial substitution of the transition metals. In this regard, Co came into focus due to its high toxicity and costs or Mn due to its electrochemical inactivity [117]. The substitution of Ni is insensible, since it provides most of the capacity to the battery cell. The choice of the dopant is attributed to the bond energy of the substituting element with oxygen, since high bonding energy is associated with improved structural stability [118]. In general, due to their electronic configuration, i.e., d-electrons, transition metals are capable of contributing to the capacity. For example, iron doped $LiNi_{0.35}Mn_{0.35}Co_{0.27}Fe_{0.03}O_2$ deliver a higher capacity when compared to its iron free counterpart [119]. In case of doping with non-transition metal elements, the capacity of the cathode active material is lowered. Figure 12a shows significant improvement upon doping of Ni-rich NMC811 cathode with various cation dopants [120]. Among tested cations, tantalum ($Ta^{5+)}$ is found to be the most effective dopant, improving the discharge capacity, capacity retention, and cycling stability at 45 °C.

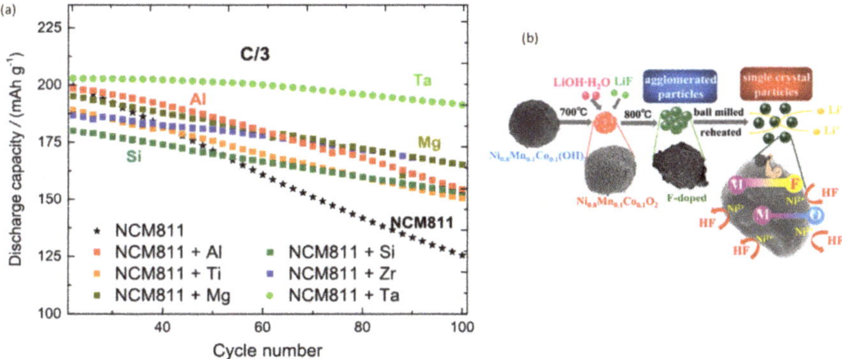

Figure 12. (a) Discharge capacities of NCM811 electrodes comprising undoped and cation doped active materials in coin cells at 45 °C, adapted from Ref. [120] with permission, copyright 2019, ACS (b) Fabrication of Fluorine doped NMC8111, reproduced from Ref. [121] with permission, Copyright 2019, Elsevier.

In addition, anion substitution is considered as one strategy, although it is predominantly adopted for Li-rich NMC materials. This involves the substitution of O^{2-} by other anions such as F^-, Cl^-, S^{2-} and so forth [19]. Among others, F^- is predominantly used as anion dopant for strengthening the binding energy between the transition-metal cations and anions, and it is attributed to its more electronegativity (3.98) compared to O^{2-} (3.44). O^{2-} substitution by F^- significantly improves the electrochemical properties of Ni-rich cathode materials, including capacity retention, rate capability, and thermal stability, and prevent the surface from attack by HF. Figure 12b depicts the synthesis scheme of fluorine doped NMC cathode material [121].

4.3. Core Shell and Gradient: Full Concentration Gradient

In general, increasing the Ni content will increase the capacity and decrease the cost, but, at the same time, decreases the structural and thermal stabilities. In view of the latter, core-shell (Figure 13a) and full concentration gradient (Figure 13b), which are also more exemplified in Figure 13c, have been proposed as enabling solutions to address the degradation phenomenon in Ni-rich cathode materials [31,122]. Hereby, a core-shell should not be confused with surface coating. For the latter, a thin, non-isostructural film (in range of Angstrom or nanometer) is coated onto the Ni-rich cathode particles, whereas a thicker, isostructural, Li-ion conductive film is applied in the case of core-shell [19].

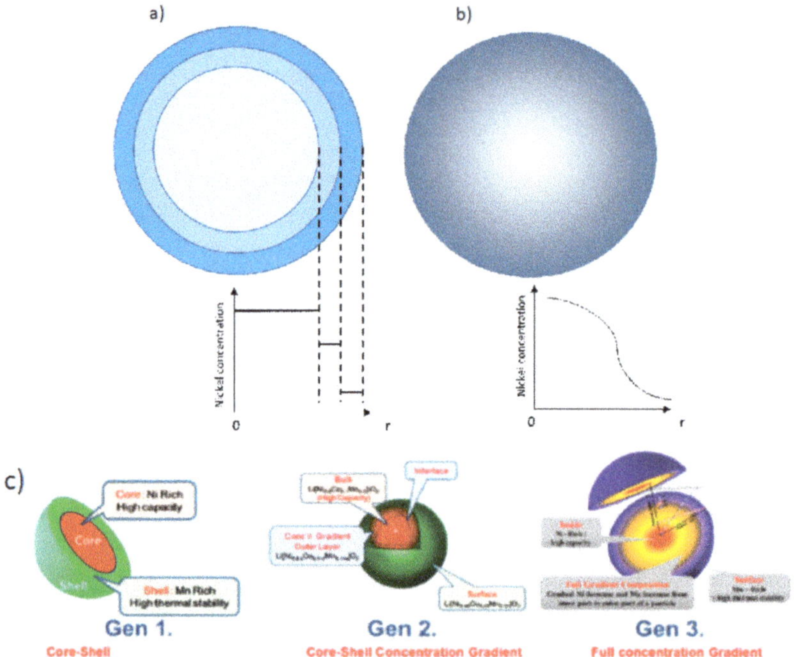

Figure 13. (**a**) Scheme of full concentration gradient through NMC particle, with Ni-rich core increasing capacity and Ni-poor shell with enhanced structural stability. (**b**) Scheme of (multilayer-) core-shell NMC particle, with Ni-rich core and Ni-poor shell(s). (**c**) (Left) Schematic drawings of a core–shell (Gen 1); (center) a Ni-rich core surrounded by a concentration gradient outer layer (Gen 2); and, (right) a FCG lithium transition metal oxide particle (Gen 3) with the nickel concentration decreasing from the center toward the outer layer and the concentration of manganese increasing accordingly. Reproduced from Ref. [122] with permission, Copyright2017, ACS.

The main target of this approach is to form a particle with a high Ni content in its core and a Ni depleted outer shell. In other words, a particle is synthesized with a structural stable shell and a high capacity core [123]. Depending on the smoothness of the concentration gradient, the modification is denoted as full concentration gradient, multilayer-core-shell, or core-shell (ordered by decreasing smoothness) [124]. Full concentration gradient (FCG) is an advanced form of core-shell structure and, accordingly, provides extra capacity, 200 mAh g^{-1} for full concentration gradient LiNi$_{0.8}$Co$_{0.1}$Mn$_{0.1}$O$_2$ vs 188 mAh g^{-1} for core-shell LiNi$_{0.8}$Co$_{0.1}$Mn$_{0.1}$O$_2$ at 4.3 V [125]. The FCG of Ni, Co, and Mn is stretched through the entire length of the particle with the compositions of Ni and Co linearly decreasing from the center and Mn smoothly increasing to the outer surface, leaving a Mn deficient interior.

When considering the synthesis route, there are differences between (multilayer-) core-shell and full concentration gradient Ni-rich cathode materials. Chen et al. [126] presented a co-precipitation process, employed and licensed by the Argonne National Laboratory, in order to obtain full concentration gradient.

During precipitation, the Ni concentration of the solution is continuously decreased by adding Ni-poor, Mn, and Co rich solutions, which results in precursor particles with a Ni-rich core and Ni-depleted outer shell. The material with concentration gradient (ca. NMC622) outperformed the heterogeneous NMC622. In case of core-shell NMC, the Ni depletion is carried out using stepwise process. At first, NMC precursor particles are obtained via co-precipitation from a Ni-rich solution. As a sufficient degree of particle growth is reached, Mn and Co are replenished, so that particle growth continues in a Ni-poor solution. Hereby, a Ni-depleted shell evolves around the Ni-rich core [127]. In both full concentration gradient and core-shell materials, the calcination temperature for lithiation must be kept relatively low to avoid transition metal diffusion and decomposition of the gradient [128].

Moreover, the considerations of cation doping and concentration gradients can be combined. Such a dopant concentration gradient is reported for aluminum [129]. Here, the inactive aluminum species stabilizes the surface of the Ni-rich NMC particle, whereas the capacity loss due to doping is diminished, since the overall amount of Al is smaller when compared to heterogeneous doping.

However, a lattice parameter mismatch is unavoidable in such materials, being linked to the difference in compositions, despite all of the beneficial features.

4.4. Surface Modification

The most upfront method to diminish that impact could be modifying the surface of the cathode particle/electrode and/or interfaces since most of the parasitic redox reactions in Ni-rich cathode materials occur at the surface and polarized interphases between the electrodes and electrolytes. Moreover, capacity fading through structural distortion commonly starts from the surface of the active cathode material and, thus, surface modifications (via surface coating) presents as one of the most sensible approaches. A great number of studies have been carried out while using different coating materials and techniques. These surface coating materials are both electrochemically and chemically inactive, enlisting metal oxides (e.g., Al_2O_3, SiO_2, etc.), fluorides (e.g., AlF_3, FeF_3 etc.), phosphates (e.g., $AlPO_4$, Li_3PO_4 etc.), lithium transition metal oxides (e.g., $LiVO_3$), conductive polymeric, and interfacial layers [130]. In general, there are two types of coatings. On one side, the non-conductive ones, where alumina is mostly the material of choice due to its high stability, and the other type is the conductive coating, i.e., Li ion conductive. However, coating always increases the resistance and thus minimization or optimization of the coating thickness is necessary. In view of this, atomic layer deposition (ALD) is one of the most widely utilized methods of interest for coating. It offers the possibility to deposit film layer by layer onto the particle, thus enabling proper control on the thickness of the film. Figure 14 shows the possible mechanism of NMC coating while using $AlPO_4$ and Al_2O_3 [131].

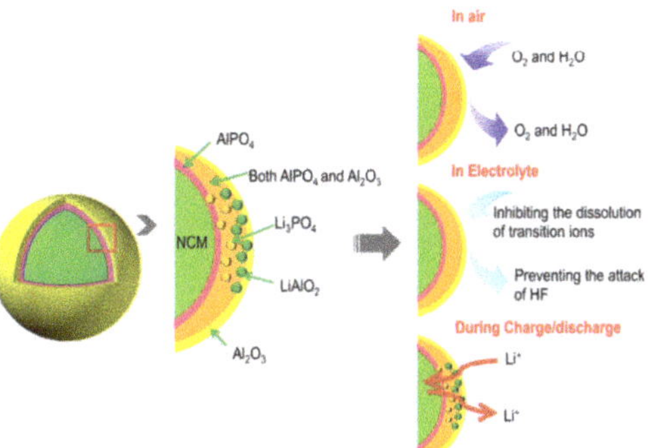

Figure 14. Possible mechanism of double-coated NMC cathode material, using Al_2O_3 and $AlPO_4$. Reproduced from Ref. [130] with permission, Copyright 2017, Elsevier.

4.5. Functional Electrolyte Additives

The use of electrolyte additives presents to be the most efficient and economic means towards solving the multifaceted hurdles that are associated with nickel rich cathode materials [75]. The additives, which are incorporated in small doses, can play multiple roles, including as cation stabilizers, reactive oxygen radicals and HF scavengers, robust cathode-electrolyte interphase (CEI) builders, surface modifiers, etc., without altering the bulk properties of the electrolyte system. Among many others, fluorinated molecules, such as Methyl (2, 2, 2-Trifluoroethyl) carbonate (FEMC) [132], bi-lithium difluoro(oxalato)borate, pentafluorophenylisocyanate, Lithium difluoro (oxalate) borate (LiDFOB) [133], lithium difluorophosphate ($Li_2PO_2F_2$), tris(2,2,2,-trifluoroethyl) phosphite (TTFP) [134], triphenylphosphine oxide boron trifluorid (TPPO-BF) [135], and other functional groups, such as bi-functional tris(trimethylsilyl) phosphate, (2-cyanoethyl) triethoxysilane, etc., are among the investigated additives and they proved to prevent transition metal dissolution and/or oxygen leaching owing to the formation of effective passivation layers both on the anode and cathode surfaces. Some additives with strong anion coordination effect, for instance tris (pentafluorophenyl) borane (TPFPB), are known to trap highly reactive oxygen species (O_2^{2-}/O_2^-) or radicals ($O_2^{\dot{Y}-}/O^{\dot{Y}-}$) that will otherwise oxidize the electrolyte and induce structural transformation [136]. Figure 15a,b show the enabling effect of sample additives (i.e., silyl-functionalized dimethoxydimethylsilane and blend of FEMC and VC) on Ni-rich cathode materials [137,138].

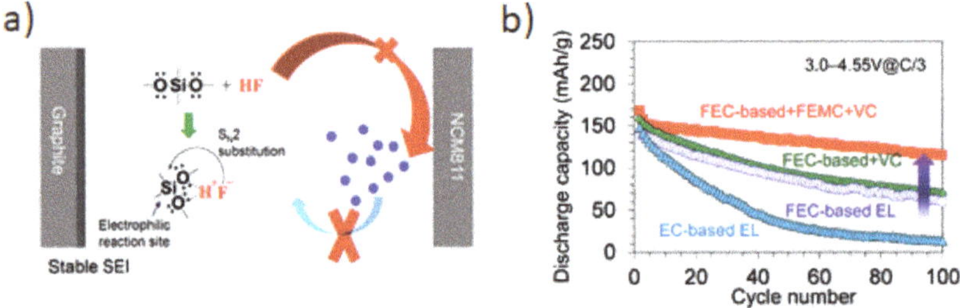

Figure 15. (**a**) Effect of silyl-functionalized dimethoxydimethylsilane additive on the surface stability of both high voltage Ni rich cathode and anode, adopted from Ref [138] with permission, Copyright 2018, Elsevier (**b**) Cycling performance, of the $LiNi_{0.5}Co_{0.2}Mn_{0.3}O_2$/Si-graphite in 1M $LiPF_6$/FEC: DEC electrolyte with out and with blended additives of FEMC-VC, obtained from Ref [137] with permission, Copyright 2016, Elsevier.

4.6. Designer Polymeric Binders

Though there is not much data regarding the utilization of tailored polymeric binders endowed with synergistic effects for Ni-rich cathode materials, i.e., with binding and filming roles, the exploration of functionalized polymeric binders could be one potential research focus to address the challenges mentioned above. A binder with special moieties could lead to the formation of a robust cathode-electrolyte interphase (CEI) and thereby prevent the drawbacks that are associated with NMC cathode materials [8,139].

5. Conclusions

The market penetration of emerging applications, such as the electro mobility and efficient integration of green energy sources into the energy mix, sturdily demands high energy density batteries, i.e., ~2.5 times more energy than the contemporary LIB can store. Advanced cathode materials are needed to meet this goal. As near-term candidates, the development of nickel (Ni)-rich cathodes have harvested huge attention from the scientific, government, and industry sectors. Ni-rich ternary-layered cathodes are characterized by high capacity, improved rate capability, lower material cost and reduced transition metal dissolution. However, the presence of various hurdles, mainly induced by the increment in Ni content, hinders the widespread commercialization of these type cathodes. These include high surface reactivity, leading to electrolyte decomposition and oxygen evolution reaction, Ni^{2+}/Li^+ mixing, phase (spine-to-rock salt type) transition, micro cracking, transition metal dissolution, and the migration of partially soluble salts to anode and thereby contaminating it etc. In response to such challenges, numerous research groups have suggested various mitigating strategies, including anion and cation doping/substitution, surface modification, deployment of role-specific electrolyte additives, polymeric binders, etc. In general, potential future research directions can be summarized, as follows:

(i)   Novel electrolyte design: The development of designer electrolyte salt anions and tailored electrolyte additives presents as one of the emerging avenues for the realization of high-voltage nickel rich cathode materials. Salt anions stable at high voltage and those do not result in the formation of reactive species are needed. On the additive side, molecules or ionic salts of multi-functionality and/or single functionality with synergistic effect are desirable.

(ii) Degradation phenomenon: The chemistry, major routes, and ultimate effects of the different degradation mechanisms require an exhaustive investigation and understanding. Of particular interest, the cathode/electrolyte interface and its dialogue with the anode and/or anode-electrolyte interface is of supreme importance and it deserves a deep characterization and, thus, comprehension while using innovative experimental and theoretical calculations. Moreover, the quantification of the different aging behaviors and understanding of the correlations among the various degradation mechanisms of Ni-rich cathode materials will be highly appreciated.

(iii) Safety appraisal: The full-scale safety profiling of high voltage and Ni-rich cathode materials is one of the stern requirement for practical operations and it needs an in-depth investigation enlisting both thermal and chemical threats making use of various arsenal characterization tools.

(iv) Advanced characterization: The use of nondestructive advanced characterization tools is required to better understand and profile the degradation and aging routes.

Author Contributions: Conceptualization, P.T. and G.G.E.; software, P.T.; writing—original draft preparation, P.T.; writing—review and editing, G.G.E. and H.J.; visualization, P.T. and G.G.E.; supervision, E.F. All authors have read and agreed to the published version of the manuscript.

Funding: This research received no external funding.

References

1.   Eshetu, G.G.; Armand, M.; Ohno, H.; Scrosati, B.; Passerini, S. Ionic Liquids as Tailored Media for the Synthesis and Processing of Energy Conversion Materials. *Energy Environ. Sci.* **2016**, *9*, 49–61. [CrossRef]

2.   Zhang, H.; Eshetu, G.G.; Judez, X.; Li, C.; Rodriguez-Martínez, L.M.; Armand, M. Electrolyte Additives for Lithium Metal Anodes and Rechargeable Lithium Metal Batteries: Progress and Perspectives. *Angew. Chem.* **2018**, *57*. [CrossRef] [PubMed]

3.   Olivetti, E.A.; Ceder, G.; Gaustad, G.G.; Fu, X. Lithium-Ion Battery Supply Chain Considerations: Analysis of Potential Bottlenecks in Critical Metals. *Joule* **2017**, *1*, 229–243. [CrossRef]

4.   Eshetu, G.G.; Mecerreyes, D.; Forsyth, M.; Zhang, H.; Armand, M. Polymeric Ionic Liquids for Lithium-Based Rechargeable Batteries. *Mol. Syst. Des. Eng.* **2019**, *4*, 294–309. [CrossRef]

5.   Ryu, H.H.; Park, K.J.; Yoon, D.R.; Aishova, A.; Yoon, C.S.; Sun, Y.K. Li[$Ni_{0.9}.Co_{0.09}W_{0.01}$]$O_2$: A New Type of Layered Oxide Cathode with High Cycling Stability. *Adv. Energy Mater.* **2019**, *9*, 1902698. [CrossRef]

6.   Winter, M.; Barnett, B.; Xu, K. Before Li Ion Batteries. *Chem. Rev.* **2018**, *118*, 11433–11456. [CrossRef]

7.   Eshetu, G.G.; Figgemeier, E. Confronting the Challenges of Next-Generation Silicon Anode-Based Lithium-Ion Batteries: Role of Designer Electrolyte Additives and Polymeric Binders. *ChemSusChem* **2019**, *12*, 2515–2539. [CrossRef]

8.   Goodenough, J.B.; Kim, Y. Spherical Carbon Particles are Used for Efficient and Stable "Metal Lithium Storage Room". *Chem. Mater.* **2010**, *22*, 587–603. [CrossRef]

9.   Armand, M. Nature Lithium Battery. *Nature* **2001**, *414*, 359–367. [CrossRef]

10. Liu, C.; Neale, Z.G.; Cao, G. Understanding Electrochemical Potentials of Cathode Materials in Rechargeable Batteries. *Mater. Today* **2016**, *19*, 109–123. [CrossRef]

11. Lee, S.H.; Lee, S.; Jin, B.S.; Kim, H.S. Optimized Electrochemical Performance of Ni Rich $LiNi_{0.91}Co_{0.06}Mn_{0.03}O_2$ Cathodes for High-Energy Lithium Ion Batteries. *Sci. Rep.* **2019**, *9*, 8901. [CrossRef] [PubMed]

12. Manthiram, A.; Song, B.; Li, W. A Perspective on Nickel-Rich Layered Oxide Cathodes for Lithium-Ion Batteries. *Energy Storage Mater.* **2017**, *6*, 125–139. [CrossRef]

13. Mizushima, K.; Jones, P.C.; Wiseman, P.J.; Goodenough, J.B. Li_xCoO_2 (0 < x <1): A new cathode material for batteries of high energy density. *Mater. Res. Bulletin* **1980**, *15*, 783–789. [CrossRef]

14. Choi, J.W.; Aurbach, D. Promise and Reality of Post-Lithium-Ion Batteries with High Energy Densities. *Nat. Rev. Mater.* **2016**, *1*, 16013. [CrossRef]

15. Schmidt, T.; Buchert, M.; Schebek, L. Investigation of the Primary Production Routes of Nickel and Cobalt Products Used for Li-Ion Batteries. *Resour. Conserv. Recycl.* **2016**, *112*, 107–122. [CrossRef]

16. Xu, B.; Qian, D.; Wang, Z.; Meng, Y.S. Recent Progress in Cathode Materials Research for Advanced Lithium Ion Batteries. *Mater. Sci. Eng. R Rep.* **2012**, *73*, 51–65. [CrossRef]

17. Zhang, W.J. Structure and Performance of $LiFePO_4$ Cathode Materials: A Review. *J. Power Sources* **2011**, *196*, 2962–2970. [CrossRef]

18. Kasnatscheew, J.; Winter, M. Thorough Look by Consideration of the Li^+ Extraction Ratio. *ACS Appl. Energy Mater.* **2019**, *2*, 7733–7737. [CrossRef]

19. Hou, P.; Yin, J.; Ding, M.; Huang, J.; Xu, X. Surface/Interfacial Structure and Chemistry of High-Energy Nickel-Rich Layered Oxide Cathodes: Advances and Perspectives. *Small* **2017**, *13*, 1–29. [CrossRef]

20. Myung, S.T.; Maglia, F.; Park, K.J.; Yoon, C.S.; Lamp, P.; Kim, S.J.; Sun, Y.K. Nickel-Rich Layered Cathode Materials for Automotive Lithium-Ion Batteries: Achievements and Perspectives. *ACS Energy Lett.* **2017**, *2*, 196–223. [CrossRef]

21. Liu, W.; Oh, P.; Liu, X.; Lee, M.J.; Cho, W.; Chae, S.; Kim, Y.; Cho, J. Nickel-Rich Layered Lithium Transition-Metal Oxide for High-Energy Lithium-Ion Batteries. *Angew. Chem.* **2015**, *54*, 4440–4457. [CrossRef] [PubMed]

22. Rozier, P.; Tarascon, J.M. Review-Li-Rich Layered Oxide Cathodes for next-Generation Li-Ion Batteries: Chances and Challenges. *J. Electrochem. Soc.* **2015**, *162*, A2490–A2499. [CrossRef]

23. Hannan, M.A.; Hoque, M.M.; Hussain, A.; Yusof, Y.; Ker, P.J. State-of-the-Art and Energy Management System of Lithium-Ion Batteries in Electric Vehicle Applications: Issues and Recommendations. *IEEE Access* **2018**, *6*, 19362–19378. [CrossRef]

24. Miao, Y.; Hynan, P.; Von Jouanne, A.; Yokochi, A. Current Li-Ion Battery Technologies in Electric Vehicles and Opportunities for Advancements. *Energies* **2019**, *12*, 1074. [CrossRef]

25. Zeng, X.; Li, M.; Abd El-Hady, D.; Alshitari, W.; Al-Bogami, A.S.; Lu, J.; Amine, K. Commercialization of Lithium Battery Technologies for Electric Vehicles. *Adv. Energy Mater.* **2019**, *1900161*, 1–25. [CrossRef]

26. Schmuch, R.; Wagner, R.; Hörpel, G.; Placke, T.; Winter, M. Performance and cost of materials for lithium-based rechargeable automotive batteries. *Nat. Energy* **2018**, *3*, 267. [CrossRef]

27. Wangda, L.; Erickson, E.M.; Arumugam, M. High-nickel layered oxide cathodes for lithium-based automotive batteries. *Nat. Energy* **2020**, *5*, 26–34. [CrossRef]

28. Ajayi, B.P.; Thapa, A.K.; Cvelbar, U.; Jasinski, J.B.; Sunkara, M.K. Atmospheric Plasma Spray Pyrolysis of Lithiated Nickel-Manganese-Cobalt Oxides for Cathodes in Lithium Ion Batteries. *Chem. Eng. Sci.* **2017**, *174*, 302–310. [CrossRef]

29. Križan, G.; Križan, J.; Dominko, R.; Gaberšček, M. Pulse Combustion Reactor as a Fast and Scalable Synthetic Method for Preparation of Li-Ion Cathode Materials. *J. Power Sources* **2017**, *363*, 218–226. [CrossRef]

30. Kim, D.; Shim, H.C.; Yun, T.G.; Hyun, S.; Han, S.M. High Throughput Combinatorial Analysis of Mechanical and Electrochemical Properties of $Li[Ni_xCo_yMn_z]O_2$ Cathode. *Extrem. Mech. Lett.* **2016**, *9*, 439–448. [CrossRef]

31. Sun, Y.K.; Chen, Z.; Noh, H.J.; Lee, D.J.; Jung, H.G.; Ren, Y.; Wang, S.; Yoon, C.S.; Myung, S.T.; Amine, K. Nanostructured High-Energy Cathode Materials for Advanced Lithium Batteries. *Nat. Mater.* **2012**, *11*, 942–947. [CrossRef] [PubMed]

32. Ren, D.; Shen, Y.; Yang, Y.; Shen, L.; Levin, B.D.A.; Yu, Y.; Muller, D.A.; Abruña, H.D. Systematic Optimization of Battery Materials: Key Parameter Optimization for the Scalable Synthesis of Uniform, High-Energy, and High Stability $LiNi_{0.6}Mn_{0.2}Co_{0.2}O_2$ Cathode Material for Lithium-Ion Batteries. *ACS Appl. Mater. Interfaces* **2017**, *9*, 35811–35819. [CrossRef] [PubMed]

33. Peralta, D.; Salomon, J.; Colin, J.F.; Boulineau, A.; Fabre, F.; Bourbon, C.; Amestoy, B.; Gutel, E.; Bloch, D.; Patoux, S. Submicronic $LiNi_{1/3}Mn_{1/3}Co_{1/3}O_2$ Synthesized by Co-Precipitation for Lithium Ion Batteries—Tailoring a Classic Process for Enhanced Energy and Power Density. *J. Power Sources* **2018**, *396*, 527–532. [CrossRef]

34. Tian, C.; Xu, Y.; Nordlund, D.; Lin, F.; Liu, J.; Sun, Z.; Liu, Y.; Doeff, M. Charge Heterogeneity and Surface Chemistry in Polycrystalline Cathode Materials. *Joule* **2018**, *2*, 464–477. [CrossRef]

35. Arai, H.; Okada, S.; Ohtsuka, H.; Ichimura, M.; Yamaki, J. Characterization and Cathode Performance of Li$_1$ − XNi$_1$ + XO$_2$ Prepared with the Excess Lithium Method. *Solid State Ion.* **1995**, *80*, 261–269. [CrossRef]

36. Oh, P.; Song, B.; Li, W.; Manthiram, A. Overcoming the Chemical Instability on Exposure to Air of Ni-Rich Layered Oxide Cathodes by Coating with Spinel LiMn$_{1.9}$Al$_{0.1}$O$_4$. *J. Mater. Chem. A* **2016**, *4*, 5839–5841. [CrossRef]

37. Cho, D.H.; Jo, C.H.; Cho, W.; Kim, Y.J.; Yashiro, H.; Sun, Y.K.; Myung, S.T. Effect of Residual Lithium Compounds on Layer Ni-Rich Li[Ni$_{0.7}$Mn$_{0.3}$]O$_2$. *J. Electrochem. Soc.* **2014**, *161*, 920–926. [CrossRef]

38. Zhang, S.S. Insight into the Gassing Problem of Li-Ion Battery. *Front. Energy Res.* **2014**, *2*, 2–5. [CrossRef]

39. Dahn, J.R.; Fuller, E.W.; Obrovac, M.; von Sacken, U. Thermal Stability of Li$_x$CoO$_2$, LixNiO$_2$ and λ-MnO$_2$ and Consequences for the Safety of Li-Ion Cells. *Solid State Ion.* **1994**, *69*, 265–270. [CrossRef]

40. Hatsukade, T.; Schiele, A.; Hartmann, P.; Brezesinski, T.; Janek, J. Origin of Carbon Dioxide Evolved during Cycling of Nickel-Rich Layered NCM Cathodes. *ACS Appl. Mater. Interfaces* **2018**, *10*, 38892–38899. [CrossRef]

41. Wang, Y.; Jiang, J.; Dahn, J.R. The Reactivity of Delithiated Li(Ni$_{1/3}$Co$_{1/3}$Mn$_{1/3}$)O$_2$, Li(Ni$_{0.8}$Co$_{0.15}$Al$_{0.05}$)O$_2$ or LiCoO$_2$ with Non-Aqueous Electrolyte. *Electrochem. Commun.* **2007**, *9*, 2534–2540. [CrossRef]

42. Jo, C.H.; Cho, D.H.; Noh, H.J.; Yashiro, H.; Sun, Y.K.; Myung, S.T. An Effective Method to Reduce Residual Lithium Compounds on Ni-Rich Li[Ni$_{0.6}$Co$_{0.2}$Mn$_{0.2}$]O$_2$ Active Material Using a Phosphoric Acid Derived Li3PO4 Nanolayer. *Nano Res.* **2015**, *8*, 1464–1479. [CrossRef]

43. Kim, U.H.; Jun, D.W.; Park, K.J.; Zhang, Q.; Kaghazchi, P.; Aurbach, D.; Major, D.T.; Goobes, G.; Dixit, M.; Leifer, N.; et al. Pushing the Limit of Layered Transition Metal Oxide Cathodes for High-Energy Density Rechargeable Li Ion Batteries. *Energy Environ. Sci.* **2018**, *11*, 1271–1279. [CrossRef]

44. Schipper, F.; Erickson, E.M.; Erk, C.; Shin, J.Y.; Chesneau, F.F.; Aurbach, D. Review-Recent Advances and Remaining Challenges for Lithium Ion Battery Cathodes I. Nickel-Rich, LiNi$_x$Co$_y$Mn$_z$O$_2$. *J. Electrochem. Soc.* **2017**, *164*, A6220–A6228. [CrossRef]

45. Li, T.; Yuan, X.-Z.; Zhang, L.; Song, D.; Shi, K.; Bock, C. *Degradation Mechanisms and Mitigation Strategies of Nickel-Rich NMC-Based Lithium-Ion Batteries*; Springer: Singapore, 2019; Volume 2018. [CrossRef]

46. Tang, M.; Yang, J.; Chen, N.; Zhu, S.; Wang, X.; Wang, T.; Zhang, C.; Xia, Y. Overall Structural Modification of a Layered Ni-Rich Cathode for Enhanced Cycling Stability and Rate Capability at High Voltage. *J. Mater. Chem. A* **2019**, *7*, 6080–6089. [CrossRef]

47. Shim, J.H.; Kim, C.Y.; Cho, S.W.; Missiul, A.; Kim, J.K.; Ahn, Y.J.; Lee, S. Effects of Heat-Treatment Atmosphere on Electrochemical Performances of Ni-Rich Mixed-Metal Oxide (LiNi$_{0.8}$Co$_{0.15}$Mn$_{0.05}$O$_2$) as a Cathode Material for Lithium Ion Battery. *Electrochim. Acta* **2014**, *138*, 15–21. [CrossRef]

48. Lim, J.M.; Hwang, T.; Kim, D.; Park, M.S.; Cho, K.; Cho, M. Intrinsic Origins of Crack Generation in Ni-Rich LiNi$_{0.8}$Co$_{0.1}$Mn$_{0.1}$O$_2$ Layered Oxide Cathode Material. *Sci. Rep.* **2017**, *7*, 39669. [CrossRef]

49. Tian, C.; Nordlund, D.; Xin, H.L.; Xu, Y.; Liu, Y.; Sokaras, D.; Lin, F.; Doeff, M.M. Depth-Dependent Redox Behavior of LiNi$_{0.6}$Mn$_{0.2}$Co$_{0.2}$O$_2$. *J. Electrochem. Soc.* **2018**, *165*, A696–A704. [CrossRef]

50. Koyama, Y.; Yabuuchi, N.; Tanaka, I.; Adachi, H.; Ohzuku, T. Solid-State Chemistry and Electrochemistry of LiCo$_{1/3}$Ni$_{1/3}$Mn$_{1/3}$O$_2$ for Advanced Lithium-Ion Batteries I. First-Principles Calculation on the Crystal and Electronic Structures. *J. Electrochem. Soc.* **2004**, *151*, A1545–A1551. [CrossRef]

51. Bak, S.M.; Hu, E.; Zhou, Y.; Yu, X.; Senanayake, S.D.; Cho, S.J.; Kim, K.B.; Chung, K.Y.; Yang, X.Q.; Nam, K.W. Structural Changes and Thermal Stability of Charged LiNi$_x$Mn$_y$Co$_z$O$_2$ Cathode Materials Studied by Combined In Situ Time-Resolved XRD and Mass Spectroscopy. *ACS Appl. Mater. Interfaces* **2014**, *6*, 22594–22601. [CrossRef]

52. Jung, R.; Metzger, M.; Maglia, F.; Stinner, C.; Gasteiger, H.A. Chemical versus Electrochemical Electrolyte Oxidation on NMC111, NMC622, NMC811, LNMO, and Conductive Carbon. *J. Phys. Chem. Lett.* **2017**, *8*, 4820–4825. [CrossRef] [PubMed]

53. Bresser, D.; Buchholz, D.; Moretti, A.; Varzi, A.; Passerini, S. Alternative Binders for Sustainable Electrochemical Energy Storage—The Transition to Aqueous Electrode Processing and Bio-Derived Polymers. *Energy Environ. Sci.* **2018**, *11*, 3096–3127. [CrossRef]

54. Doberdò, I.; Löffler, N.; Laszczynski, N.; Cericola, D.; Penazzi, N.; Bodoardo, S.; Kim, G.-T.; Passerini, S. Enabling Aqueous Binders for Lithium Battery Cathodes—Carbon Coating of Aluminum Current Collector. *J. Power Sources* **2014**, *248*, 1000–1006. [CrossRef]

55. Du, Z.; Rollag, K.M.; Li, J.; An, S.J.; Wood, M.; Sheng, Y.; Mukherjee, P.P.; Daniel, C.; Wood, D.L. Enabling Aqueous Processing for Crack-Free Thick Electrodes. *J. Power Sources* **2017**, *354*, 200–206. [CrossRef]

56. Ahmadi, L.; Young, S.B.; Fowler, M.; Fraser, R.A.; Achachlouei, M.A. A Cascaded Life Cycle: Reuse of Electric Vehicle Lithium-Ion Battery Packs in Energy Storage Systems. *Int. J. Life Cycl. Assess.* **2017**, *22*, 111–124. [CrossRef]

57. Pavoni, F.H.; Sita, L.E.; dos Santos, C.S.; da Silva, S.P.; da Silva, P.R.C.; Scarminio, J. $LiCoO_2$ Particle Size Distribution as a Function of the State of Health of Discarded Cell Phone Batteries. *Powder Technol.* **2018**, *326*, 78–83. [CrossRef]

58. Yang, Y.; Xu, S.; He, Y. Lithium Recycling and Cathode Material Regeneration from Acid Leach Liquor of Spent Lithium-Ion Battery via Facile Co-Extraction and Co-Precipitation Processes. *Waste Manag.* **2017**, *64*, 219–227. [CrossRef]

59. Yang, Y.; Huang, G.; Xu, S.; He, Y.; Liu, X. Thermal Treatment Process for the Recovery of Valuable Metals from Spent Lithium-Ion Batteries. *Hydrometallurgy* **2016**, *165*, 390–396. [CrossRef]

60. Zheng, X.; Gao, W.; Zhang, X.; He, M.; Lin, X.; Cao, H.; Zhang, Y.; Sun, Z. Spent Lithium-Ion Battery Recycling—Reductive Ammonia Leaching of Metals from Cathode Scrap by Sodium Sulphite. *Waste Manag.* **2017**, *60*, 680–688. [CrossRef]

61. Mohtat, P.; Nezampasandarbabi, F.; Mohan, S.; Siegel, J.B.; Stefanopoulou, A.G.; James, A.; Uddin, K.; Chouchelamane, G.H.; Suttman, A.; Gong, X.; et al. Aging Phenomena for Lithium-Ion Batteries. *J. Power Sources* **2017**, *147*, 98–103. [CrossRef]

62. Waldmann, T.; Iturrondobeitia, A.; Kasper, M.; Ghanbari, N.; Aguesse, F.; Bekaert, E.; Daniel, L.; Genies, S.; Gordon, I.J.; Löble, M.W.; et al. Review—Post-Mortem Analysis of Aged Lithium-Ion Batteries: Disassembly Methodology and Physico-Chemical Analysis Techniques. *J. Electrochem. Soc.* **2016**, *163*, A2149–A2164. [CrossRef]

63. Wu, Y.; Saxena, S.; Xing, Y.; Wang, Y.; Li, C.; Yung, W.K.C.; Pecht, M. Analysis of Manufacturing-Induced Defects and Structural Deformations in Lithium-Ion Batteries Using Computed Tomography. *Energies* **2018**, *11*, 925. [CrossRef]

64. Li, J.; Du, Z.; Ruther, R.E.; An, S.J.; David, L.A.; Hays, K.; Wood, M.; Phillip, N.D.; Sheng, Y.; Mao, C.; et al. Toward Low-Cost, High-Energy Density, and High-Power Density Lithium-Ion Batteries. *JOM* **2017**, *69*, 1484–1496. [CrossRef]

65. Mohanty, D.; Hockaday, E.; Li, J.; Hensley, D.K.; Daniel, C.; Wood, D.L. Effect of Electrode Manufacturing Defects on Electrochemical Performance of Lithium-Ion Batteries: Cognizance of the Battery Failure Sources. *J. Power Sources* **2016**, *312*, 70–79. [CrossRef]

66. Mohanty, D.; Li, J.; Born, R.; Maxey, L.C.; Dinwiddie, R.B.; Daniel, C.; Wood, D.L. Non-Destructive Evaluation of Slot-Die-Coated Lithium Secondary Battery Electrodes by in-Line Laser Caliper and IR Thermography Methods. *Anal. Methods* **2014**, *6*, 674–683. [CrossRef]

67. Birkl, C.R.; Roberts, M.R.; McTurk, E.; Bruce, P.G.; Howey, D.A. Degradation Diagnostics for Lithium Ion Cells. *J. Power Sources* **2017**, *341*, 373–386. [CrossRef]

68. Dahn, H.M.; Smith, A.J.; Burns, J.C.; Stevens, D.A.; Dahn, J.R. User-Friendly Differential Voltage Analysis Freeware for the Analysis of Degradation Mechanisms in Li-Ion Batteries. *J. Electrochem. Soc.* **2012**, *159*, A1405–A1409. [CrossRef]

69. Gabrisch, H.; Yi, T.; Yazami, R. Transmission Electron Microscope Studies of $LiNi_{1/3}Mn_{1/3}Co_{1/3}O_2$ before and after Long-Term Aging at 70 °C. *Electrochem. Solid-State Lett.* **2008**, *11*, 7–12. [CrossRef]

70. Mu, L.; Lin, R.; Xu, R.; Han, L.; Xia, S.; Sokaras, D.; Steiner, J.D.; Weng, T.C.; Nordlund, D.; Doeff, M.M.; et al. Oxygen Release Induced Chemomechanical Breakdown of Layered Cathode Materials. *Nano Lett.* **2018**, *18*, 3241–3249. [CrossRef]

71. Jung, S.K.; Gwon, H.; Hong, J.; Park, K.Y.; Seo, D.H.; Kim, H.; Hyun, J.; Yang, W.; Kang, K. Understanding the Degradation Mechanisms of $LiNi_{0.5}Co_{0.2}Mn_{0.3}O_2$ Cathode Material in Lithium Ion Batteries. *Adv. Energy Mater.* **2014**, *4*, 1–7. [CrossRef]

72. Li, J.; Downie, L.E.; Ma, L.; Qiu, W.; Dahn, J.R. Study of the Failure Mechanisms of $LiNi_{0.8}Mn_{0.1}Co_{0.1}O_2$ Cathode Material for Lithium Ion Batteries. *J. Electrochem. Soc.* **2015**, *162*, A1401–A1408. [CrossRef]

73. Jung, R.; Metzger, M.; Maglia, F.; Stinner, C.; Gasteiger, H.A. Oxygen Release and Its Effect on the Cycling Stability of $LiNi_xMn_yCo_zO_2$ (NMC) Cathode Materials for Li-Ion Batteries. *J. Electrochem. Soc.* **2017**, *164*, A1361–A1377. [CrossRef]

74. Nam, K.W.; Yoon, W.S.; Yang, X.Q. Structural Changes and Thermal Stability of Charged $LiNi_{1/3}Co_{1/3}Mn_{1/3}O_2$ Cathode Material for Li-Ion Batteries Studied by Time-Resolved XRD. *J. Power Sources* **2009**, *189*, 515–518. [CrossRef]

75. Young, D.; Park, I.; Shin, Y.; Seo, D.; Kang, Y.; Doo, S.; Koh, M. Ni-Stabilizing Additives for Completion of Ni-Rich Layered Cathode Systems in Lithium-Ion Batteries: An Ab Initio Study. *J. Power Sources* **2019**, *418*, 74–83. [CrossRef]

76. Zhang, Y.; Katayama, Y.; Tatara, R.; Giordano, L.; Yu, Y.; Fraggedakis, D.; Sun, J.; Maglia, F.; Jung, R.; Bazant, M.Z.; et al. Revealing Electrolyte Oxidation via Carbonate Dehydrogenation on Ni-Based Oxides in Li-Ion Batteries by in Situ Fourier Transform Infrared Spectroscopy. *Energy Environ. Sci.* **2019**. [CrossRef]

77. Lin, F.; Markus, I.M.; Nordlund, D.; Weng, T.C.; Asta, M.D.; Xin, H.L.; Doeff, M.M. Surface Reconstruction and Chemical Evolution of Stoichiometric Layered Cathode Materials for Lithium-Ion Batteries. *Nat. Commun.* **2014**, *5*, 3529. [CrossRef]

78. Kim, J.M.; Chung, H.T. Role of Transition Metals in Layered $Li[Ni,Co,Mn]O_2$ under Electrochemical Operation. *Electrochim. Acta* **2004**, *49*, 3573–3580. [CrossRef]

79. Kondrakov, A.O.; Schmidt, A.; Xu, J.; Geßwein, H.; Mönig, R.; Hartmann, P.; Sommer, H.; Brezesinski, T.; Janek, J. Anisotropic Lattice Strain and Mechanical Degradation of High- and Low-Nickel NCM Cathode Materials for Li-Ion Batteries. *J. Phys. Chem. C* **2017**, *121*, 3286–3294. [CrossRef]

80. Li, X.; Qiao, Y.; Guo, S.; Xu, Z.; Zhu, H.; Zhang, X.; Yuan, Y.; He, P.; Ishida, M.; Zhou, H. Direct Visualization of the Reversible O^{2-}/O^- Redox Process in Li-Rich Cathode Materials. *Adv. Mater.* **2018**, *30*, 2–7. [CrossRef]

81. Nation, L.; Li, J.; James, C.; Qi, Y.; Dudney, N.; Sheldon, B.W. In Situ Stress Measurements during Electrochemical Cycling of Lithium-Rich Cathodes. *J. Power Sources* **2017**, *364*, 383–391. [CrossRef]

82. Ko, D.S.; Park, J.H.; Park, S.; Ham, Y.N.; Ahn, S.J.; Park, J.H.; Han, H.N.; Lee, E.; Jeon, W.S.; Jung, C. Microstructural Visualization of Compositional Changes Induced by Transition Metal Dissolution in Ni-Rich Layered Cathode Materials by High-Resolution Particle Analysis. *Nano Energy* **2019**, *56*, 434–442. [CrossRef]

83. Li, Y.; Cheng, X.; Zhang, Y.; Zhao, K. Recent Advance in Understanding the Electro-Chemo-Mechanical Behavior of Lithium-Ion Batteries by Electron Microscopy. *Mater. Today Nano* **2019**, *7*, 100040. [CrossRef]

84. Ruan, Y.; Song, X.; Fu, Y.; Song, C.; Battaglia, V. Structural Evolution and Capacity Degradation Mechanism of $LiNi_{0.6}Mn_{0.2}Co_{0.2}O_2$ Cathode Materials. *J. Power Sources* **2018**, *400*, 539–548. [CrossRef]

85. Besli, M.M.; Xia, S.; Kuppan, S.; Huang, Y.; Metzger, M.; Shukla, A.K.; Schneider, G.; Hellstrom, S.; Christensen, J.; Doeff, M.M.; et al. Mesoscale Chemomechanical Interplay of the $LiNi_{0.8}Co_{0.15}Al_{0.05}O_2$ Cathode in Solid-State Polymer Batteries. *Chem. Mater.* **2019**, *31*, 491–501. [CrossRef]

86. Nara, H.; Morita, K.; Mukoyama, D.; Yokoshima, T.; Momma, T.; Osaka, T. Impedance Analysis of $LiNi_{1/3}Mn_{1/3}Co_{1/3}O_2$ Cathodes with Different Secondary-Particle Size Distribution in Lithium-Ion Battery. *Electrochim. Acta* **2017**, *241*, 323–330. [CrossRef]

87. Zhang, S.S. Understanding of Performance Degradation of $LiNi_{0.80}Co_{0.10}Mn_{0.10}O_2$ Cathode Material Operating at High Potentials. *J. Energy Chem.* **2020**, *41*, 135–141. [CrossRef]

88. Nowak, S.; Winter, M. The Role of Cations on the Performance of Lithium Ion Batteries: A Quantitative Analytical Approach. *Acc. Chem. Res.* **2018**, *51*, 265–272. [CrossRef]

89. Hwang, S.; Kim, S.Y.; Chung, K.Y.; Stach, E.A.; Kim, S.M.; Chang, W. Determination of the Mechanism and Extent of Surface Degradation in Ni-Based Cathode Materials after Repeated Electrochemical Cycling. *APL Mater.* **2016**, *4*. [CrossRef]

90. Zhao, W.; Zheng, J.; Zou, L.; Jia, H.; Liu, B.; Wang, H.; Engelhard, M.H.; Wang, C.; Xu, W.; Yang, Y.; et al. High Voltage Operation of Ni-Rich NMC Cathodes Enabled by Stable Electrode/Electrolyte Interphases. *Adv. Energy Mater.* **2018**, *8*. [CrossRef]

91. Xiong, D.J.; Ellis, L.D.; Li, J.; Li, H.; Hynes, T.; Allen, J.P.; Xia, J.; Hall, D.S.; Hill, I.G.; Dahn, J.R. Measuring Oxygen Release from Delithiated $LiNixMnyCo_{1-x-y}O_2$ and Its Effects on the Performance of High Voltage Li-Ion Cells. *J. Electrochem. Soc.* **2017**, *164*, A3025–A3037. [CrossRef]

92. Hu, E.; Wang, X.; Yu, X.; Yang, X.Q. Probing the Complexities of Structural Changes in Layered Oxide Cathode Materials for Li-Ion Batteries during Fast Charge-Discharge Cycling and Heating. *Acc. Chem. Res.* **2018**, *51*, 290–298. [CrossRef]

93. An, S.J.; Li, J.; Daniel, C.; Mohanty, D.; Nagpure, S.; Wood, D.L. The State of Understanding of the Lithium-Ion-Battery Graphite Solid Electrolyte Interphase (SEI) and Its Relationship to Formation Cycling. *Carbon N. Y.* **2016**, *105*, 52–76. [CrossRef]

94. Shkrob, I.A.; Kropf, A.J.; Marin, T.W.; Li, Y.; Poluektov, O.G.; Niklas, J.; Abraham, D.P. Manganese in Graphite Anode and Capacity Fade in Li Ion Batteries. *J. Phys. Chem. C* **2014**, *118*, 24335–24348. [CrossRef]
95. Gilbert, J.A.; Shkrob, I.A.; Abraham, D.P. Transition Metal Dissolution, Ion Migration, Electrocatalytic Reduction and Capacity Loss in Lithium-Ion Full Cells. *J. Electrochem. Soc.* **2017**, *164*, A389–A399. [CrossRef]
96. Yeon, N.; Yim, T.; Ho, J.; Yu, J.; Lee, Z. Microstructural Study on Degradation Mechanism of Layered Electron Microscopy. *J. Power Sources* **2016**, *307*, 641–648. [CrossRef]
97. Guéguen, A.; Streich, D.; He, M.; Mendez, M.; Chesneau, F.F.; Novák, P.; Berg, E.J. Decomposition of LiPF6 in High Energy Lithium-Ion Batteries Studied with Online Electrochemical Mass Spectrometry. *J. Electrochem. Soc.* **2016**, *163*, A1095–A1100. [CrossRef]
98. Freiberg, A.T.S.; Roos, M.K.; Wandt, J.; De Vivie-Riedle, R.; Gasteiger, H.A. Singlet Oxygen Reactivity with Carbonate Solvents Used for Li-Ion Battery Electrolytes. *J. Phys. Chem. A* **2018**, *122*, 8828–8839. [CrossRef]
99. Waag, W.; Käbitz, S.; Sauer, D.U. Experimental Investigation of the Lithium-Ion Battery Impedance Characteristic at Various Conditions and Aging States and Its Influence on the Application. *Appl. Energy* **2013**, *102*, 885–897. [CrossRef]
100. Mu, L.; Yuan, Q.; Tian, C.; Wei, C.; Zhang, K.; Liu, J.; Pianetta, P.; Doeff, M.M.; Liu, Y.; Lin, F. Propagation Topography of Redox Phase Transformations in Heterogeneous Layered Oxide Cathode Materials. *Nat. Commun.* **2018**, *9*. [CrossRef]
101. Radin, M.D.; Hy, S.; Sina, M.; Fang, C.; Liu, H.; Vinckeviciute, J.; Zhang, M.; Whittingham, M.S.; Meng, Y.S.; Van der Ven, A. Narrowing the Gap between Theoretical and Practical Capacities in Li-Ion Layered Oxide Cathode Materials. *Adv. Energy Mater.* **2017**, *7*, 1–33. [CrossRef]
102. Solchenbach, S.; Metzger, M.; Egawa, M.; Beyer, H.; Gasteiger, H.A. Quantification of PF5 and POF3 from Side Reactions of LiPF6 in Li-Ion Batteries. *J. Electrochem. Soc.* **2018**, *165*, A3022–A3028. [CrossRef]
103. Pierre-Etienne, C.; David, P.; Mikael, C.; Pascal, M. Impact of Morphological Changes of $LiNi_{1/3}Mn_{1/3}Co_{1/3}O_2$ on Lithium-Ion Cathode Performances. *J. Power Sources* **2017**, *346*, 13–23. [CrossRef]
104. Chen, Z.; Wang, J.; Chao, D.; Baikie, T.; Bai, L.; Chen, S.; Zhao, Y.; Sum, T.C.; Lin, J.; Shen, Z. Hierarchical Porous $LiNi_{1/3}Co_{1/3}Mn_{1/3}O_2$ Nano-/Micro Spherical Cathode Material: Minimized Cation Mixing and Improved Li^+ Mobility for Enhanced Electrochemical Performance. *Sci. Rep.* **2016**, *6*, 25771. [CrossRef]
105. Wang, Y.; Roller, J.; Maric, R. Morphology-Controlled One-Step Synthesis of Nanostructured $LiNi_{1/3}Mn_{1/3}Co_{1/3}O_2$ Electrodes for Li-Ion Batteries. *ACS Omega* **2018**, *3*, 3966–3973. [CrossRef]
106. Li, J.; Li, H.; Stone, W.; Weber, R.; Hy, S.; Dahn, J.R. Synthesis of Single Crystal $LiNi_{0.5}Mn_{0.3}Co_{0.2}O_2$ for Lithium Ion Batteries. *J. Electrochem. Soc.* **2017**, *164*, A3529–A3537. [CrossRef]
107. Li, H.; Li, J.; Ma, X.; Dahn, J.R. Synthesis of Single Crystal $LiNi_{0.6}Mn_{0.2}Co_{0.2}O_2$ with Enhanced Electrochemical Performance for Lithium Ion Batteries. *J. Electrochem. Soc.* **2018**, *165*, A1038–A1045. [CrossRef]
108. Li, J.; Li, H.; Stone, W.; Glazier, S.; Dahn, J.R. Development of Electrolytes for Single Crystal NMC532/Artificial Graphite Cells with Long Lifetime. *J. Electrochem. Soc.* **2018**, *165*, A626–A635. [CrossRef]
109. Sharifi-Asl, S.; Chen, G.; Croy, J.; Balasubramanian, M.; Shahbazian-Yassar, R. Aberration-Corrected Scanning Transmission Electron Microscopy of Single Crystals and Chemically-Gradient NMC Cathodes. *Microsc. Microanal.* **2018**, *24*, 1536–1537. [CrossRef]
110. Xie, H.; Du, K.; Hu, G.; Peng, Z.; Cao, Y. The Role of Sodium in $LiNi_{0.8}Co_{0.15}Al_{0.05}O_2$ Cathode Material and Its Electrochemical Behaviors. *J. Phys. Chem. C* **2016**, *120*, 3235–3241. [CrossRef]
111. Pouillerie, C.; Croguennec, L.; Delmas, C. Modifications Observed Upon Cycling. *Solid State Ion.* **2000**, *132*, 15–29. [CrossRef]
112. Jang, Y.I.; Huang, B.; Wang, H.; Maskaly, G.R.; Ceder, G.; Sadoway, D.R.; Chiang, Y.M.; Liu, H.; Tamura, H. Synthesis and Characterization of $LiAl_yCO_{1-y}O_2$ and $LiAl_yNi_{1-y}O_2$. *J. Power Sources* **1999**, *81*, 589–593. [CrossRef]
113. Croguennec, L.; Suard, E.; Willmann, P.; Delmas, C. Structural and Electrochemical Characterization of the $LiNi_{1-y}Ti_yO_2$ Electrode Materials Obtained by Direct Solid-State Reactions. *Chem. Mater.* **2002**, *14*, 2149–2157. [CrossRef]
114. Zhang, X.; Jiang, W.J.; Mauger, A.; Gendron, F.; Julien, C.M. Minimization of the Cation Mixing in $Li_{1+x}(NMC)_{1-x}O_2$ as Cathode Material. *J. Power Sources* **2010**, *195*, 1292–1301. [CrossRef]
115. Nayak, P.K.; Erickson, E.M.; Schipper, F.; Penki, T.R.; Munichandraiah, N.; Adelhelm, P.; Sclar, H.; Amalraj, F.; Markovsky, B.; Aurbach, D. Review on Challenges and Recent Advances in the Electrochemical Performance of High Capacity Li- and Mn-Rich Cathode Materials for Li-Ion Batteries. *Adv. Energy Mater.* **2018**, *8*, 1–16. [CrossRef]

116. Wang, R.; Qian, G.; Liu, T.; Li, M.; Liu, J.; Zhang, B.; Zhu, W.; Li, S.; Zhao, W.; Yang, W.; et al. Tuning Li-Enrichment in High-Ni Layered Oxide Cathodes to Optimize Electrochemical Performance for Li-Ion Battery. *Nano Energy* **2019**, *62*, 709–717. [CrossRef]

117. Tian, C.; Lin, F.; Doeff, M.M. Electrochemical Characteristics of Layered Transition Metal Oxide Cathode Materials for Lithium Ion Batteries: Surface, Bulk Behavior, and Thermal Properties. *Acc. Chem. Res.* **2018**, *51*, 89–96. [CrossRef]

118. Eilers-Rethwisch, M.; Winter, M.; Schappacher, F.M. Synthesis, Electrochemical Investigation and Structural Analysis of Doped Li[Ni$_{0.6}$Mn$_{0.2}$Co$_{0.2-x}$M$_x$]O$_2$ (x = 0, 0.05; M = Al, Fe, Sn) Cathode Materials. *J. Power Sources* **2018**, *387*, 101–107. [CrossRef]

119. Mi, C.; Han, E.; Li, L.; Zhu, L.; Cheng, F.; Dai, X. Effect of Iron Doping on LiNi$_{0.35}$Co$_{0.30}$Mn$_{0.35}$O$_2$. *Solid State Ion.* **2018**, *325*, 24–29. [CrossRef]

120. Weigel, T.; Schipper, F.; Erickson, E.M.; Susai, F.A.; Markovsky, B.; Aurbach, D. Structural and Electrochemical Aspects of LiNi$_{0.8}$Co$_{0.1}$Mn$_{0.1}$O$_2$ Cathode Materials Doped by Various Cations. *ACS Energy Lett.* **2019**, *4*, 508–516. [CrossRef]

121. Zhao, Z.; Huang, B.; Wang, M.; Yang, X.; Gu, Y. Facile Synthesis of Fluorine Doped Single Crystal Ni-Rich Cathode Material for Lithium-Ion Batteries. *Solid State Ion.* **2019**, *342*, 115065. [CrossRef]

122. Konarov, A.; Myung, S.T.; Sun, Y.K. Cathode Materials for Future Electric Vehicles and Energy Storage Systems. *ACS Energy Lett.* **2017**, *2*, 703–708. [CrossRef]

123. Lin, F.; Nordlund, D.; Li, Y.; Quan, M.K.; Cheng, L.; Weng, T.C.; Liu, Y.; Xin, H.L.; Doeff, M.M. Metal Segregation in Hierarchically Structured Cathode Materials for High-Energy Lithium Batteries. *Nat. Energy* **2016**, *1*, 1–8. [CrossRef]

124. Hou, P.; Zhang, H.; Zi, Z.; Zhang, L.; Xu, X. Core-Shell and Concentration-Gradient Cathodes Prepared via Co-Precipitation Reaction for Advanced Lithium-Ion Batteries. *J. Mater. Chem. A* **2017**, *5*, 4254–4279. [CrossRef]

125. Sun, Y.K.; Myung, S.T.; Park, B.C.; Amine, K. Synthesis of Spherical Nano- To Microscale Core-Shell Particles Li[(Ni$_{0.8}$Co$_{0.1}$Mn$_{0.1}$)$_{1-x}$(Ni$_{0.5}$Mn$_{0.5}$)$_x$]O$_2$ and Their Applications to Lithium Batteries. *Chem. Mater.* **2006**, *18*, 5159–5163. [CrossRef]

126. Chen, X.; Li, D.; Mo, Y.; Jia, X.; Jia, J.; Yao, C.; Chen, D.; Chen, Y. Cathode Materials with Cross-Stack Structures for Suppressing Intergranular Cracking and High-Performance Lithium-Ion Batteries. *Electrochim. Acta* **2018**, *261*, 513–520. [CrossRef]

127. Qiu, Z.; Zhang, Y.; Huang, X.; Duan, J.; Wang, D.; Nayaka, G.P.; Li, X.; Dong, P. Beneficial Effect of Incorporating Ni-Rich Oxide and Layered over-Lithiated Oxide into High-Energy-Density Cathode Materials for Lithium-Ion Batteries. *J. Power Sources* **2018**, *400*, 341–349. [CrossRef]

128. Li, G.; Qi, L.; Xiao, P.; Yu, Y.; Chen, X.; Yang, W. Effect of Precursor Structures on the Electrochemical Performance of Ni-Rich LiNi$_{0.88}$Co$_{0.12}$O$_2$ Cathode Materials. *Electrochim. Acta* **2018**, *270*, 319–329. [CrossRef]

129. Hou, P.; Li, F.; Sun, Y.; Pan, M.; Wang, X.; Shao, M.; Xu, X. Improving Li$^+$ Kinetics and Structural Stability of Nickel-Rich Layered Cathodes by Heterogeneous Inactive-Al^{3+} Doping. *ACS Sustain. Chem. Eng.* **2018**, *6*, 5653–5661. [CrossRef]

130. Chen, Z.; Qin, Y.; Amine, K.; Sun, Y.K. Role of Surface Coating on Cathode Materials for Lithium-Ion Batteries. *J. Mater. Chem.* **2010**, *20*, 7606–7612. [CrossRef]

131. Zhao, R.; Liang, J.; Huang, J.; Zeng, R.; Zhang, J.; Chen, H.; Shi, G. Improving the Ni-Rich LiNi$_{0.5}$Co$_{0.2}$Mn$_{0.3}$O$_2$ Cathode Properties at High Operating Voltage by Double Coating Layer of Al$_2$O$_3$ and AlPO$_4$. *J. Alloys Compd.* **2017**, *724*, 1109–1116. [CrossRef]

132. Co, L.; Battery, M.O.; Lee, Y.; Nam, K.; Hwang, E.; Kwon, Y.; Kang, D.; Kim, S.; Song, S. Interfacial Origin of Performance Improvement and Fade for 4.6 V. *J. Phys. Chem. C* **2014**, *118*, 10631–10639.

133. Gao, H.; Maglia, F.; Lamp, P.; Amine, K.; Chen, Z. Mechanistic Study of Electrolyte Additives to Stabilize High-Voltage Cathode-Electrolyte Interface in Lithium-Ion Batteries. *ACS Appl. Mater. Interfaces* **2017**, *9*, 44542–44549. [CrossRef] [PubMed]

134. He, M.; Su, C.C.; Peebles, C.; Feng, Z.; Connell, J.G.; Liao, C.; Wang, Y.; Shkrob, I.A.; Zhang, Z. Mechanistic Insight in the Function of Phosphite Additives for Protection of LiNi$_{0.5}$Co$_{0.2}$Mn$_{0.3}$O$_2$ Cathode in High Voltage Li-Ion Cells. *ACS Appl. Mater. Interfaces* **2016**, *8*, 11450–11458. [CrossRef]

135. Nie, M.; Madec, L.; Xia, J.; Hall, D.S.; Dahn, J.R. Some Lewis Acid-Base Adducts Involving Boron Trifluoride as Electrolyte Additives for Lithium Ion Cells. *J. Power Sources* **2016**, *328*, 433–442. [CrossRef]

136. Liu, Z.; Chai, J.; Xu, G.; Wang, Q.; Cui, G. Functional Lithium Borate Salts and Their Potential Application in High Performance Lithium Batteries. *Coord. Chem. Rev.* **2015**, *292*, 56–73. [CrossRef]
137. Nguyen, D.T.; Kang, J.; Nam, K.M.; Paik, Y.; Song, S.W. Understanding Interfacial Chemistry and Stability for Performance Improvement and Fade of High-Energy Li-Ion Battery of LiNi$_{0.5}$Co$_{0.2}$Mn$_{0.3}$O$_2$//Silicon-Graphite. *J. Power Sources* **2016**, *303*, 150–158. [CrossRef]
138. Jang, S.H.; Jung, K.; Yim, T. Silyl-Group Functionalized Organic Additive for High Voltage Ni-Rich Cathode Material. *Curr. Appl. Phys.* **2018**, *18*, 1345–1351. [CrossRef]
139. Shi, Y.; Zhou, X.; Yu, G. Material and Structural Design of Novel Binder Systems for High-Energy, High-Power Lithium-Ion Batteries. *Acc. Chem. Res.* **2017**, *50*, 2642–2652. [CrossRef]

MDPI
St. Alban-Anlage 66
4052 Basel
Switzerland
Tel. +41 61 683 77 34
Fax +41 61 302 89 18
www.mdpi.com

Batteries Editorial Office
E-mail: batteries@mdpi.com
www.mdpi.com/journal/batteries